# Rapeseed and Canola Oil

# RAPESEED AND CANOLA OIL
## Production, Processing, Properties and Uses

Edited by

FRANK D. GUNSTONE
Professor Emeritus
University of St Andrews and
Honorary Research Professor
Scottish Crop Research Institute
Dundee

**Blackwell** Publishing

**CRC Press**

© 2004 by Blackwell Publishing Ltd

Editorial offices:
Blackwell Publishing Ltd, 9600 Garsington Road, Oxford OX4 2DQ, UK
  *Tel*: +44 (0)1865 776868
Blackwell Publishing Asia Pty Ltd, 550 Swanston Street, Carlton, Victoria 3053, Australia
  *Tel*: +61 (0)3 8359 1011

ISBN 1-4051-1625-0

Published in the USA and Canada (only) by
CRC Press LLC
2000 Corporate Blvd., N.W.
Boca Raton, FL 33431, USA
Orders from the USA and Canada (only) to
CRC Press LLC

USA and Canada only:
ISBN 0-8493-2364-9

The right of the Author to be identified as the Author of this Work has been asserted in accordance with the Copyright, Designs and Patents Act 1988.

All rights reserved. No part of this publication may be reproduced, stored in a retrieval system, or transmitted, in any form or by any means, electronic, mechanical, photocopying, recording or otherwise, except as permitted by the UK Copyright, Designs and Patents Act 1988, without the prior permission of the publisher.

This book contains information obtained from authentic and highly regarded sources. Reprinted material is quoted with permission, and sources are indicated. Reasonable efforts have been made to publish reliable data and information, but the author and the publisher cannot assume responsibility for the validity of all materials or for the consequences of their use.

**Trademark notice:** Product or corporate names may be trademarks or registered trademarks, and are used only for identification and explanation, without intent to infringe.

First published 2004

British Library Cataloguing-in-Publication Data
A catalogue record for this book is available from the British Library

Library of Congress Cataloging-in-Publication Data
A catalog record for this book is available from the Library of Congress

Set in 10.5/12 pt Times
by Integra Software Services Pvt. Ltd, Pondicherry, India
Printed and bound in Great Britain
by MPG Books, Bodmin, Cornwall

The publisher's policy is to use permanent paper from mills that operate a sustainable forestry policy, and which has been manufactured from pulp processed using acid-free and elementary chlorine-free practices. Furthermore, the publisher ensures that the text paper and cover board used have met acceptable environmental accreditation standards.

For further information on Blackwell Publishing, visit our website:
www.blackwellpublishing.com

# Contents

| | |
|---|---|
| **Contributors** | x |
| **Preface** | xi |

## 1 Rapeseeds and rapeseed oil: agronomy, production, and trade — 1
### E.J. BOOTH and F.D. GUNSTONE

| | | |
|---|---|---|
| 1.1 | Oilseed rape in context | 1 |
| 1.2 | Major developments in variety types | 2 |
| 1.3 | Crop establishment | 3 |
| 1.4 | Fertiliser requirement | 6 |
| 1.5 | Crop protection | 7 |
| | 1.5.1 Crop protection – weeds | 7 |
| | 1.5.2 Crop protection – pests | 8 |
| | 1.5.3 Crop protection – diseases | 9 |
| 1.6 | Maturity and harvesting | 10 |
| 1.7 | Production and trade for oilseeds and oil | 10 |
| 1.8 | Rapeseed | 11 |
| 1.9 | Rapeseed oil | 15 |
| | References | 15 |

## 2 Extraction and refining — 17
### E.J. BOOTH

| | | |
|---|---|---|
| 2.1 | Introduction | 17 |
| 2.2 | Oil extraction steps | 17 |
| | 2.2.1 Pre-treatment | 17 |
| |     2.2.1.1 Seed cleaning | 17 |
| |     2.2.1.2 Tempering | 18 |
| |     2.2.1.3 Dehulling | 19 |
| |     2.2.1.4 Flaking | 19 |
| |     2.2.1.5 Conditioning | 20 |
| | 2.2.2 Mechanical extraction | 21 |
| |     2.2.2.1 Oil extraction based solely on mechanical methods | 22 |
| |     2.2.2.2 Oil settling and filtering | 23 |
| | 2.2.3 Solvent extraction | 23 |
| |     2.2.3.1 Solvent recovery | 24 |
| |     2.2.3.2 Desolventising – toasting | 25 |
| |     2.2.3.3 Alternative solvents for oil extraction | 26 |
| | 2.2.4 Composition of crude oil | 26 |
| 2.3 | Refining | 27 |
| | 2.3.1 Degumming | 28 |
| | 2.3.2 Physical refining | 29 |
| | 2.3.3 Alkali refining | 29 |

|     |        | 2.3.4 Bleaching | 30 |
| --- | --- | --- | --- |
|     |        | 2.3.5 Winterisation | 31 |
|     |        | 2.3.6 Deodourisation | 31 |
|     | 2.4    | Biorefining – an alternative oilseed processing method | 32 |
|     | References |  | 35 |

## 3 Chemical composition of canola and rapeseed oils    37
### W.M.N. RATNAYAKE and J.K. DAUN

|     |     |     |     |
| --- | --- | --- | --- |
| 3.1 | Brief history of the development of rapeseed oils |  | 37 |
| 3.2 | Development of specialty types of rapeseed oils |  | 41 |
| 3.3 | Minor fatty acids |  | 44 |
| 3.4 | Triacylglycerols of rapeseed and canola oils |  | 48 |
| 3.5 | Minor lipid components |  | 59 |
|     | 3.5.1 Sterols |  | 59 |
|     | 3.5.2 Tocopherols |  | 63 |
|     | 3.5.3 Carotenoids |  | 66 |
|     | 3.5.4 Waxes |  | 66 |
|     | 3.5.5 Polar lipids |  | 67 |
| 3.6 | Chlorophyll |  | 68 |
| 3.7 | Sulfur and sulfur-containing compounds |  | 70 |
| 3.8 | Minerals |  | 71 |
| 3.9 | Conclusions |  | 72 |
| References |  |  | 73 |

## 4 Chemical and physical properties of canola and rapeseed oil    79
### DÉRICK ROUSSEAU

|     |     |     |     |
| --- | --- | --- | --- |
| 4.1 | Introduction |  | 79 |
| 4.2 | Chemical properties |  | 79 |
|     | 4.2.1 Saponification value |  | 80 |
|     | 4.2.2 Iodine value |  | 80 |
|     | 4.2.3 Oxidative stability |  | 81 |
|     |        | 4.2.3.1 Mechanism | 81 |
|     |        | 4.2.3.2 Susceptibility to oxidation | 84 |
|     |        | 4.2.3.3 Peroxide value | 85 |
|     |        | 4.2.3.4 Thiobarbituric acid (TBA) test | 85 |
|     |        | 4.2.3.5 *p*-Anisidine | 85 |
|     |        | 4.2.3.6 Conjugated dienes | 86 |
|     |        | 4.2.3.7 Chromatography | 86 |
|     |        | 4.2.3.8 Electron-spin resonance | 86 |
|     |        | 4.2.3.9 Sensory analysis | 86 |
| 4.3 | Physical properties |  | 87 |
|     | 4.3.1 Relative density |  | 87 |
|     | 4.3.2 Viscosity |  | 88 |
|     | 4.3.3 Surface and interfacial tension |  | 89 |
|     | 4.3.4 Refractive index |  | 90 |
|     | 4.3.5 Specific heat: heat of fusion or crystallisation |  | 90 |
|     | 4.3.6 Heat of combustion |  | 91 |
|     | 4.3.7 Smoke, flash and fire point |  | 91 |
|     | 4.3.8 Solubility |  | 92 |

|       |         | 4.3.9    | Cold test                                               | 92  |
|-------|---------|----------|---------------------------------------------------------|-----|
|       |         | 4.3.10   | Spectroscopic properties                                | 92  |
|       |         | 4.3.11   | Melting behaviour, polymorphism and crystal structure   | 93  |
|       |         |          | 4.3.11.1  Unsaturation level                            | 93  |
|       |         |          | 4.3.11.2  Acyl chain length                             | 94  |
|       |         |          | 4.3.11.3  Fatty acid isomers                            | 94  |
|       |         |          | 4.3.11.4  Positional distribution                       | 94  |
|       | 4.4     | Modification strategies                                            | 97  |
|       |         | 4.4.1    | Hydrogenation                                           | 97  |
|       |         | 4.4.2    | Interesterification                                     | 101 |
|       | References                                                                    | 105 |

# 5 High erucic oil: its production and uses — 111
## C. TEMPLE-HEALD

|       |         |                                                                      |     |
|-------|---------|----------------------------------------------------------------------|-----|
| 5.1   | Introduction                                                                  | 111 |
| 5.2   | Crucifer oilseeds                                                             | 111 |
|       | 5.2.1   | *Brassica napus* (HERO)                                              | 111 |
|       |         | 5.2.1.1  HEAR agronomy                                               | 113 |
|       | 5.2.2   | Crambe abyssinica                                                    | 115 |
|       |         | 5.2.2.1  Crambe agronomy                                             | 115 |
|       | 5.2.3   | Mustard rapeseed                                                     | 116 |
| 5.3   | Processing of HEAR oils                                                       | 117 |
|       | 5.3.1   | Batch processes                                                      | 118 |
|       | 5.3.2   | Continuous splitting processes                                       | 118 |
|       | 5.3.3   | Other splitting processes                                            | 119 |
| 5.4   | Downstream processing of the split HEAR fatty acids                           | 120 |
|       | 5.4.1   | Fractional distillation                                              | 120 |
|       | 5.4.2   | Dry or melt crystallisation                                          | 121 |
| 5.5   | Quality problems associated with processing HEAR oils                         | 121 |
|       | 5.5.1   | Downstream processing problems                                       | 122 |
| 5.6   | Meal quality                                                                  | 122 |
| 5.7   | Users and producers of erucic acid                                            | 124 |
| 5.8   | Uses of erucic acid                                                           | 126 |
| 5.9   | Genetic modification of HEAR crops                                            | 128 |
| 5.10  | Ideal crop for industrial users                                               | 129 |
| References                                                                            | 129 |

# 6 Food uses and nutritional properties — 131
## BRUCE E. McDONALD

|       |         |                                                                      |     |
|-------|---------|----------------------------------------------------------------------|-----|
| 6.1   | Introduction                                                                  | 131 |
| 6.2   | Food uses                                                                     | 132 |
|       | 6.2.1   | Salad oils, salad dressings and mayonnaise                           | 132 |
|       | 6.2.2   | Margarine                                                            | 132 |
|       | 6.2.3   | Other uses                                                           | 133 |
| 6.3   | Nutritional properties                                                        | 134 |
| 6.4   | Dietary fat and cardiovascular disease                                        | 134 |
| 6.5   | Effect of canola oil on plasma cholesterol and lipoproteins                   | 135 |
|       | 6.5.1   | Studies with normolipidemic subjects                                 | 135 |
|       | 6.5.2   | Studies with hyperlipidemic subjects                                 | 137 |

|  |  |  |  |
|---|---|---|---|
| | 6.5.3 | Potential effect of phytosterols in canola oil on plasma cholesterol levels | 140 |
| | 6.5.4 | Effect of canola oil intake on lipid peroxidation | 140 |
| | 6.5.5 | Canola oil and thrombogenesis | 142 |
| | 6.5.6 | Effect of canola oil on fatty acid composition of plasma and platelet phospholipids | 143 |
| | 6.5.7 | Effect of canola oil on clotting time and factors involved in clot formation | 146 |
| | 6.5.8 | Canola oil and cardiac arrhythmia | 147 |
| 6.6 | The Lyon Diet Heart Study: the canola oil connection | | 147 |
| 6.7 | Summary | | 149 |
| | References | | 149 |

# 7 Non-food uses  154
## KERR WALKER

| | | | |
|---|---|---|---|
| 7.1 | Introduction | | 154 |
| 7.2 | Biodiesel | | 154 |
| | 7.2.1 | Biodiesel feedstocks | 155 |
| | 7.2.2 | Production of biodiesel | 158 |
| | 7.2.3 | Fuel characteristics | 160 |
| | 7.2.4 | Emissions | 160 |
| | 7.2.5 | Economics of biodiesel production | 163 |
| | 7.2.6 | Biodiesel market opportunities | 166 |
| | 7.2.7 | Biodiesel production in Europe | 166 |
| 7.3 | Lubricants | | 167 |
| | 7.3.1 | Hydraulic fluids | 169 |
| | 7.3.2 | Future market potential | 171 |
| | 7.3.3 | Greases | 171 |
| | 7.3.4 | Mould release agents | 172 |
| | 7.3.5 | Motor and gear oils | 173 |
| | 7.3.6 | Metal working fluids | 174 |
| | 7.3.7 | Chainsaw oil | 176 |
| 7.4 | Surfactants | | 177 |
| 7.5 | Paints and inks | | 179 |
| 7.6 | Polymers | | 181 |
| | References | | 182 |

# 8 Potential and future prospects for rapeseed oil  186
## CHRISTIAN MÖLLERS

| | | | |
|---|---|---|---|
| 8.1 | Introduction | | 186 |
| 8.2 | Oil content and triacylglycerol structure | | 188 |
| 8.3 | Modification of the $C_{18}$ fatty acid composition | | 189 |
| | 8.3.1 | Development of rapeseed with modified linolenic acid (18:3) content | 189 |
| | 8.3.2 | Development of rapeseed with an increased oleic acid (18:1) content | 191 |
| | 8.3.3 | Development of high oleic acid/low linolenic acid oilseed rape | 192 |
| | 8.3.4 | Development of high stearic acid (18:0) oilseed rape | 193 |
| 8.4 | Low saturated fatty acids | | 193 |
| 8.5 | Medium and short chain fatty acids | | 194 |
| 8.6 | Gamma linolenic acid | | 196 |
| 8.7 | Long chain polyunsaturated fatty acids | | 197 |
| 8.8 | High erucic acid | | 198 |

| | | | |
|---|---|---|---|
| 8.9 | Miscellaneous unusual fatty acids | | 200 |
| 8.10 | Minor bioactive constituents | | 202 |
| | 8.10.1 | Polar lipids | 202 |
| | 8.10.2 | Tocopherols | 203 |
| | 8.10.3 | Sterols | 206 |
| | 8.10.4 | Carotenoids | 208 |
| | 8.10.5 | Chlorophyll | 210 |
| 8.11 | Conclusions and outlook | | 211 |
| | References | | 212 |

**List of acronyms**     **218**
**Index**     **220**

# Contributors

**Elaine J. Booth** — Scottish Agricultural College, Craibstone Estate, Bucksburn, Aberdeen AB21 9YA, UK

**James K. Daun** — Grain Research Laboratory, Canadian Grain Commission, 1404-303 Main Street, Winnipeg MB, R3C 3G8, Canada

**Frank D. Gunstone** — Professor Emeritus, University of St Andrews and Honorary Research Professor, Scottish Crop Research Institute, Invergowrie, Dundee DD2 5DA, UK

**Bruce E. McDonald** — Manitoba Health Research Council, P127-720 Bannatyne Avenue, Winnipeg MB, R3E 0W3, Canada

**Christian Möllers** — Institut für Pflanzenbau und Pflanzenzüchtung, Universität Göttingen, Von Siebold Strasse 8, D-37075 Göttingen, Germany

**W.M. Nimal Ratnayake** — Nutrition Research Division, Food Directorate, Health Canada, Postal Locator 2203C, Ottawa ON, K1A 0L2, Canada

**Dérick Rousseau** — School of Nutrition, Ryerson University, 350 Victoria Street, Toronto ON, M5B 2K3, Canada

**Clare Temple-Heald** — Croda Chemicals Europe Ltd, Oak Road, Clough Road, Hull, East Yorkshire HU6 7PH, UK

**Kerr C. Walker** — Scottish Agricultural College, Craibstone Estate, Bucksburn, Aberdeen AB21 9YA, UK

# Preface

In 2002 we published *Vegetable Oils in Food Technology*, with chapters on ten major oils and several minor oils. In further discussion between the publisher and the editor, it was agreed that there was a case for developing one of these chapters into a more detailed treatment. That agreement has culminated in the present volume devoted to rapeseed oil.

There are several reasons why rapeseed oil (also called canola oil) merits this fuller treatment. Brassica oilseeds have been grown and used by humans for thousands of years. The oil was rich in erucic acid and was used industrially as a lubricant, an illuminant and an animal feed as well as, in some countries, a food oil. The situation changed in the 1960s when Canadian researchers developed varieties of rapeseed that were low in erucic acid and produced a healthy food oil of enhanced value. Apart from linseed, this is the only oilseed crop of any significance that can be grown in Northern climes. Today it is grown extensively in Canada and in northern Europe. The plant remains the subject of intensive research by both traditional seed breeding methods and modern biotechnological means to improve its agronomy and the quality of both the oil and the seed meal.

Rapeseed is now the second largest oilseed crop after soybean, and the third largest vegetable oil after soybean oil and palm oil, and it is therefore an important contributor to the annual supply of vegetable oils required to meet increasing demand – particularly from developing countries. Oilseed rape is grown extensively in India and in China, as well as in Canada and northern Europe.

The first two chapters are devoted to agronomy, production of refined oil, and trade matters. Chapters 3 and 4 are concerned with the chemical composition of rapeseed oil and its physical and chemical properties. Then follow three chapters on the use of the high-erucic acid oil, and the food and non-food uses of the greater volume of low-erucic acid oil. Nutritional properties are included in the chapter on food uses. The final chapter is devoted to the potential and prospects of rapeseed (canola) oil.

The book is directed primarily at the producers and processors of fatty oils, together with the users in the food and oleochemical industries. These will include oils and fats chemists and technologists, chemical engineers in the oil processing industry, nutritionists and seed technologists.

I wish to thank the authors for their willingness to contribute to this book, and for the high quality of their chapters. I also acknowledge the generous help and advice that I have received from Graeme MacKintosh and David McDade of Blackwell Publishing Ltd.

Frank D. Gunstone

# 1 Rapeseeds and rapeseed oil: agronomy, production, and trade

E.J. Booth and F.D. Gunstone

## 1.1 Oilseed rape in context

Brassica oilseeds have been grown by humans for thousands of years and are one of the few edible oilseeds capable of being grown in cool temperate climates. They are closely related to the condiment mustards used for flavouring and for their reputed medicinal properties. Records indicate early cultivation of vegetable forms of the crop in India in 1500 BP (Prakash, 1980), and in China more than 1000 BP (Li, 1980). Cultivation extended across Europe in the Middle Ages, and by the fifteenth century rapeseed was grown in the Rhineland as a source of lamp oil and also for cooking fat (Heresbach, 1570). Demand for rapeseed oil grew significantly in the developed world during the twentieth century with concurrent improvements in agronomic techniques, processing methods, and varieties.

There are a number of rapeseed species grown in the major oilseed rape-producing areas of the world. *Brassica rapa* (turnip rape) is the most cold-hardy species and currently accounts for half the area grown in western Canada, largely because of its earlier maturity. Ecotypes of this species are also grown in the Indian subcontinent. *B. napus* (swede rape) is the most commonly grown rapeseed in Europe, Canada, and China. Both spring- and winter-sown varieties are available, although winter types dominate due to their higher yields in favourable growing conditions. *B. juncea* is well adapted to drier growing conditions and is widely grown in northern India and China, where brown- and yellow-seeded varieties are available. This species has been introduced into parts of Australia and is used in mainstream breeding programmes of *B. napus* for drier climates. *B. carinata* is less widely grown than the other species and is largely restricted to Ethiopia and the surrounding countries in East Africa.

Oilseed rape provides a convenient alternative for cereal-based agricultural systems as it is broad leaved and can be grown as a break crop for a continuous run of cereals. Minimal investment in new machinery is required as the bulk of oilseed cultivation operations can be conducted with existing cereal equipment. The timing of work required for oilseed rape throughout the season allows arable work peaks to be spread throughout the year.

Oilseed rape has beneficial effects for the following crops in the rotation. Its deep rooting tap root opens up the soil and can improve soil structure, particularly of clay soil, and break up compacted subsurface layers of soil. Nutrient residues left after the crop has been harvested improve the fertility of the soil with subsequent benefits for the following crops. Yields of wheat crops following oilseed rape can typically yield around 35% more than in a continuous cereal sequence (Wibberley, 1996). The breakdown of glucosinolates from Brassica residues left in the soil may also have a biocidal effect and aid control of crop pests and soil diseases.

## 1.2 Major developments in variety types

Considerable changes to quality aspects of both rapeseed oil and meal rapeseed have been achieved by plant breeding efforts in the latter part of the twentieth century. There were concerns, mainly in the Western world, about the high levels of erucic acid in rapeseed oil and the perception that this was linked to heart disease in animals. Breeding programmes in Canada produced the first low erucic acid rapeseed varieties in the 1960s. This characteristic has now been introduced to *B. napus*, *B. rapa*, and *B. juncea*. Low erucic varieties of *B. juncea* now exist, but the bulk of this crop is still high erucic and high erucic acid oil from different rapeseed types continues to be accepted for human consumption in Asia.

The rapeseed meal left after oil extraction, is useful as a high protein animal feed. Quantities which can be fed are however limited, primarily due to the presence of sulfur-bearing compounds known as glucosinolates. High intakes cause problems of palatability due to the hot mustard-like taste of the glucosinolate by-products and can be associated with goitrogenic, liver, and kidney abnormalities and fertility problems of livestock. The low glucosinolate trait from the variety Bronowksi was successfully incorporated into spring varieties of *B. napus* and *B. rapa*, and later into winter varieties of *B. napus*, as a result of Canadian breeding efforts in the 1970s. It has been proved to be more difficult to incorporate the low glucosinolate character into *B. juncea* and efforts to achieve this are still ongoing.

These breeding developments led to the production of rapeseed low in both erucic acid in the oil and glucosinolates in the meal, the so-called *double-low* varieties. The name canola was established with the licensing of the first double-low variety in Canada in 1974. The canola trademark is held by the Canola Council in Canada and may be permitted for use to describe rapeseed with less than 2% erucic acid in the oil and less than 30 µM/g glucosinolates in the meal (Anon., 2003). The term *canola* is used in many English-speaking countries such as the USA and Australia, but in the UK *rapeseed* continues to be used in reference to both double-low types and other quality types such as high erucic acid rapeseed for industrial purposes.

A major objective of oilseed rape breeding in recent years has been the production of hybrids, due to the observed advantages of heterosis in terms of yield and vigour of the three main species, *B. napus*, *B. rapa*, and *B. juncea*. During hybrid production, it is necessary to induce male sterility in one parent to ensure that cross-pollination occurs and that self-pollination is avoided. This has been proved to be far from straightforward in oilseed rape. The Polima system of hybridisation has been used in China, Australia, and Canada, however, this system has been associated with problems in achieving sterility in all conditions and has also reduced petal size affecting pollination. The Ogu-INRA cms system, applying cytoplasmic male sterility discovered in radish, is the principal means of commercial production of *B. napus* varieties in Europe. A disadvantage of this system is that the gene for restoration of fertility is closely linked to that for high glucosinolates content which means that it is difficult to achieve sufficiently low glucosinolate levels in varieties.

The initial difficulties in producing hybrid varieties led to the introduction of variety associations within Europe. These were composed of a non-restored, male sterile-hybrid seed fraction plus a pollinating fraction, typically in the ratio of 80:20. Variety associations were widely grown for a period during the 1990s, but the area devoted to them is now declining. Hybrids with fully restored fertility developed using the Male Sterile Lembke system have become available and further hybrid technology systems are being developed.

Oilseed rape has also proved amenable to genetic modification and many breeding programmes are underway to modify traits and introduce new quality characteristics, principally with *B. napus*. Oilseed rape, or canola, is the third most widely grown commercial genetically modified crop after soybean and maize with an area equivalent to that of cotton (Walker *et al.*, 2000). Tolerance to one of the several broad-spectrum herbicides was one of the first characteristics to be incorporated in crops using genetic modification, and oilseed rape varieties tolerant to either glyphosate or glufosinate have been developed. These are widely grown in North America, but no genetically modified oilseed rape has yet been accepted for commercial cultivation in Europe. Rapeseed modified to produce lauric acid within the oil, produced through genetic modification, has also been grown commercially in North America but this is now discontinued. Further developments with Brassica breeding are being facilitated by new techniques in biotechnology, especially in genetic fingerprinting.

## 1.3 Crop establishment

Agronomic practices vary from country to country along with species, variety, and prevailing market conditions, but there are general common principles which are outlined here. Rapeseed can tolerate a wide range of soil pH levels ranging from 5.5 to 8.0, enabling cultivation on slightly more acidic soil than other crops including barley, beans, and sugar beet. Seeds should be drilled into a fine, firm, moist, and well-structured seedbed aiming to encourage rapid and uniform germination and establishment and in dry conditions can be vulnerable to lack of water. Once seed has imbibed water, soil temperature is the main factor affecting speed of germination and proportion of seeds producing viable plants. At temperatures of 21–25°C, germination can take place in only one day, but at temperatures of 2°C, germination can take 11–14 days.

Optimum date of sowing varies according to the latitude and the date of onset of winter. In northern European countries such as the UK and Denmark, early sowing of winter varieties is critical for satisfactory establishment and the optimum sowing date is generally in the latter half of August, whereas in the south of Europe, e.g. southern France and Germany, sowing dates can be extended until early September. The aim in all cases is to produce plants that are sufficiently large to withstand the rigours of winter through either direct frost kill or frost heave. A guide for the northern part of the UK is to achieve at least five true leaves and reasonably vigorous plants by December. Another reference suggests plants with 6–8 leaves prior to the onset of severe weather (Auld *et al.*, 1983). The potential for good winter hardiness was determined to be at the optimum when the stem base was 5–16 mm in diameter and stem elongation was less than 30 mm in western Canada (Topinka *et al.*, 1991).

For spring varieties, optimum sowing dates relate to both soil conditions and conditions predicted for harvest. In northern climates, timing of spring crop establishment should make best use of the short season and enable appropriate timing of maturation and harvest before the onset of inclement weather. In warmer conditions, crop establishment may be dictated by available soil moisture.

Optimum plant populations will vary according to the date of sowing, method of crop establishment, soil fertility, and method of harvesting. Several issues relating to the variety also influence seed rates, including whether the variety is of winter- or spring-sown type, conventional or hybrid, and in some cases, particular varietal agronomic characteristics.

The rape plant has a remarkable capacity to respond to external factors and is considered to be *plastic* in terms of its yield components. Similar yields can be obtained from a wide range of plant populations. Indeed, Ogilvy (1984) showed that yield varied by only 10% of the maximum over the full range of seed rates tested of 3–12 kg/ha. Very high seed rates tend to result in high plant populations with many plants competing with each other early in development and producing tall, thin-stemmed crops prone to lodging prior to harvest. Conversely, low seed rates lead to low plant populations producing open crops slow to form a crop canopy. These crops tend to be susceptible to crop pests and produce shorter, thick-stemmed crops. However, plants sown at low seed rates are able to compensate by producing more branches and more pods per plant. Mendham and Salisbury (1995) reported that increased branching and pod production per plant allowed the population to be reduced from 200 to 40 plants/m$^2$ with less than 20% loss in yield, provided sufficient water was available to allow compensatory growth. Plant populations of less than 20 plants/m$^2$ resulted in greater reductions in yield, but plant density of less than 10 was required to result in a yield of less than 50% of the control. A drawback with lower plant populations is that each plant in the thinner crop will produce more branches leading to a prolonged maturation time as pods on different branches will ripen at different timings.

Standard advice in the UK has been to sow at 120 seeds/m$^2$ for conventional varieties (equivalent to approximately 6–7 kg/ha), leaving a considerable margin for winter kill. Optimal plant population in the spring was found to be 80–100 plants/m$^2$. There is now some evidence to suggest that less-dense crop canopies are beneficial to yield as they minimise shading, encouraging better light penetration to lower pods and consequently improved pod and seed retention with reduced lodging. There is a move to lower seed rates or even to consider reducing crop-canopy density of very thick plants in the spring by mowing (Lunn and Spink, 2000). The work indicates that an open crop canopy should be aimed for, with as few as 25–50 plants/m$^2$ established, for crops which are sown in good conditions at the optimum sowing time.

The recommended seed rate for winter hybrids was set lower than conventional types at 70 plants/m$^2$ for many varieties. For the variety association types, this reflected the need to avoid out-competing the pollinator fraction within the crop stand, which is relied upon for pollination. For restored hybrids, the low seed rate relates to a need to reduce the expenditure on more costly hybrid seed technology. Seed merchants are now selling certain conventional varieties on an area pack basis and are recommending lower seed rates for varieties with weaker straw in order to encourage thinner crops with thicker-stemmed individual plants and avoid lodging problems.

Spring oilseed rape has a shorter growing season than winter oilseed rape and tends to produce fewer branches. Recommended seed rates are hence slightly higher than winter oilseed rape. In the UK, recommended plant populations are in the region of 150 plants/m$^2$ for conventional *B. napus* varieties, generally requiring 7–8 kg seeds per ha. Plant populations for hybrid varieties can be reduced slightly due to expected higher vigour, but as seed size tends to be larger, this may not result in a reduction in seed rate per hectare.

Choice of crop establishment method will vary depending on factors such as machinery and labour availability, and economics of the system including cost of inputs and value of outputs. Broadcasting of the seed requires a slightly higher seed rate as the seed will be vulnerable to variable moisture availability and attack from predators such as slugs and pigeons. Seed drilling using a cereal drill may be the most popular method, but adjustments to seed rate and depth are required. Precision drilling where seed is placed at a constant depth at regular intervals may have advantages, but is slower than conventional drilling and will require good weather conditions at drilling. The use of direct drilling without ploughing and with minimal use of cultivation is the subject of rekindled interest in northern Europe in an attempt to reduce costs. Risks of crop failure will be increased, but establishment may be successful where conditions are favourable. In China, the practice of transplanting seedlings from seedbeds to fields is used, but would be associated with unacceptably high labour costs in many regions.

## 1.4 Fertiliser requirement

Oilseed rape is responsive to fertiliser and a crop yielding 3 t/ha will require nitrogen application input of 150–210 kg/ha (Pouzet, 1995). For winter oilseed rape in European conditions, application of nitrogen in autumn can result in greater foliage growth in early sown crops but rarely an increase in yield, and consequently only a low rate of nitrogen, if any, is required. Nitrogen application in the autumn is vulnerable to leaching in winter leading to an increase of nitrogen in watercourses; hence lower applications at this time can reduce environmental concerns. In the spring, oilseed rape requires nitrogen earlier than other crops as it begins spring growth very early, before soil mineralisation takes place. The economic rate of nitrogen application in the spring can be assessed by consideration of the price of nitrogen and the expected value of rapeseed utilising knowledge of the crop's response to nitrogen.

Spring-sown oilseed rape requires nitrogen later in the spring compared to winter oilseed rape coinciding with more active mineralisation rates. This factor, along with the lower expected yield of spring oilseed rape, leads to a lower requirement for applied nitrogen and typically 120 kg/ha is applied in the UK with current nitrogen prices and rapeseed values.

Crop requirement for phosphate is not very high (about 60 kg/ha for winter oilseed rape), and, being relatively immobile in the soil, can be applied to the seedbed. Oilseed rape needs large amounts of potash as more than 200 kg/ha is mobilised by the plant. Much of this will be supplied from soil reserves, and the application rate for moderate potash soils where straw of the previous crop is removed will be in the region of 50 kg/ha.

Oilseed rape has a high requirement for sulfur and with environmental clean-up of power stations reducing atmospheric supply of sulfur, deficiency of this nutrient causing chlorosis of the leaves and reduction of yield has been more widely seen since the mid-1980s. Sulfate will be mineralised from soil reserves, but soil supply will be small and like nitrate, supply is difficult to predict due to leaching. Rates for sulfur application may be based on sulfur levels in the soil and knowledge of sulfur atmospheric deposition. Recommended rates vary in the major European oilseed rape growing countries from 10 to 30 kg sulfur/ha in the UK, to an average of 30 kg sulfur/ha in France, 30–40 kg sulfur/ha in Denmark and 20–50 kg sulfur/ha in Germany (Walker and Dawson, 2002).

The rapeseed crop has a high requirement for water and is particularly sensitive to drought at germination and establishment. However, irrigation is not widely used in the areas of western Canada and the Mediterranean where facilities are available.

## 1.5 Crop protection

### 1.5.1 Crop protection – weeds

The occurrence of weeds within the oilseed rape crop can cause a range of significant problems which are responsible for considerable growing costs. Indeed, weeds are acknowledged as the most important limiting factors in Canada (Orson, 1995). Weeds cause crop protection problems in a number of different ways. First they cause direct yield losses through competition for light, nutrients, and space. In rotation with cereals, volunteer plants from the previous cereal crops are particularly competitive. Secondly, their presence can be associated with weed seeds in the harvested seed samples. Weeds can also interfere with harvesting, especially if swathing is used (see Section 1.6) and where weeds remaining green delay drying of the crop.

In some parts of the world, weeds are controlled through cultural means alone, whilst throughout Europe and Canada control of weeds is frequently achieved by a combination of agronomic practice and use of herbicides. Cultural methods include use of rotation so that weeds giving greatest problems in Brassicas are controlled in different parts of the rotation. Stubble cleaning by cultivation and use of stale seedbeds, where the ground is cultivated and the seedbed prepared to encourage germination of weeds before a further cultivation kills germinating weeds, just prior to sowing the crop. Inversion tillage will also contribute to weed control. Fields known to have suffered a high shedding loss from a previous cereal crop and likely to produce many volunteers should be avoided. Ensuring that the oilseed crop is competitive is an important factor of weed control. Use of timely sowing at the optimum seed rate on good seedbed conditions and adequate disease and pest control will all contribute towards a competitive crop.

In terms of chemical weed control, non-selective herbicides such as glufosinate ammonium, glyphosate, and paraquat are used prior to or shortly after drilling to control weeds which have emerged after primary cultivation. Pre-drilling herbicides such as trifluralin are commonly used and can provide cheap weed control. However, they have a limited weed species spectrum and some resistance has been noted in Canada. Pre-emergence residual soil acting herbicides such as metazachlor offer a better range of weed control but require moist soil for optimum activity. Post-emergence selective herbicides are commonly used for control of annual grass weeds.

Varieties tolerant to a broad-spectrum herbicide have become very popular in North America, with 76% of the Canadian crops cultivated in 1999 being herbicide tolerant (Harker *et al.*, 2000). Two of these types, glufosinate- and glyphosate-tolerant varieties, are produced using genetic modification, with the third, imidazolinone tolerant, being derived from conventional breeding techniques. Herbicide-tolerant types may offer advantages in certain circumstances, for instance, in the control of Brassica weeds or where dry conditions prevent optimal activity of conventional herbicides.

*1.5.2  Crop protection – pests*

Insect pests of oilseed rape are principally Crucifer specialists and many of these species use glucosinolates which are found throughout Brassica plants as attractants for feeding or ovipositional stimuli. They are important limiting factors to the production of Brassica oilseeds and their control may form a significant portion of costs. There is a wide range of pest species associated with Brassicas and these may be partitioned into pests which affect the crop at establishment, during growth, and in the run-up to harvest.

At establishment, flea beetles (*Phyllotreta* spp.) may attack spring crops and cause typical *shot-holing* of the cotyledons and leaves. They are a problem particularly in warm, dry conditions. Slugs can also cause significant damage at the cotyledon and early leaf stage and are associated with wet conditions on heavy soils. Problems with slugs may be exacerbated by minimal cultivation techniques where crop trash persists from the previous crop.

During growth a range of pest problems may occur. Cabbage-stem flea beetle (*Psylliodes chrysocephala*) is one of the most important pests on winter-sown oilseed rape in Europe (Ekbom, 1995). The larvae of the beetle eat into the leaf stalk and base of the plant during autumn and winter causing in some cases the plant to die during winter. Several species of aphid can cause damage. When aphid populations are high, direct damage due to sucking can occur and *Myzus persicae* can also act as a virus vector for beet western yellow virus particularly in the autumn. Larvae from a range of *Lepidoptera* species can feed on leaves, but damage is frequently limited particularly as the crop matures.

As the crop approaches harvest, a further range of pests occur. Pollen beetles (*Meligethes* spp.) are the most significant pests of oilseed rape in Scandinavia (Nilsson, 1987) and are the most common pests of the Scottish oilseed rape crop (Evans, 2001). Adult beetles feed on pollen from buds and flowers, also eggs are laid inside the buds and larvae eat stamens causing abortion of the bud and pod. They are more important for spring-sown crops, as the later flowering of this type coincides more closely with oviposition. Seed weevil (*Ceuthorhynchus assimilis*), in Europe and the North America, and pod midge (*Dasinaura brassicae*), in Europe, only lay eggs into the pods, the larvae from both species feeding on the developing seeds. The presence of pod midge larvae can cause splitting of pods.

For several pest species such as cabbage-stem flea beetle and pollen beetle, economic thresholds for number of pests present should be used in order to both achieve optimal control of the pest and also minimise environmental consequences of insecticide application. Effective and regular monitoring of crops throughout the season is necessary. Cultivation of crops in fields next to the previous season's oilseed rape and growing winter- and spring-sown types next to each other should be avoided in order to reduce problems caused by pests such as pollen beetle and cabbage-stem weevil, which can readily migrate from field to field (Evans, 2001). Better knowledge of factors stimulating or attracting insects and also deterring pests is developing and this should enable more targeted plant breeding in the future.

*1.5.3 Crop protection – diseases*

Oilseed rape is affected by a range of diseases and the importance of these vary from area to area. Sclerotinia stem rot (*Sclerotinia sclerotiorum*) and stem canker (*Leptosphaeria maculans*), also known as black leg, are the major diseases of oilseed rape in Canada (Rimmer and Buckwaldt, 1995). They are also commonly found in Europe, where there are also a number of additional important diseases. Verticillium wilt is a particular problem in Sweden and Germany, light leaf spot (*Pyrenopeziza brassicae*) in northern parts of Europe and clubroot (*Plasmodiophora brassicae*) in Scandinavian countries. In China, sclerotinia stem rot is found to be a major disease and viral disease also causes substantial yield losses. Alternaria (*Alternaria brassicae*) causes problems for *B. rapa* in northern parts of India and simultaneous infection of white rust (*Albugo candida*) and downy mildew (*Perenospora parasitica*) is common on both *B. rapa* and *B. juncea* in this region.

Control of disease has involved a range of strategies. It has been possible to utilise natural variation to breed varieties, particularly *B. napus*, with improved resistance to a number of diseases including light leaf spot and stem canker. For other diseases like alternaria, stem rot, and white rust, availability of resistant lines is more limited. Cultural control methods, particularly rotation, are important means of controlling diseases such as clubroot and sclerotinia, and good agronomic practice will limit the number of susceptible crops in the rotation. Agro-chemical use may also be part of the control of Brassica diseases; seed treatments and foliar applications are both routinely used for the control of alternaria in India, and fungicide spray programmes are also utilised to control light leaf spot and stem rot in Europe.

## 1.6 Maturity and harvesting

Oilseed rape is regarded as mature when all the seeds have turned black and the moisture content of the seeds is less than 15%. Harvesting too early, before seeds have matured sufficiently, may increase chlorophyll levels in the oil thereby reducing quality. For storage of the harvested crop, a moisture content of 9% is required and artificial drying will be needed if the seed is above this level. The crop is at risk of high losses due to the small size of the seeds and also because of the growth habit of the crop, with differing maturity of individual branches of the plants particularly in northern areas where there is a slow maturation phase.

Direct combining with no harvest pre-treatment may be possible in certain circumstances where the crop reaches an even and early maturity. In many cases however, either swathing or desiccation prior to combining is needed. Swathing is used for the bulk of the Canadian crop (Pouzet, 1995) and much of the European crop, and involves cutting the stem of the crop and leaving the upper part of the plants to mature and dry on the stubble platform. This method is well suited to areas at risk of high winds which could cause unacceptable losses in a standing crop. Desiccation using a range of different chemical desiccants including glyphosate, diquat, or glufosinate ammonia allows the seed to mature in the still standing crop. This latter method may be most suited to areas prone to damp conditions at harvest when air movement through the crop is restricted.

## 1.7 Production and trade for oilseeds and oil

The following discussion on production and trade for rapeseeds and rapeseed oil is based on the data presented in Tables 1.1–1.6 collated from Oil World Annuals for 2002 and 2003, and from Revised Oil World 2020.

The figures in Tables 1.2–1.5 cover the seven-year period from 1996/97 to 2002/03 with some forecasts for 2003/04. The production of seed increased steadily over many years and reached a maximum of 42.6 million tonnes in

**Table 1.1** Average production (million tonnes) of rapeseeds and rapeseed oil for selected five-year periods and comparison with totals for 10 oilseeds and 17 oils and fats

| 5-year period | 10 seeds | Rapeseed | 17 oils and fats | Rapeseed oil |
| --- | --- | --- | --- | --- |
| 1990/91–1994/95 | 232.4 | 27.29 (11.7%) | 86.8 | 9.66 (11.1%) |
| 1995/96–1999/00 | 280.1 | 35.49 (12.7%) | 105.1 | 12.64 (12.0%) |
| 2000/01–2004/05* | 336.6 | 42.27 (12.6%) | 126.5 | 15.34 (12.1%) |
| 2005/06–2009/10* | 381.7 | 48.27 (12.6%) | 146.7 | 17.72 (12.1%) |

* Forecasts.
*Source*: Mielke (2002b).

1999/00. This resulted from a peak in the area devoted to rapeseed production and from a high average yield. In particular, it reflects a large production from the high yielding European countries. Since then, while yields have been maintained, the area under cultivation has fallen and this has reduced production. This fall is probably linked with the low prices for all oils and fats during the three-year period from 1999/00 to 2001/02. In 2001/02, rapeseed production was 11.4% of the total production from ten major oilseeds and rapeseed oil production was 11.4% of the total production from 17 oils and fats. In 2002/03 both these figures were 10.1%.

The figures in Table 1.1 cover production over the past ten years and predictions for the next ten years for rapeseed and rapeseed oil. Both the seeds and the oil are expected to maintain a level of around 12% of the increasing totals for seeds and for oils and fats.

## 1.8 Rapeseed

Table 1.2 contains summarising data for rapeseeds and rapeseed oil. Normally about 95% of the seed is crushed to give oil (39%) and seed meal (60%). Material is exported either as seed for local crushing or as oil. Over the seven-year period, exported seeds have been 12–21% of total seed production, and exported oil has varied between 9 and 17% of total oil production. Expressing exported seeds as oil equivalent and adding this to exported oil, the total exports are 28–38% of total oil. The balance (62–72%) represents oil consumed in the countries in which the seed is grown. Imports are not discussed at this stage but total imports have to be virtually the same as total exports.

**Table 1.2** World totals (million tonnes) for production and exports of rapeseeds and rapeseed oil from 1996/97 to 2002/03 with forecasts for 2003/04

| Year | 1996/97 | 1997/98 | 1998/99 | 1999/00 | 2000/01 | 2001/02 | 2002/03 | 2003/04 |
|---|---|---|---|---|---|---|---|---|
| **Rapeseed** | | | | | | | | |
| Production | 31.00 | 33.11 | 36.13 | 42.56 | 37.53 | 36.66 | 32.53 | 37.22 |
| Harvest area[a] | 22.0 | 23.3 | 24.8 | 27.2 | 24.7 | 23.9 | 22.5 | 25.0 |
| Yield[b] | 1.41 | 1.42 | 1.46 | 1.56 | 1.52 | 1.53 | 1.44 | 1.49 |
| Crushing | 29.81 | 31.50 | 32.55 | 37.18 | 35.76 | 34.23 | 31.56 | 34.30 |
| Exports[c] | 3.79 | 4.96 | 7.54 | 8.93 | 7.53 | 5.48 | 4.84 | – |
| **Rapeseed oil** | | | | | | | | |
| Production | 11.48 | 12.22 | 12.68 | 14.52 | 13.98 | 13.47 | 12.47 | 13.55 |
| Consumption | 11.52 | 12.12 | 12.71 | 14.41 | 14.20 | 13.66 | 12.54 | 13.50 |
| Exports[c] | 1.80 | 2.11 | 1.92 | 1.94 | 1.40 | 1.29 | 1.11 | 1.30 |

[a] million hectares.
[b] tonnes/hectare.
[c] Imports are not cited – they are virtually the same as exports.
*Source:* Mielke (2002a).

**Table 1.3** Major producing, exporting, and importing countries of rapeseeds in million tonnes from 1996/97 to 2002/03

| Year | 1996/97 | 1997/98 | 1998/99 | 1999/00 | 2000/01 | 2001/02 | 2002/03 |
|---|---|---|---|---|---|---|---|
| **Rapeseed production** | | | | | | | |
| World | 31.00 | 33.11 | 36.13 | 42.62 | 37.53 | 36.66 | 32.53 |
| EU-15 | 7.34 | 8.73 | 9.52 | 11.47 | 8.95 | 8.87 | 9.34 |
| Germany | 2.18 | 2.87 | 3.39 | 4.28 | 3.59 | 4.16 | 3.87 |
| France | 2.90 | 3.50 | 3.73 | 4.47 | 3.48 | 2.87 | 3.32 |
| UK | 1.41 | 1.53 | 1.57 | 1.74 | 1.16 | 1.16 | 1.47 |
| Cent Europe | 1.30 | 1.53 | 2.02 | 2.77 | 2.23 | 2.62 | 2.26 |
| Czech Rep. | 0.53 | 0.56 | 0.68 | 0.93 | 0.84 | 0.97 | 0.71 |
| Poland | 0.45 | 0.59 | 1.10 | 1.13 | 0.96 | 1.06 | 0.99 |
| Canada | 5.06 | 6.39 | 7.64 | 8.80 | 7.21 | 5.15 | 3.95 |
| USA | 0.22 | 0.35 | 0.71 | 0.62 | 0.92 | 0.91 | 0.71 |
| China | 9.20 | 9.58 | 8.30 | 10.13 | 11.38 | 11.32 | 10.53 |
| India | 6.30 | 4.65 | 5.00 | 5.10 | 3.75 | 4.85 | 3.70 |
| Australia | 0.64 | 0.86 | 1.76 | 2.43 | 1.78 | 1.63 | 0.79 |
| **Rapeseed exports** | | | | | | | |
| World | 3.79 | 4.96 | 7.54 | 8.90 | 7.53 | 5.48 | 4.83 |
| EU-15 | 0.50 | 0.69 | 1.33 | 1.19 | 0.34 | 0.55 | 0.78 |
| Cent Europe | 0.18 | 0.24 | 0.74 | 0.98 | 0.76 | 0.70 | 0.57 |
| Canada | 2.55 | 3.17 | 3.71 | 4.29 | 4.48 | 2.46 | 2.50 |
| Australia | 0.31 | 0.57 | 1.35 | 2.00 | 1.41 | 1.36 | 0.47 |
| **Rapeseed imports** | | | | | | | |
| World | 3.81 | 4.30 | 7.62 | 9.18 | 7.63 | 5.64 | 4.78 |
| EU-15 | 0.36 | 0.32 | 1.06 | 0.84 | 1.04 | 0.75 | 0.72 |
| Mexico | 0.50 | 0.68 | 0.85 | 0.98 | 0.86 | 0.96 | 0.60 |
| China | – | 0.32 | 2.52 | 3.58 | 2.36 | 0.77 | 0.15 |
| Japan | 2.00 | 2.09 | 2.17 | 2.23 | 2.18 | 2.09 | 2.02 |
| Pakistan | 0.03 | 0.02 | 0.15 | 0.50 | 0.30 | 0.37 | 0.53 |
| Bangladesh | 0.12 | 0.20 | 0.20 | 0.28 | 0.21 | 0.18 | 0.17 |

*Source*: Mielke (2002a).

**Table 1.4** Yields of rapeseed (tonnes/hectare) from 1996/97 to 2002/03

| Year | 1996/97 | 1997/98 | 1998/99 | 1999/00 | 2000/01 | 2001/02 | 2002/03 |
|---|---|---|---|---|---|---|---|
| World | 1.41 | 1.42 | 1.46 | 1.56 | 1.52 | 1.53 | 1.44 |
| EU-15 | 2.80 | 3.11 | 3.08 | 3.22 | 2.98 | 2.97 | 3.03 |
| Cent Europe | 1.84 | 2.06 | 2.30 | 2.15 | 2.13 | 2.38 | 2.13 |
| Canada | 1.47 | 1.32 | 1.41 | 1.58 | 1.48 | 1.37 | 1.34 |
| USA | 1.55 | 1.39 | 1.62 | 1.46 | 1.50 | 1.54 | 1.37 |
| China | 1.37 | 1.48 | 1.27 | 1.47 | 1.52 | 1.51 | 1.37 |
| India | 0.92 | 0.73 | 0.83 | 0.84 | 0.75 | 0.90 | 0.72 |
| Australia | 1.52 | 1.26 | 1.50 | 1.27 | 1.31 | 1.46 | 1.07 |

*Source*: Mielke (2002a).

**Table 1.5** Major producing, exporting, and importing countries of rapeseed oil in million tonnes from 1996/97 to 2002/03

| Year | 1996/97 | 1997/98 | 1998/99 | 1999/00 | 2000/01 | 2001/02 | 2002/03 |
|---|---|---|---|---|---|---|---|
| **Rapeseed oil production** | | | | | | | |
| World | 11.48 | 12.22 | 12.68 | 14.52 | 13.98 | 13.47 | 12.40 |
| EU-15 | 2.93 | 3.46 | 3.54 | 3.84 | 3.53 | 3.67 | 3.57 |
| Germany | 1.36 | 1.67 | 1.68 | 1.88 | 1.76 | 1.85 | 1.82 |
| France | 0.38 | 0.46 | 0.46 | 0.54 | 0.52 | 0.59 | 0.60 |
| UK | 0.61 | 0.67 | 0.64 | 0.62 | 0.57 | 0.59 | 0.57 |
| Cent Europe | 0.58 | 0.55 | 0.64 | 0.62 | 0.64 | 0.66 | 0.68 |
| Czech Rep. | 0.19 | 0.18 | 0.18 | 0.16 | 0.22 | 0.22 | 0.19 |
| Poland | 0.32 | 0.29 | 0.37 | 0.33 | 0.34 | 0.32 | 0.36 |
| Canada | 1.14 | 1.37 | 1.26 | 1.27 | 1.22 | 0.97 | 0.88 |
| USA | 0.17 | 0.26 | 0.28 | 0.31 | 0.33 | 0.30 | 0.28 |
| China | 3.01 | 3.11 | 3.67 | 4.72 | 4.67 | 4.27 | 3.74 |
| Japan | 0.84 | 0.86 | 0.90 | 0.92 | 0.90 | 0.87 | 0.83 |
| India | 2.17 | 1.86 | 1.43 | 1.66 | 1.59 | 1.69 | 1.37 |
| Australia | 0.11 | 0.12 | 0.15 | 0.15 | 0.15 | 0.15 | 0.13 |
| **Rapeseed oil exports** | | | | | | | |
| World | 1.80 | 2.11 | 1.92 | 1.91 | 1.40 | 1.29 | 1.11 |
| EU-15 | 0.60 | 0.77 | 0.71 | 0.61 | 0.20 | 0.36 | 0.29 |
| Canada | 0.67 | 0.71 | 0.74 | 0.79 | 0.74 | 0.56 | 0.51 |
| USA | 0.13 | 0.16 | 0.12 | 0.13 | 0.08 | 0.12 | 0.07 |
| **Rapeseed oil imports** | | | | | | | |
| World | 1.64 | 2.10 | 2.04 | 1.96 | 1.38 | 1.32 | 1.13 |
| USA | 0.50 | 0.50 | 0.50 | 0.53 | 0.54 | 0.50 | 0.43 |
| Ex-USSR | 0.14 | 0.28 | 0.22 | 0.20 | 0.10 | 0.07 | 0.06 |
| **Rapeseed oil consumption** | | | | | | | |
| World | 11.52 | 12.12 | 12.71 | 14.43 | 14.20 | 13.66 | 12.54 |
| EU-15 | 2.43 | 2.63 | 2.80 | 3.21 | 3.42 | 3.34 | 3.32 |
| Cent Europe | 0.60 | 0.56 | 0.60 | 0.67 | 0.64 | 0.69 | 0.70 |
| Canada | 0.56 | 0.63 | 0.61 | 0.60 | 0.54 | 0.46 | 0.42 |
| USA | 0.54 | 0.59 | 0.64 | 0.70 | 0.83 | 0.71 | 0.65 |
| Mexico | 0.25 | 0.31 | 0.42 | 0.42 | 0.40 | 0.40 | 0.33 |
| China | 3.26 | 3.45 | 3.91 | 4.68 | 4.70 | 4.34 | 3.77 |
| India | 2.13 | 2.02 | 1.62 | 1.83 | 1.65 | 1.74 | 1.43 |
| Japan | 0.85 | 0.86 | 0.90 | 0.93 | 0.92 | 0.89 | 0.86 |

*Source*: Mielke (2002a).

Table 1.3 gives details of the production of rapeseed and its exports and imports by country/region. In most years, China is the dominant producer and in some years additional supplies of seed are imported into that country for local crushing. Recent falls in imports probably reflect the tightness of supply of rapeseed compared with competing supplies of soybeans, soybean oil, and palm oil. EU-15 is the second largest producer of seeds, especially in Germany, France, and UK. Europe is a net importer of oils and fats and is anxious to improve native supplies of oils and fats from rape, sunflower, and olive, which

**Table 1.6** Consumption of rapeseed oil in selected countries as a % of total consumption of oils and fats in those countries during 2001 and 2002

|  | 2001 | 2002 |
|---|---|---|
| Czech Rep. | 44.6 | 45.1 |
| Poland | 33.8 | 35.3 |
| Canada | 33.4 | 29.8 |
| UK | 32.7 | 28.9 |
| Japan | 31.7 | 30.5 |
| Germany | 30.2 | 31.1 |
| France | 23.2 | 22.1 |
| China | 21.8 | 18.0 |
| EU-15 | 17.8 | 17.4 |
| Mexico | 15.7 | 12.5 |
| India | 14.1 | 11.8 |
| USA | 4.8 | 4.4 |

*Source*: Mielke (2002a).

are themselves obtained from different countries within Europe. Rapeseed oil is important in the European economy and there is much research being undertaken to improve both quality and quantity (Leckband *et al.*, 2002). Low erucic rapeseed oil was first developed in Canada where seeds meeting the defined standards for erucic acid and glucosinolates were called canola (Section 1.2). Production in Canada grew rapidly and because of its modest population (31 million) it became a significant exporter of both seeds and oil. Interest in canola has spilled over into USA. Production of seed is increasing at the same time as there is a significant import of oil. India grows important amounts of rape (mainly as mustard). The levels fluctuate with local conditions (mainly climatic) and it is all consumed locally. Rape also grows successfully in Central Europe, especially in Poland and the Czech Republic. Exports from this region are increasing. Rapeseed is also grown in Australia and, because this is another country with a small population (19–20 million), seed is available for export, particularly to China and Japan. Recent decline in production is due to a fall in the area under cultivation and due to the vagaries of climate.

Table 1.4 provides information on yields of rapeseed in the major producing countries. It is obvious that these vary markedly. The highest yields are produced in EU-15 (around 3 t/ha), followed by Central Europe (above 2 t/ha), Canada, USA, China, and Australia (each about 1.5 t/ha), and India (below 1 t/ha). The marked differences in yield in the various areas of production have an effect on average world yields. Increasing production in EU-15 and Central Europe will raise world yields and contrariwise, increasing production in India will pull world yields down. These changes in numbers occur without any underlying change of yield in individual countries. They reflect only a change in the total mix.

Exports of seeds over the six years covered in Table 1.3 have ranged from 3.8 to 8.9 million tonnes. This has come predominantly from Canada (generally about 50% of that country's production) and also from Australia and Central Europe.

Japan, with little or no indigenous supplies of oilseeds, has been a regular importer of around 2 million tonnes of seeds for local crushing. China has also been a significant, but less regular, importer of rapeseed and EU-15 and Mexico each import almost 1 million tonne each year.

## 1.9 Rapeseed oil

Production of rapeseed oil rose sharply for a number of years and peaked in 1999/00 at 14.5 million tonnes since when it has fallen back somewhat. Details for a seven-year period are given in Table 1.5. The major producers are China (30.2% of total production in 2002/03), EU-15 (28.8%), India (11.0%), and Canada (7.1%).

Exports of rapeseed oil have fallen recently and are now less than 10% of total rapeseed oil production. Only Canada and EU-15 are now significant exporters of the oil. European exports have probably fallen because of the increasing use of this commodity to make biodiesel (rapeseed oil methyl esters) (see Chapter 7). Capacity to make the esters was greatly increased in the years of big harvests and the declining harvests which followed led to a cut in exports in order to meet the oleochemical demand. Imports of oil are very widely spread with the USA making the largest demand.

The final part of Table 1.5 shows the major consumers of rapeseed oil. The largest are now China (30.1% of world production in 2002/03), EU-15 (26.5%), and India (11.4%). The situation in Japan is interesting. Japan grows no rapeseed but imports and extracts the seed it needs for its own use, all at levels that are remarkably steady over the six years covered in the tables. Table 1.6 shows how important a commodity rapeseed oil is in those countries in which it is grown. There are six countries in which rapeseed oil exceeds 30% of the total oil used in those countries.

Fry (2001) has reported that for the quarter century, 1976–2000, the trend growth rate for rapeseed oil was 7.3% resulting from increases of 4.4% in harvested area and 2.4% in oil yield.

## References

Anon. (2003) Canola standards and regulations. <Canola-council.org/pubs/standards.html>.

Auld, D.L., Bettis, B.L. and Dial, M.J. (1983) Planting date and cultivar effect on winter rape production. *Agronomy Journal*, **76**, 197–200.

Ekbom, B. (1995) Insect pests. In: *Brassica Oilseeds, Production and Utilization* (eds D. Kimber and D.I. McGregor). CAB International, Wallingford, UK, pp. 141–152.

Evans, A. (2001) Pests of oilseed rape: a Scottish perspective. *SAC Technical Note* T511.

Fry, J. (2001) The world's oils and fat needs in the 21st century: lessons from the 20th century. Lecture presented at a meeting of the Oils and Fats Group of the Society of Chemical Industry in Hull, England.

Harker, K.N., Blackshaw, R.E., Kirkland, K.J., Derkson, D.A. and Wall, D. (2000) Herbicide-tolerant canola: weed control and yield comparisons in western Canada. *Canadian Journal of Plant Science*, **80**, 647–654.

Heresbach, K. (1570) Rei rustica libri quator. Cologne (translated by G. Markam, 1631, London).

Leckband, G., Radess, H. and Frauen, M. (2002) NAPUS 2000 – a research project to produce functional foods from rapeseed. *Lipid Technology Newsletter*, **8**, 79–83.

Li, C.S. (1980) Classification and evolution of mustard crops (*Brassica juncea*) in China. *Cruciferae Newsletter*, **5**, 33–36.

Lunn, G. and Spink, J. (2000) Effective oilseed rape canopies. *HGCA Topic Sheet*, No. 37, Summer 2000.

Mendham, N.J. and Salisbury, P.A. (1995) Physiology: crop development, growth and yield. In: *Brassica Oilseeds, Production and Utilization* (eds D. Kimber and D.I. McGregor). CAB International, Wallingford, UK, pp. 11–64.

Mielke, T. (2002a) Oil World Annuals 2002 and 2003, ISTA Mielke GmbH, Hamburg.*

Mielke, T. (2002b) The Revised Oil World 2020: Supply, Demand, and Prices, ISTA Mielke GmbH, Hamburg.*

*ISTA Mielke GmbH of Hamburg, Germany, produces weekly, monthly, annual, and occasional issues devoted to the production and use of 12 oilseeds, 17 oils and fats, and 10 oilmeals.

Nilsson, C. (1987) Yield losses in summer rape caused by pollen beetles (*Meligethes* spp.). *Swedish Journal of Agricultural Research*, **17**, 105–111.

Ogilvy, S.E. (1984) The influence of seed rate on population structure and yield of winter oilseed rape. *Aspects of Applied Biology*, **6**, 59–66.

Orson, J. (1995) Weeds and their control. In: *Brassica Oilseeds, Production and Utilization* (eds D. Kimber and D.I. McGregor). CAB International, Wallingford, UK, pp. 93–109.

Pouzet, A. (1995) Agronomy. In: *Brassica Oilseeds, Production and Utilization* (eds D. Kimber and D.I. McGregor). CAB International, Wallingford, UK, pp. 11–64.

Prakash, S. (1980) Cruciferous oilseeds in India. In: *Brassica Crops and Wild Allies* (eds S. Tsunoda, K. Hinata and C. Gomez-Campo), Japan Scientific Press, Tokyo, Japan, pp. 151–163.

Rimmer, S.R. and Buckwaldt, L. (1995) Diseases. In: *Brassica Oilseeds, Production and Utilization* (eds D. Kimber and D.I. McGregor). CAB International, Wallingford, UK, pp. 111–140.

Topinka, A.R.C., Downey, R.K. and Rakow, G.F.W. (1991) Effect of agronomic practices on the overwintering of winter canola in southern Alberta. *Proceedings 8th International Rapeseed Congress* (ed. D.I. McGregor), Saskatoon, Canada, pp. 665–670.

Walker, K.C. and Dawson, C.J. (2002) Sulphur fertiliser recommendations in Europe. *The International Fertiliser Society*, UK, Proceedings 506, 20pp.

Walker, R.L., Walker, K.C. and Booth, E.J. (2000) What does the future hold for GM crops? *Aspects of Applied Biology*, **62**, Farming systems for the new Millennium, 173–180.

Wibberley, J. (1996) A brief history of rotations, economic considerations and future directions. *Aspects of Applied Biology*, **47**, 1–10.

# 2 Extraction and refining
E.J. Booth

## 2.1 Introduction

Rapeseed is grown principally for the oil which is extracted from the seed and has both food and industrial applications. The solid residue or meal left after oil extraction is, however, also an important product as a high protein animal feedstuff and processing takes into account both the oil and the meal.

All oilseed extraction processes have certain objectives in common. They aim to maximise the oil yield, minimise damage to the oil and meal and minimise undesirable impurities in both fractions. There are a range of scales and methods of processing undertaken. Very small-scale mechanical extraction systems may press a few kilograms of seed per day and are perhaps more common in the developing world. Large-scale commercial processes use mechanical and possibly also solvent processing methods to ensure maximum oil extraction and are capable of crushing up to several thousand tonnes of seed per day.

In this chapter, extraction refers to the removal of the crude oil from the seed and treatments undertaken on both oil and meal in this process. The most commonly used commercial means of oil extraction are described with new technologies discussed where appropriate. The processing of this crude oil to make it suitable for consumption as liquid oil is also described. Refining describes the removal of impurities from the oil to give refined oil. Further processing of the liquid oil into a modified product is dealt with in another chapter.

## 2.2 Oil extraction steps

Current large-scale commercial rapeseed oil extraction or crushing involves a number of steps including seed cleaning, optional tempering and dehulling, flaking, conditioning, mechanical extraction by pre-pressing and extrusion, and/or expansion, frequently followed by solvent extraction (Fig. 2.1).

### 2.2.1 Pre-treatment

#### 2.2.1.1 Seed cleaning
Seed cleaning is necessary to remove foreign matter such as portions of the plant, weed seeds, other grains, soil material and dust before processing. Treatment typically consists of three basic steps: aspiration, screen separation to

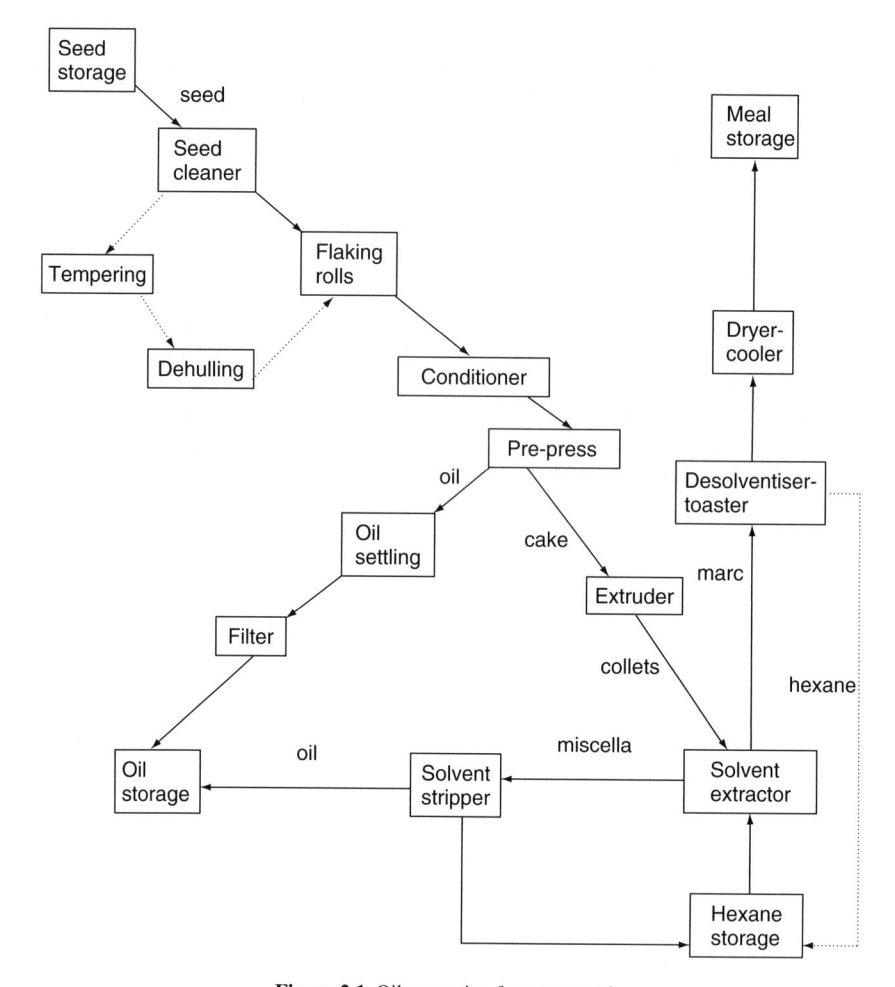

**Figure 2.1** Oil extraction from rapeseed.

remove oversized particles and screen separation to remove undersized particles. In general, rapeseed is cleaned to less than 2.5% dockage before processing (Unger, 1990).

### 2.2.1.2 Tempering

In many extraction plants, and particularly in colder climates, the cleaned seed is tempered, or heated, to 30–40°C prior to processing. This helps to avoid the problem of shattering in the flaking unit associated with cold seed, which can lead to irregular sizes of flakes thereby reducing extraction efficiency and producing fine particles which are difficult to remove. Tempering is carried out either by indirect application of heat in a rotary kiln with steam-heated tubes or by direct hot air contact which can be carried out in units similar to grain driers. A retention time of 30–45 minutes is used in combination with gentle heating to uniformly and thoroughly heat all rapeseeds. Seed moisture adjustment may also be carried out at this stage by regulating the airflow through the rotary kiln.

### 2.2.1.3 Dehulling

Removal of the hull of rapeseed is expensive due to the small size of the seeds. The removal of hull produces meal with lower fibre and higher protein content. Dehulling reduces the carry-over of some impurities such as hull pigments to the crushing process and reduces subsequent processing costs. It is, however, estimated that dehulled meal would have to attract a premium of 15% to make dehulling viable on a commercial basis (Carr, 1995) and dehulling presently has limited economic viability. Dehulling is achieved by using mechanical or pneumatic impact separation of the hull from the seed and air aspiration and/or fluidised bed sorter to remove the hulls.

### 2.2.1.4 Flaking

Cleaned rapeseed is flaked to facilitate oil extraction by rolling to break open the seed coat and rupture a portion of the oil cells. Lipid particles are dispersed throughout the intact rapeseed surrounded by cell membranes or cell walls. Rupture of the cell walls allows lipid particles to migrate to the outer surface of the flake where they can be separated from the solid residue. The rupturing of the cell wall also allows solvent (in the later processes of extraction) to penetrate the seed material, to dissolve and progressively dilute the viscous lipid portions which can then flow out of the cell structure to the outer surfaces of the flake. The large surface area of the flake allows easy contact of the lipid particles for efficient oil extraction.

Flaking is achieved by passing the seed through one or two pairs of cast iron rollers, typically 500–800 mm in diameter and 1000–1500 mm long. These revolve on large swivel-suspension roller bearings with one roll operated at 2–5% higher rpm so that the roll surfaces wipe each other and shear the seeds into flakes. To ensure that flakes are evenly distributed along the complete width of the flaking rollers, vibrating feeders are used to spread the seed across the rollers. The separation distance between rollers is precisely adjusted to give the desired flake thickness. Scrapers on the rollers prevent the relatively high oil content flakes sticking to the surface of the rollers. The thickness of the flakes is important for the efficiency of the oil extraction and a flake thickness of 0.30–0.38 mm generally gives good results (Carr, 1995). Thin flakes lead to better oil extraction as distances of diffusion of solvent and oil out of the flake are reduced. However, flakes thinner than 0.20 mm are very fragile and small particles may contaminate the oil and be difficult to remove during oil filtration. Flakes thicker than 0.4 mm may be associated with lower oil yield.

Some plants carry out flaking in two successive steps with the first roll cracking the seed and providing flakes of 0.4–0.7 mm thickness (Unger, 1990). The second operation flakes the seed to a thickness of 0.2–0.3 mm. This two-stage operation reduces the chance of any seed passing over the ends of the rolls and remaining whole, which would impede subsequent oil extraction. Limiting the thickness reduction in the first rolling step can counteract the resilience associated with rapeseed and ensure production of good-quality flakes. The flaking operation is carried out on a continuous basis and flaking mill units are capable of processing up to 450 t/day. It is important to ensure that subsequent conditioning occurs immediately to prevent the need for storage of flakes.

### 2.2.1.5 Conditioning

Thermal treatment, known as conditioning or cooking, is required to help breakdown the oil bodies within the seed. Heating the rapeseed flakes to 75–85°C serves a number of important functions. It ruptures the remaining oil cells and promotes coalescing of minute lipid particles to larger oil droplets. This change can be observed by comparing the uncooked flakes, which do not have an oily appearance, with the cooked flakes which look and feel oily. Oil viscosity is also reduced allowing oil to be separated from solids in the following processing.

Cooking also allows the moisture content of the flakes to be adjusted to that required during the subsequent mechanical oil extraction phase. A moisture content of 5–6.5% is needed for feedstock to many screw presses.

Additionally, conditioning achieves deactivation of enzymes within the seeds. Myrosinase is of particular interest as, in the presence of moisture, it is responsible for the breakdown of glucosinolates within the seed to products such as isothiocyanates and nitriles which are harmful when fed to animals. Glucosinolate breakdown can also lead to a release of the sulfur to the oil, which can compromise the efficiency of the nickel catalyst used during subsequent hydrogenation processing. Lipases, enzymes responsible for the breakdown of triacylglycerols and phospholipids, are also deactivated during cooking.

Conditioning is carried out using either drum- or stack-type conditioners. Drum conditioners give higher heat transfer rates, but may result in damage to the fragile flakes. A stack conditioner vessel consists of a series of 4–8 vertical, cylindrical, steam-heated steel kettles. Each kettle is 30–50 cm in diameter and 50–70 cm high. Conditioner capacities can range from 50 to 1000 t/day. Flaked seed enters the top of the conditioner. Myrosinase is destroyed in this top kettle and a rapid heating to above 80°C along with careful moisture control at 6–10% is necessary for optimum heat inactivation (Carr, 1995). Blades, called sweeps, are fixed to a vertical shaft and are positioned above each tray. These, along with a chute and gate mechanism, mix and push seed down the conditioner from kettle to kettle, maintaining a uniform depth of flakes on each tray. Flakes entering the conditioner are rapidly heated to 80°C and then maintained at 80–105°C during the conditioning phase. Excessive heat of over 100°C for extended periods is avoided as this can lead to protein damage. Careful heating, however, can improve the quality of the meal by increasing protein availability.

Cell wall degrading enzymes, namely cellulases, hemicellulases and pectinases are now considered as a form of seed pre-treatment in conventional rapeseed processing (Derksen et al., 1994). The enzymes lead to increased permeability of cell walls and encourage release of oil. As a result, solvent extraction times can be halved and mechanical pressing efficiency increased to 90%. Costs are still however too high for this pre-treatment to be viable.

### 2.2.2 Mechanical extraction

It is common practice for rapeseed to undergo an initial mechanical extraction to produce press cake with an oil content of below 20%, followed by solvent extraction to remove the bulk of the remaining oil. Pre-pressing aims to maximise oil removal from the flaked seeds and produce press cake of acceptable quality, by compressing small flakes into a press cake which will facilitate good solvent contact and percolation in the extractor (Unger, 1990).

Conditioned seed flakes are fed into a series of low-pressure continuous screw presses or expellers consisting of a rotating screw shaft within a cylindrical barrel and cage unit lined with bars. The flat bars, made of hardened steel, are set edgeways around the periphery of the cage and are positioned to allow cake solids to be retained in the barrel whilst the oil flows out of the barrel between the carefully spaced gaps of the bars. Pressure and heat are developed within the barrel by the rotating screw shaft working against an adjustable choke partially constricting the discharge of the cake from the barrel. This method prevents the use of excessive power, pressure and temperature.

Cake discharged from the pre-presser should be spongy, permeable and resistant to disintegration on its way to the solvent extractor. Moisture content of the cake is also important to achieve the correct consistency for solvent extraction. Consistencies which are either too dry and granular or too wet and sloppy should be avoided and the optimum moisture content should be 4–5%. Diffusion of solvent in the solvent extractor is greatly influenced by cake thickness that should be between 3.2 and 4.8 mm (Carr, 1995). Percolation of solvent down through the cake bed will be affected by size and thickness of cake fragments. Therefore to ensure that fragments of optimum size enter the solvent extractor, attention must be paid to cake thickness and durability. Oil content of flaked, conditioned seed can be reduced from 42% to approximately 16% using this process and units can press up to 350 t of flakes per day.

The use of an additional mechanical pressing stage, extrusion, has become more frequent over the last 15 years and is now incorporated into many rapeseed crushing plants (Buhr, 1990). Extrusion aims to facilitate enhanced solvent extraction by creating material which is more porous, allowing better performance in solvent extraction compared to press cake derived from conventional processing (Pickard, 2001). The process restructures the cake to increase bulk density, improves oil extractability, inactivates enzymes, removes free liquids/oils and fats from their solid components and cooks the protein constituent for satisfactory agglomeration.

The material will enter the extruder after exiting the pre-presser. Extruders consist of a barrel with a rotating shaft fitted with flights. Steam is added and the cake material is mixed by the action of breaker screws along the length of the barrel. Pressure, created by the injected steam and friction from the rotating shaft, raises the temperature. At the end of the barrel the material is pressed against small die openings in the end plate. The sudden release of pressure during discharge causes the material to *expand*, producing segments of porous material known as collets. The swelling of the collets liberates the oil with the vapourising moisture and creates pores within the collets. Once the water is vapourised, the oil is reabsorbed by the collets, but can be readily extracted in a solvent extractor due to the pore structure.

There is interest in using extruders in place of expellers due to their lower capital costs and also higher throughputs. Seed flakes are not required to be as thin as for expellers, as a greater percentage of oil cells can be ruptured thermally due to the high temperatures used in extrusion. Less oil will be removed however. After extrusion alone, the oil content of flakes will have been reduced from 42 to 25–30%.

*2.2.2.1 Oil extraction based solely on mechanical methods*

Some, particularly smaller-scale processing plants, use only mechanical extraction, without solvent extraction, to extract oil from rapeseed. This is less efficient than utilisation of solvent extraction and recoveries of oil are lower. Improvements can be made by the use of high-powered mechanical pressing, resulting in cake with a residue of 10% oil. A further press may extract another 1–2% oil. It should be noted that this requires a high power cost to achieve a high throughput and also necessitates setting up a large number of pressing machines which are technically demanding. Heat treatment during mechanical processing can increase the efficiency of oil yield from solely mechanical pressing, to leave a residue of 6–10% oil in the meal. However, the content of undesirable compounds such as phospholipids, pigments and sterols in the resulting oil is also increased by heat treatment.

The relatively high oil content of meal extracted by mechanical means alone, compared to solvent-extracted meal, may suppress the value of the meal if it is sold solely as a protein supplement to animal feed. Combined with the loss in oil extraction efficiency this makes mechanical extraction less attractive to crushers. Some processors have achieved recognition of the additional energy content of resulting meal due to higher oil residue and may obtain a higher meal value.

*2.2.2.2 Oil settling and filtering*

Oil liberated through the expeller contains around 3% solid matter. This is removed by gravity-settling in a screening tank for approximately 3 hours, with the temperature maintained at approximately 66°C. Settled solids are known as *foots* and arise as a result of pressure within the expeller on the conditioned flakes. The amount of foots can be minimised by optimisation of flaking and conditioning. Foots may be continuously dredged off and re-cycled to the conditioner for further pressing, or they may be re-pressed in a separate foots press. Oil derived from the foots press is re-cycled to the screenings tank for resettlement of suspended small solid particles known as *fines*. Remaining suspended fines in the oil can be removed by either filtration or centrifugation of the oil continuously drawn off the unfiltered oil tank. Oil from the extruder stage is also subject to settling and filtration processes. Cake exits the expeller at 83–94°C with 15–18% residual oil. It is processed through the cake sizer to reduce its size before entering the solvent-extraction plant.

*2.2.3 Solvent extraction*

Solvent extraction aims to remove as much oil as possible while minimising solvent loss. The most frequently used solvent over many decades is n-hexane because of its ready availability, oil solubility, water mixing behaviour, boiling point and heat of evaporation. It is, however, associated with several disadvantages compared to other oil extraction methods:

- Equipment required is more expensive, mechanical maintenance is more costly and power requirement is high.
- Hexane is highly flammable and in solvent-extraction plants, it is used in quantities in excess of 50 000 L, with a significant proportion at or near its boiling point. It is associated with fire and explosion risks and every year a solvent-extraction plant explodes somewhere in the world (Anon., 1998). Location in a separate building away from high temperature facilities is usually necessary.
- Due to the dusty nature of low oil content meal, there are also dust-explosion risks.

Nevertheless, where the oil content of the starting material is relatively high, as in rapeseed, it is found that at high volumes (e.g. 3000 t/day) use of solvent extraction is more economic allowing extraction of a greater proportion of oil.

At low volumes of seed (less than 150 t/day) it may be cheaper in terms of capital costs to utilise solely mechanical oil extraction (Booth *et al.*, 1993).

On application to the seed material, solvent dissolves the trapped oil and separates it from the solids as miscella (oil plus solvent). Most extractors operate countercurrently and continuously, moving the press cake and miscella in opposite directions. After sufficient intermixing, the extracted meal leaves the solvent extractor unit at one end and miscella at the other. Either a basket-percolation type extractor or a shallow-bed loop extractor can be used. For both types it is important to ensure that the cake is positioned to have full contact with the hexane and also that there is sufficient retention time to allow the large volumes of solvent to wash and separate the miscella with minimum solvent carry-over in the discharged solids.

With the basket-based extractor, solvent or miscella is flooded through the cake in five to eight stages, with each stage containing a higher solvent: oil ratio. The solvent percolates by gravity through the cake, saturating cake fragments. The miscella has a lower viscosity than oil alone, promoting the diffusion of oil into the miscella solution and then onto the surface of the cake. It is washed away by the flow of miscella, flowing out of the base of the basket, to be pumped to the next basket of cake. Immediately before discharge from the solvent extractor, the meal will be washed with pure solvent.

In the shallow-bed extractor, cake is conveyed on a perforated steel belt through the length of the extractor to be treated with a number of miscella and solvent sprays. Pure solvent is introduced just prior to the cake discharge section and this is circulated countercurrent to the flow of the cake via miscella pumps. The miscella becomes progressively richer in oil as it moves around the solvent extractor and is discharged at the cake inlet, where it contains a high proportion of oil. Hexane-saturated meal leaving the solvent extractor after the pure-solvent wash is known as marc and contains less than 1% oil.

Raised operating temperatures of the solvent extractor will encourage rapid extraction of oil but temperatures are limited to 50–55°C due to the vapour pressure of n-hexane. Higher temperatures will lead to an unacceptably high quantity of solvent vapour which must be recovered and re-used. The flammability of hexane gives rise to safety concerns and the need for careful management and plant maintenance. Before hexane is introduced into the system, and after a shutdown, many plants will purge all vessels of air using an inert gas to reduce the risk of fire or explosion.

*2.2.3.1 Solvent recovery*

Miscella discharged from the extractor contains approximately 25% oil, requiring separation from crude oil for further processing and hexane for re-utilisation. Oil is freed from the miscella by processing through a series of stills, stripping columns and condensers. After filtering and cooling, the hexane is discharged from the solvent-extraction plant for storage or further treatment. Air and vapours released during solvent extraction are scrubbed to remove traces of solvent before they can be discharged from the plant. Gravity separation is used to recover further solvent from this material, for further use in the solvent-extraction process. Losses of only 3.0 litres of solvent per tonne of rapeseed processed are consistently achieved in well-run plants (Unger, 1990).

*2.2.3.2 Desolventising – toasting*
Meal exiting the solvent extractor contains 30–35% solvent which must be removed before it is used as an animal feed. Solvent is removed from the meal by evaporation in a desolventiser-toaster, which also dries and crisps the meal.

Transport of meal in a closed conveyor system to the desolventiser-toaster is necessary to avoid premature loss of solvent. The desolventiser-toaster is an enclosed vessel consisting of a vertical stack of cylindrical gas-tight pans, which are each steam heated from the base. Meal enters this unit at approximately 57°C and is heated to 105°C, to drop by gravity from tray to tray via automatic doors. A rotating sweep arm above each tray ensures heat transfer, prevents extracted flakes from sticking to the tray and mixes extracted cake above the trays.

The solvent is gradually volatilised and recovered for further use. The addition of steam to the meal displaces the hexane absorbed by the protein, and this process continues in each tray progressively removing the solvent. Further heating can be applied to improve desolventising if necessary. The temperatures used in successive trays permit drying and toasting of the meal. At the end of the process, the meal is discharged to a drier–cooler at a temperature of approximately 100°C and a moisture content of 10–12%. At this point the meal is virtually solvent-free and has a moisture content of 15–18% with a lipid content of 1%. It is then dried to a moisture content of 8–10%, cooled and milled for delivery to feed manufacturers.

The use of high temperature and moisture in this process further inactivates any remaining myrosinase and lowers glucosinolate content through thermal decomposition. Nevertheless, it is important to maintain careful control of temperatures, moisture content and retention time in different parts of the oil extraction process to minimise amino acid damage and protect protein quality.

Environmental regulations are now demanding ever more stringent outcomes on solvent recovery and the additional retention time in the desolventiser-toaster necessary for stripping solvent to more exacting standards may compromise protein quality. Development of a new technology to increase steam density combined with slotted screen trays in the desolventiser-toaster offers scope to allow meal to be more effectively stripped of solvent within conventional retention times (Kemper, 2000).

### 2.2.3.3 Alternative solvents for oil extraction

With the drawbacks associated with hexane as a solvent for oilseed processing, much effort has been expended on the development of alternative solvents. Ideally such an alternative should have a high solvent power and selectivity for triacylglycerols, be easily removed from meal and oil, have a low flammability, be stable, non-reactive and non-toxic and have high purity (Derksen *et al.*, 1994). Many solvents have been evaluated for suitability but all have been associated with certain problems. Isopropanol has attracted attention. It has health and environmental advantages, having lower toxicity and lower flammability. However, it is also associated with several disadvantages which have limited further developments of its use in commercial extraction. Isopropanol has lower solvent power for triacylglycerols, a higher heat of vapourisation and higher boiling point than hexane, requiring a higher energy input. It also readily mixes with water needing careful control of seed moisture content to maintain solvent power.

Carbon dioxide under supercritical conditions of temperature (over 31°C) and pressure (over 73 bar) is also of interest as an alternative solvent. This option has the benefit of a lack of toxicity, ease of solvent recovery from miscella and meal, and non-flammability. Supercritical $CO_2$ is highly selective for triacylglycerols, minimising content of undesirable compounds such as phospholipids, pigments and sterols and reducing refining requirement. Use of supercritical $CO_2$ is limited by cost and difficulties in adoption to continuous extraction procedures preferred for large-scale processing.

### 2.2.4 Composition of crude oil

Crude oil obtained from the extraction of rapeseed consists of approximately 98% triacylglycerols, or esters resulting from the combination of one molecule of glycerol with three molecules of fatty acids. Phospholipids (also known as gums), free fatty acids, pigments, sterols, waxes, meal, oxidised materials, moisture and dirt account for the remaining 2%. Oil refining has the objectives of removing the impurities and maximising the yield of neutral oil, and also minimising both damage to oil quality and reduction of natural tocopherol antioxidant content. Phospholipids, free fatty acids and pigments are of greatest concern to the oil refiner, and techniques have been developed for straightforward removal of most impurities. Phospholipids are removed by degumming or alkali refining, free fatty acids and odour/flavour components by physical refining, and pigments by bleaching (Fig. 2.2). Waxes may be removed by winterisation and deodourisation may also be used to remove odour and flavour.

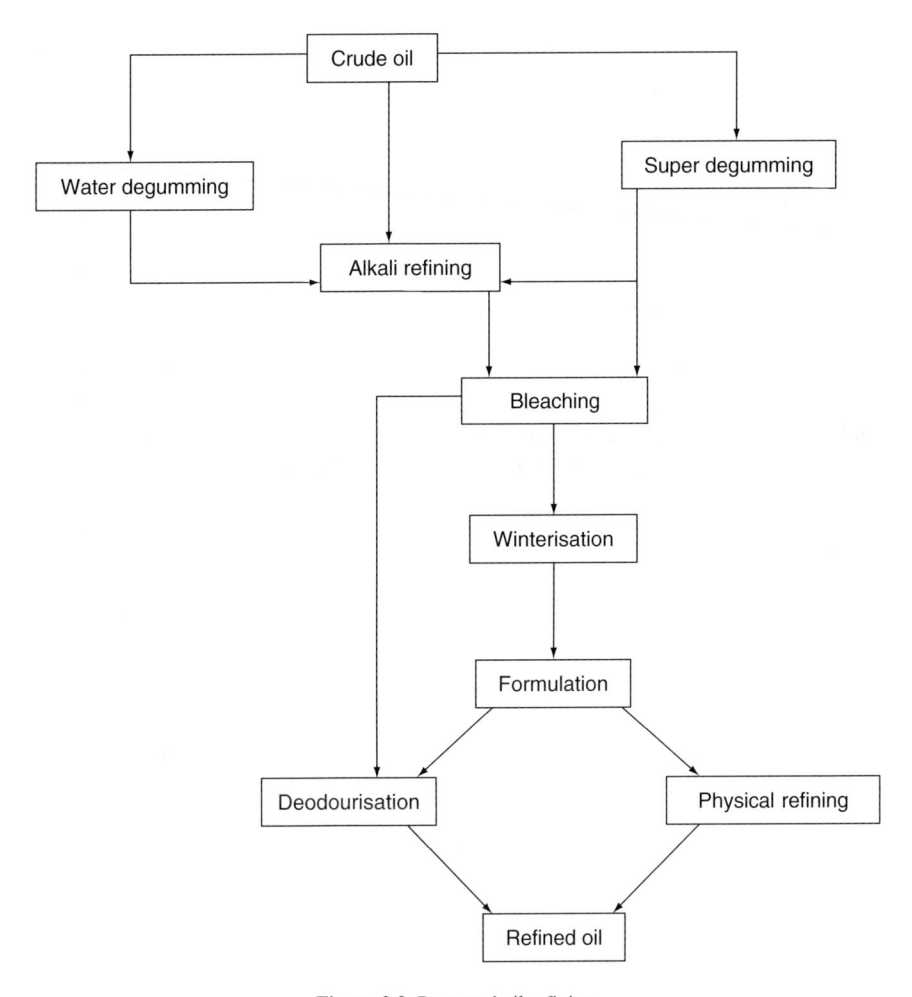

**Figure 2.2** Rapeseed oil refining.

## 2.3 Refining

The removal of phospholipids has been a major refining challenge over the years (Carr, 1995). A phospholipid molecule consists of one molecule of glycerol, two molecules of various fatty acids and phosphoric acid. Crude rapeseed oil contains approximately 1.25% phospholipid or 500 ppm phospholipid measured as phosphorus (Mag, 2001). Soluble phospholipids are likely to form emulsions, causing processing problems through incurring further costs due to chemical refining losses, or by settling as a sludge during storage. It is therefore necessary to remove them. To prevent settling during long periods of storage or transport, there are advantages in the removal of phospholipids just after extraction rather than in the refinery. Phospholipids were conventionally removed using alkali refining, but developments with water and acid degumming (also known as superdegumming), in combination with physical refining to remove free fatty acids, now provide an alternative. Several new technical approaches are also under development.

## 2.3.1 Degumming

It is possible to remove some phospholipids by water degumming as they are hydratable, but non-hydratable forms, which mostly exist in combination with calcium, magnesium or iron cations, must undergo a pre-treatment with acid.

In water degumming, hydratable phospholipids are converted to hydrated gums, which are insoluble in oil and can be separated using centrifugal action. Lecithin (phosphatidylcholine) and cephalin (phosphatidylethanolamine) are the main phospholipids found in rapeseed oil, and both are hydratable. Lecithin (including cephalin) can be obtained as a by-product from rapeseed oil, and although dark in colour, may be produced if market conditions allow. In the water degumming process, crude oil at 80°C is mixed with a small quantity of water or steam (2–4%) for 10–30 minutes. All traces of water are subsequently removed by drying discharged oil under vacuum before it is pumped to storage tanks. This oil will now contain 100–250 ppm non-hydratable phospholipids measured as phosphorus (Mag, 2001), which may be marketed as crude degummed oil or may undergo further processing.

The other phospholipids, consisting of non-hydratable types, require alternative means of removal, namely acid degumming or alkali refining. In acid degumming, the remaining phospholipids are converted to the hydratable form before centrifugation by the addition of either phosphoric, citric or malic acid. The acids act as sequestering agents to break the cation complex within the non-hydratable phospholipids, resulting in oil-insoluble metals, salts and hydratable phospholipids. Hydration with water then makes the phospholipids oil-insoluble.

In acid degumming, incoming oil is heated to 70°C and 0.1% of 75% phosphoric acid is mixed with the oil for approximately 20 minutes. Up to 2% of water is incorporated and the resulting hydrated gums are removed by centrifuging. If the oil is to be stored prior to bleaching, it must be vacuum dried. Residual concentrations are in the order of 25 ppm phospholipid measured as phosphorus (Mag, 2001).

Further developments in degumming include the addition of enough dilute caustic soda to partially neutralise the phosphoric acid, but insufficient to form soap. The bulk of the gums is then removed in the first separator and the remaining fine phospholipid particles are removed by subsequent hot-water treatment and centrifuging. This can give a residual phosphorus level of less than 10 ppm. Extracted gums can be blended back into the meal where they help to reduce dustiness and increase its metabolisable energy content (Pickard, 2001).

*2.3.2 Physical refining*

Physical refining has developed relatively recently and can be used particularly for acid degummed oil. Oil must be well degummed before entry into this process, which can then be more economical. Physical refining uses steam distillation to remove free fatty acids, flavour and some colour from oils low in phospholipid.

At the start of the process, acid degummed oil is mixed with 0.05–0.1% of 85% phosphoric acid at 110°C under vacuum (Carr, 1995). Oil is then mixed with approximately 1.0% acid-activated bleaching clay for 10–15 minutes. The oil is separated from spent clay and impurities, including precipitated phospholipids, adsorbed chlorophyll and some carotenoids, and is then pumped through a deaerator to a deodouriser–deacidifier for approximately 1 hour at 260°C and under 1–3 mm mercury vacuum. This has the purpose of removing most of the free fatty acids, but also deodourises the oil, volatilising the odour/flavour impurities by the injection of steam. The oil is finally cooled and filtered for storage. As deodourisation is normally the last refining process to be carried out, physical refining may be delayed until other processes such as oil modification have been undertaken.

For quality oils, it is important that oil has been degummed to remove metals and phospholipids before physical refining takes place. The use of physical refining has a number of advantages compared to chemical refining; it can lower the loss of neutral oil in by-products, reduce the number of purification operations and eliminate acidulation problems associated with the soapstock by-product produced by alkali refining. Physical refining may therefore be associated with lower capital and operating costs and with lower environmental problems arising from large volumes of effluent.

*2.3.3 Alkali refining*

Alkali refining offers benefits over physical refining where the quality of seed entering the refining plant is variable. Seed quality will vary significantly from year to year due to climate and agronomic practices in terms of impurities affecting colour, appearance, odour, taste and keeping quality. Alkali refining is therefore frequently used as the main process used by refiners, or is available for conditions when physical refining cannot meet product specifications.

Crude or degummed oil is treated with a solution of phosphoric acid to help precipitate phospholipids, before being continuously mixed with dilute sodium hydroxide (caustic soda) solution and heated to neutralise free fatty acids and any excess phosphoric acid and precipitate phospholipids. Centrifuges are used to continuously separate the soapstock from oil and any small amounts of residual soap are removed by mixing with hot, soft water and by further centrifugation. Water is removed by passing through a continuous vacuum drying stage.

It is important to provide the alkali refining process with a uniform flow of homogeneous material, so moderate agitation in the oil tanks and accurate metering is required. Food-grade phosphoric acid at 75% and a rate of 100–500 ppm is used. The strength of sodium hydroxide solution used varies according to the free fatty acid and phospholipid content (Carr, 1995). The concentration necessary is determined by using a specific gravity hydrometer calibrated in degrees Baumé (Bé) and is usually between 15 and 18 Bé for rapeseed oil. Different temperatures and mixing times are favoured by refiners, with European processors using a short mix method with the crude oil heated to between 85 and 105°C and rapid delivery to the centrifuge at 90°C. Oil is reheated to between 85 and 90°C and 10–20% softened water is added, before the second centrifuging and drying. The soap phase produced from this process can be added to the meal or it can be acidulated and used as a feed ingredient.

### 2.3.4 Bleaching

Bleaching aims to remove the carotenoid and chlorophyll pigments to produce oil colour of an acceptable quality. Minor impurities such as soaps, residual phospholipids, metals and oxidation products are also removed at this stage, improving the organoleptic quality and oxidative stability of the final oil.

In the bleaching process, acid-treated bentonite clay is activated by heat treatment to adsorb the pigments and other contaminants. The process is undertaken under continuous vacuum to protect the oil from oxidation at high bleaching temperatures. As the oil enters the bleacher, it is heated to approximately 100°C before contacting the acid-activated bleaching clay. The addition of bleaching clay is metered to achieve a constant value of 0.5–2.0% bleaching clay to oil ratio and flow rates are adjusted to give a contact time of 10–20 minutes. Oil that has been subject to alkali refining tends to require less bleaching earth, as the caustic soda used reacts with colour pigments and other impurities (Kellens, 2001). Bleached oil is continuously pumped from the bottom of the bleacher into a filter to separate the spent clay. The filter used is usually a hermetic leaf filter with stainless-steel mesh elements. The clay is blown with steam in order to dry it within the filter. The filter can be opened and the clay detached by vibration.

Batch bleaching is the conventional means of undertaking this process and it is still used by many refiners. Continuous vacuum bleaching is more efficient at deaeration than batch vacuum bleaching and is therefore more effective in protecting the oil from oxidation. Heat recovery through the continuous process is possible giving potential energy savings. Multi-step bleaching processes are being developed to reduce bleaching earth consumption. Here the bleaching earth is added in two consecutive stages with filtration after each stage. Due to increasingly stringent regulations applied to disposal of bleaching clays, efforts are ongoing to reduce solid bleaching waste.

### 2.3.5 Winterisation

Traces of wax (esters of fatty acids and long-chain alcohols) can form in the oil in certain growing seasons. These small concentrations of wax (100–200 ppm) can produce cloudiness in oil, particularly when the oil is stored at low temperatures. Where there are demands for perfectly clear oils at both ambient and refrigerator temperatures, low-temperature winterisation is carried out to precipitate and remove waxes before bottling the oil. Waxes tend to solidify at low temperatures and can be removed as solid particles by low-temperature filtration.

Oil is rapidly cooled, initially by a heat exchanger. Further slow cooling to between 0 and 19°C for around 4 hours allows crystallisation to be complete. Subsequent filtration may be improved by adding a filter aid to the oil on entry to the process. This reduces the wax content to less than 50 ppm, no longer producing a visible haze. Due to the relatively small amounts of solids present, most winterisation processes are continuous.

### 2.3.6 Deodourisation

Deodourisation is the final refining step for edible oils. Its main purpose is to remove compounds from the oil typical of the seed from which it was derived. Odour and flavour impurities are more volatile than triacylglycerols enabling the use of steam distillation for their removal. Deodourisation also destroys peroxides in the oil, removes aldehydes and ketones or other volatile products arising from atmospheric oxidation and destroys the relatively unstable carotenoid pigments, thereby reducing oil colour. Processes such as bleaching and hydrogenation form traces of compounds associated with odour and flavour. Deodourisation serves to physically refine the oil as an alternative to alkali refining. Any blending of liquid oils with base stocks containing solids, necessary to meet required melting behaviour, is carried out before deodourisation takes place.

Deodourisation is undertaken as a continuous or semi-continuous process in vertical stainless-steel vessels filled with five to seven shallow trays. Continuous deodourisation is designed for larger operations and has advantages of lower utility consumption, shorter processing time, reduced labour and high heat recovery. Oil, deaerated under high vacuum, is delivered to the top steam-heated tray, before moving down to the second tray where it is heated to around 240–260°C. It then flows through two deodourising trays countercurrent to a superheated, stripping steam flow. The oil is retained in each tray for 20 minutes and a vacuum is maintained to improve volatilisation of impurities and product oil quality. The oil is cooled in a water-chilled cooling tray before filtering and transfer to a storage tank, from where it is ready for packaging and end use.

Tocopherol losses and *trans* isomerisation are now perceived as important factors associated with deodourisation. Tocopherols are natural antioxidants, found in higher quantities in oils that are highly unsaturated. Part of the tocopherols is lost during deodourisation as a result of steam distillation or due to thermal degeneration. Deodouriser distillates now attract a market for tocopherol production (vitamin E). Highly polyunsaturated oils are particularly sensitive to *trans* isomerisation, especially the linolenic acid present in rapeseed oil. *Trans* isomerisation is favoured by high temperatures during deodourisation, with *trans* isomer formation being negligible below 220°C and rising to nearly exponential levels above 240°C (Kellens, 2001).

## 2.4 Biorefining – an alternative oilseed processing method

The biorefinery concept was developed as an environmentally friendly processing technique to separate and refine whole crops into their botanical and chemical components. It is intended that the biorefinery would be situated in the same locality as the site of production of the crops to permit short transportation of biomass. The process is based on aqueous extraction, utilising *Green Chemistry* to separate biomolecules occurring in oil and protein-rich crops (Gylling, 1993). Relatively mild methods are used to enable enzyme-processing steps to release native products and ensure that their functional properties are preserved. It is also hoped to eliminate waste products.

The biorefinery process for rapeseed was initially developed by Novo Nordisk and the Chemistry Department of the Royal Veterinary and Agricultural University in Copenhagen in the late 1980s (Anon., 2000). Biorefining is considered to offer potential as a supplement and alternative to traditional rapeseed processing. It has the potential to add value to a range of products produced from rapeseed for use as food, animal feed and also non-food uses. The process has been developed to pilot scale on the island of Bornholm, Denmark.

The biorefinery process involves inactivation of myrosinases, use of cell wall degrading enzymes on milled seed suspended in water and centrifugations resulting in four fractions: oil, protein-rich meal, syrup and hulls. Processing using the biorefinery technique can be divided into three stages: pre-treatment, release of oil and purification of products (Fig. 2.3).

The pre-treatment stage inactivates the enzymes myrosinases, lipoxygenases and lipases present in rapeseed. These enzymes may otherwise promote the

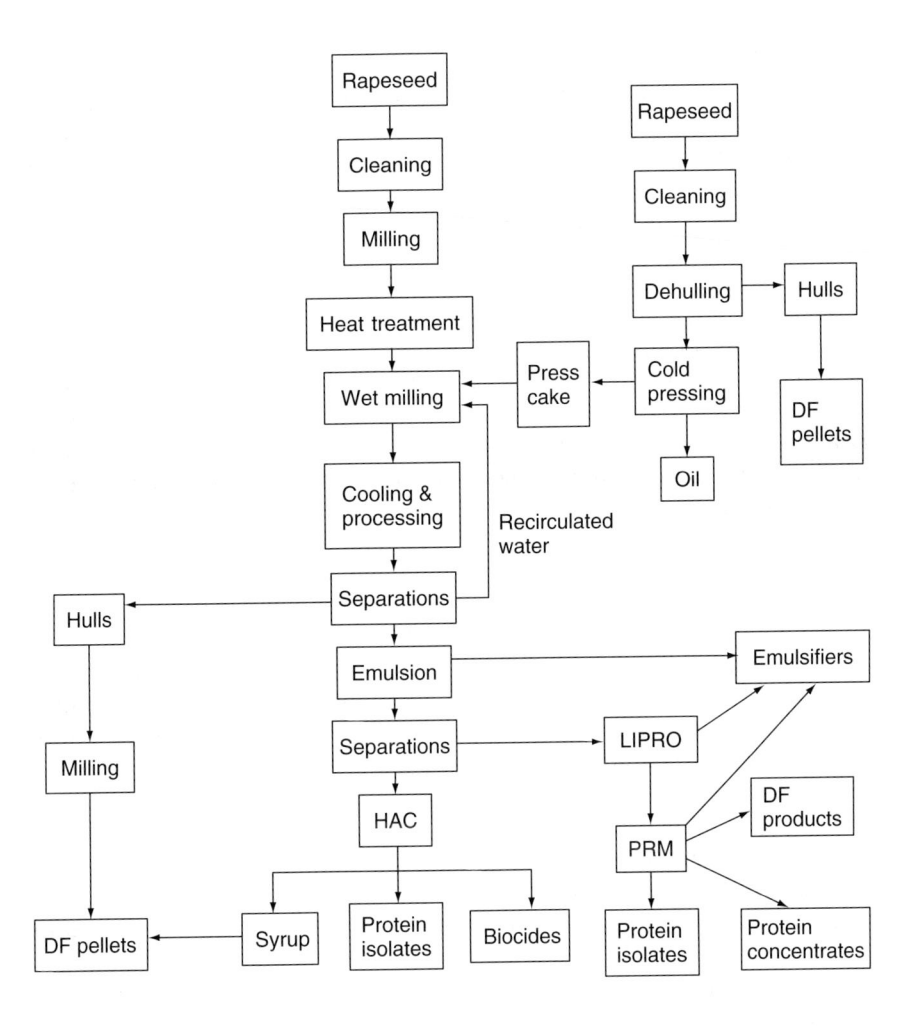

```
                    Rapeseed                    Rapeseed
                        │                           │
                    Cleaning                    Cleaning
                        │                           │
                     Milling                   Dehulling ──→ Hulls
                        │                           │            │
                Heat treatment                   Cold            │
                        │         Press       pressing           ↓
                   Wet milling ←── cake ←────    │             DF
                        │                        │            pellets
                 Cooling &                      Oil
                 processing     Recirculated
                        │          water
                   Separations
      ┌─────────────────┤
   Hulls            Emulsion ──────────────────→ Emulsifiers
      │                 │                           ↑
   Milling         Separations ────→ LIPRO ─────────┤
      │                 │                │          │
      │                HAC               │       DF products
      │          ┌──────┼──────┐        PRM ──┬────────┐
      ↓          ↓      ↓      ↓         │    ↓        ↓
  DF pellets  Syrup  Protein  Biocides   │  Protein  Protein
                     isolates            │  isolates concentrates
```

HAC     Hydrophilic and amphiphilic compounds
LIPRO   Lipophilic protein in complex with amphiphilic lipids
PRM     Protein-rich meal
DF      Dietary fibre

**Figure 2.3** Basic processing scheme for the biorefining of oilseed rape (Anon., 2000).

hydrolysis or oxidation of the intended high-quality oil, protein and other products from biorefining. After the seed is cleaned, it is dry milled with a hammer mill and enzyme degradation is effected by heating with water to 85–99°C for 20 minutes (Jensen *et al.*, 1990).

Oil release is assisted by degradation of cell walls using the enzyme strain SP-311 based on *Aspergillus niger* and developed by Novo Nordisk. This enzyme has catalytic activity towards hydrolysis of pectic substances, hemicellulose, cellulose and other cell wall constituents. In the oil release stage, the suspension is allowed to cool to 50°C in order to protect the native biomolecules. The suspension is wet milled, the pH adjusted to 4.5 and the SP-311 enzyme added. Enzymic treatment proceeds for approximately 4 hours.

Separation and purification of products follow this process. Decanting removes the hulls. Washing and centrifuging steps separate the oil, protein and other functional biomolecules. Spray drying is the last process used to produce final isolates from the protein-rich meal (PRM) and syrup from the hydrophilic and amphiphilic compounds (HAC) fraction.

It is also possible to enter press cake into the biorefinery process and for it to proceed only through the separation stages. Here the rapeseed would first be dehulled and cold pressed before biorefining. Oil is recovered at a yield of 35% seed dry matter (Bagger *et al.*, 2003), up to 8% lower than the conventional solvent-extraction process. The oil has particular quality features associated with the mild conditions of processing. Native antioxidants are preserved within the oil giving it a resistance towards autoxidation (Anon., 2000). Another advantage of biorefinery-produced oil is the much reduced phospholipid content compared to conventional processes, with lecithins from biorefined oil being present at 0.03% compared to 1.8% for conventionally produced oil (Anon., 1998) and with low levels of other unwanted constituents. It was concluded that further oil refining was largely unnecessary (Bagger *et al.*, 2003).

Biorefining permits the separation of glucosinolates, the principal constituents limiting the amount of meal that can be fed to animals, from the protein fraction. This gives a significant advantage to the meal compared to the meal from conventionally processed rapeseed. The use of protein to produce protein-based films is also under development. Dietary fibre (DF) within rapeseed protein products creates problems for protein yield and work has been concentrated on this to increase the quality of protein concentration. Techniques have been developed to produce fibre pellets from hulls and concentrated syrup after isolation of protein, oil and glucosinolates as a ruminant feedstuff. Utilisation of these products eliminates the production of waste from the process.

Due to the mild conditions of processing, only limited glucosinolate degradation occurs. Glucosinolates are known to have biocidal activity and their concentration in the HAC fraction during biorefining gives potential for convenient isolation for biocidal application. Tests of glucosinolate material from the biorefinery process have indicated activity in controlling a number of fungal pathogens and pests, although much further work is required to develop products with consistent rates of activity (Anon., 2000).

At present biorefining of rapeseed is restricted to pilot plant scale. An assessment considering commercial scale-up indicated that a biorefinery processing 100 000 t seeds per year had a higher requirement of electricity, steam and particularly process water, but a lower requirement of cooling water (Anon., 1993; cited by Anon., 1998). Labour requirement was similar for both. Avoidance of use of petrochemical solvent extraction was considered an advantage for ease of operation and environmental aspects. Financial comparisons including capital costs indicated that the biorefinery process was considerably more expensive.

Improved prices in recognition of product benefits are required in order to make biorefining financially attractive. Since these products are not yet commercially available, their true value in the market place can only be estimated at present. One study considered the viability of a relatively modest plant, using average market prices for conventional products and calculated substitution prices for specific biorefined products, which would not be available from conventional processing (Pedersen and Gylling, 2001). Value of oil is assumed to be similar to refined oil, due to its improved composition. The protein isolate was compared to casein and the protein concentrate to wheat gluten. The hulls and DFs were considered to be of the same value as wheat bran, the carbohydrates were assumed to compare with molasses and the emulsifiers to diacylglycerols. No comparable product could be identified for the glucosinolate biocides and a conservative value was placed on this biorefinery derivative. Based on the assumptions outlined, estimates indicate that biorefining could be profitable at an annual capacity of 8600 t rapeseed per year.

## References

Anon. (1993) Technical information, Novo Nordisk, NJ-930126 KiJa.

Anon. (1998) Aqueous enzymatic extraction of oil from rapeseeds, Denmark, 1991–1994. Manufacture of food products and beverages. Environmental management centre, international cleaner production information clearinghouse. http://www.emcentre.com/unepweb/tec_case/food_15/process/p16.htm

Anon. (2000) FAIR CT95-0260 High quality oils, proteins and bioactive products for food and non-food purposes based on biorefining of Cruciferous oilseed crops. *Final Report*, EU Project.

Bagger, C.L., Sorensen, H., Sorensen, J.C. and Sorensen, S. (2003) Biorefining, the soft processing alternative. *Proceedings of the 11th GCIRC International Rapeseed Congress*, Copenhagen, Denmark, 650.

Booth, E., Cook, P., Entwistle, G., Stott, A. and Walker, K. (1993) Oilseed rape processing in Scotland; the options for unconventional rape products. *SAC Economic Report* No. 41. The Scottish Agricultural College, Aberdeen, 90pp.

Buhr, N. (1990) Mechanical pressing. In: *Edible Fats and Oil Processing: Basic Principles and Modern Practices* (ed. D. Erickson). Champaign, Illinois: American Oil Chemists' Society.

Carr, R.A. (1995) Processing the seed and oil. In: *Brassica Oilseeds; Production and Utilization* (eds Kimber D. and D.I. McGregor). CAB International, Oxon, UK, pp. 267–290.

Derksen, J.T.P., Muuse, B.G. and Cuperus F.P. (1994) Processing of novel oil crops and seed oils. In: *Designer Oil Crops; Breeding, Processing and Biotechnology* (ed. D.J. Murphy). VCH Verlagsgesellschaft mbH, Weinheim, Federal Republic of Germany, pp. 253–281.

Gylling, M. (1993) The whole crop biorefinery project, midterm assessment – concept, status, perspectives. *The Bioraf Denmark Foundation*.

Jensen, S.K., Olsen, H.S. and Sorensen, H. (1990) Aqueous enzymatic processing of rapeseed for production of high quality products. In: *Canola and Rapeseed, Production, Chemistry, Nutrition and Processing Technology* (ed. F. Shahidi). Van Nostrand Reinhold, New York, pp. 331–343.

Kellens, M. (2001) *Current Developments in Oil Refining Technology*. De Smet Group, Antwerp, Belgium.

Kemper, T.G. (2000) Innovations in meal desolventising. *AOCS annual meeting and expo*, San Diego, California.

Mag, T. (2001) Canola Seed and Oil Processing. Canola Council of Canada. http://www.canola-council.org

Pedersen, S.M. and Gylling, M. (2001) The economics of producing quality oils, proteins and bioactive products for food and non-food purposes based on biorefining. Ministeriet for Fodevarer, Landbrug og Fiskeri, Statens Jordbrugs-og Fiskeriokonomiske Institut Working Paper no. 04/2001.

Pickard, M.D. (2001) Canola meal feed industry guide, section I, processing of canola seed for quality meal. Canola Council of Canada. http://www.canola-council.org/pubs/processing.html

Unger, E.H. (1990) Commercial processing of canola and rapeseed: crushing and oil extraction. In: *Canola and Rapeseed, Production, Chemistry, Nutrition and Processing Technology* (ed. F. Shahidi). Van Nostrand Reinhold, New York, pp. 235–249.

# 3 Chemical composition of canola and rapeseed oils
W.M.N. Ratnayake and J.K. Daun

## 3.1 Brief history of the development of rapeseed oils

Canola and rapeseed are common names used to describe crops including the species *Brassica napus* L., *Brassica rapa* L. (formerly *Brassica campestris* L.), *Brassica juncea* L. Czern. (Bengtsson *et al.*, 1972). Several different subspecies of each type are grown in Canada and other regions of the world, especially both winter and summer types of *B. napus* and *B. rapa*. The Sarson type of *B. rapa* is widely grown in India.

Fats and oils are chiefly characterized by their fatty acid composition. Rapeseed, like more than two-thirds of the members of the Cruciferae family, is characterized by the presence of significant amounts of fatty acids with chain lengths longer than $C_{18}$, the principal fatty acid being erucic acid (*cis*-22:1(*n*-9)). Traditional rapeseed oils produced in Canada (Table 3.1) had erucic acid levels ranging from the low 20% for *B. rapa* cultivars to about 40% for *B. napus* types. These oils were also characterized by significant amounts of oleic, linoleic and eicosenoic fatty acids. Concerns about the nutritional intake of erucic acid were raised in the mid-1950s (Sauer and Kramer, 1983).

During the 1960s, researchers actively worked to develop varieties of rapeseed with low levels of erucic acid. By the 1970s health authorities moved to restrict the level of erucic acid in rapeseed oil consumed by humans and at that time low erucic acid rapeseed (LEAR) was introduced. In Canada, the erucic acid limit was first set at a maximum of 5% but this was lowered to 2% maximum in the early 1980s and industry standards today are aiming at a maximum of 1% erucic acid (Daun and Adolphe, 1997).

Researchers were also able to reduce the level of glucosinolates, another antinutritional component of rapeseed, by the mid-1970s and the new product was given the name canola (Daun, 1984). The term canola has now gained acceptance in many parts of the world. The fatty acid composition of canola oil is similar to that of LEAR oil (Table 3.1). The oil is characterized by a high proportion of oleic acid, usually close to 60% with significant amounts of linoleic and α-linolenic acids. This coupled with a very low level of saturated fatty acids (6–7%) has made canola oil desirable from a nutritional point of view.

There is a natural range in fatty acid composition among different high and low erucic acid rapeseed oils. The variation is due to a combination of environmental and variety effects (Jonsson, 1975; Holmes and Bennet, 1979). This range has been recognized by Codex Alimentarius Commission (2001) (Table 3.2).

**Table 3.1** Development of fatty acid composition of rapeseed types that have been or are in current commercial production. Data from analysis of authentic samples from the collection at the Canadian Grain Commission, Grain Research Laboratory. Samples collected from plant breeders or from the Canadian Grain Commission, Harvest Survey of Canola

| Date | Species | Cultivar | 12:0 | 14:0 | 16:0 | 16:1 | 18:0 | 18:1 | 18:2 | 18:3 | 20:0 | 20:1 | 20:2 | 22:0 | 22:1 | 24:0 | 24:1 |
|---|---|---|---|---|---|---|---|---|---|---|---|---|---|---|---|---|---|
| | | | | | | | | Traditional rapeseed | | | | | | | | | |
| 1940 | B. napus | Argentine type | ND | 0.1 | 3.2 | 0.2 | 1.1 | 15.9 | 14.0 | 8.8 | 0.7 | 10.1 | 0.7 | 0.6 | 42.1 | 0.2 | 1.0 |
| 1961 | B. napus | Nugget | ND | 0.1 | 3.0 | 0.3 | 1.6 | 20.8 | 12.2 | 6.4 | 1.0 | 13.0 | 0.5 | 0.6 | 38.1 | 0.2 | 0.9 |
| 1964 | B. rapa | Echo | ND | 0.1 | 2.3 | 0.2 | 1.1 | 32.5 | 17.3 | 9.9 | 0.5 | 10.7 | 0.6 | 0.4 | 23.0 | 0.2 | 1.4 |
| | | | | | | | | Low erucic acid rapeseed | | | | | | | | | |
| 1968 | B. napus | Oro | ND | 0.1 | 4.1 | 0.3 | 2.0 | 63.7 | 18.2 | 7.6 | 0.7 | 1.5 | 0.1 | 0.4 | 0.3 | 0.2 | 0.3 |
| 1971 | B. rapa | Span | ND | 0.1 | 3.2 | 0.3 | 1.2 | 55.4 | 21.9 | 10.3 | 0.4 | 2.6 | 0.2 | 0.2 | 3.1 | 0.2 | 0.4 |
| 1973 | B. napus | Midas | ND | 0.1 | 4.0 | 0.3 | 1.6 | 59.3 | 21.0 | 10.7 | 0.6 | 1.3 | 0.1 | 0.4 | 0.1 | 0.2 | 0.2 |
| | | | | | | | | Low erucic acid and low glucosinolate rapeseed (canola) | | | | | | | | | |
| 1974 | B. napus | Tower | ND | 0.1 | 4.0 | 0.3 | 1.5 | 58.0 | 21.7 | 11.3 | 0.5 | 1.4 | 0.1 | 0.3 | 0.1 | 0.2 | 0.2 |
| 1980 | B. rapa | Tobin | ND | 0.1 | 3.2 | 0.2 | 1.4 | 57.5 | 21.9 | 12.9 | 0.5 | 1.1 | 0.1 | 0.2 | 0.2 | 0.2 | 0.3 |
| 1977 | B. napus | Jet Neuf | ND | 0.1 | 5.0 | 0.4 | 1.7 | 51.8 | 19.2 | 7.3 | 0.6 | 5.7 | 0.2 | 0.3 | 6.6 | 0.2 | 0.2 |
| 1996 | B. napus | 46A65 | ND | 0.1 | 3.6 | 0.3 | 2.1 | 66.4 | 17.9 | 6.6 | 0.7 | 1.4 | 0.1 | 0.4 | 0.0 | 0.2 | 0.1 |
| 1997 | B. rapa | Reward | ND | 0.1 | 3.3 | 0.3 | 1.6 | 59.6 | 20.9 | 11.5 | 0.5 | 1.0 | 0.1 | 0.3 | 0.2 | 0.1 | 0.2 |

|  |  |  |  |  |  |  |  |  |  |  |  |  |  |  |  |  |
|---|---|---|---|---|---|---|---|---|---|---|---|---|---|---|---|---|
|  |  |  |  |  |  |  |  |  |  |  |  | High erucic acid rapeseed |  |  |  |  |
| 1975 | *B. rapa* (sarson) | R-500 | ND | ND | 1.6 | 0.2 | 0.8 | 9.2 | 12.2 | 9.6 | 0.7 | 2.3 | 0.3 | 1.1 | 57.5 | 0.5 | 1.8 |
| 1982 | *B. napus* | Reston | ND | ND | 3.0 | 0.3 | 1.1 | 17.1 | 13.6 | 8.5 | 0.7 | 8.9 | 0.6 | 0.5 | 42.8 | 0.2 | 1.2 |
| 1989 | *B. napus* | Hero | ND | ND | 2.8 | 0.2 | 1.1 | 13.9 | 11.9 | 7.9 | 0.9 | 6.2 | 0.4 | 0.8 | 50.8 | 0.3 | 1.1 |
| 1999 | *B. napus* | Millennium 03 | ND | ND | 2.6 | 0.2 | 0.8 | 11.5 | 11.6 | 8.9 | 0.7 | 6.1 | 0.4 | 0.7 | 54.0 | 0.2 | 1.0 |
|  |  |  |  |  |  |  |  | Specialty rapeseed and canola |  |  |  |  |  |  |  |  |
| 1987 | *B. napus* | Stellar | ND | 0.1 | 4.0 | 0.2 | 1.9 | 62.1 | 25.5 | 2.5 | 0.7 | 1.3 | 0.1 | 0.4 | 0.1 | 0.2 | 0.2 |
| 1994 | *B. napus* | Allons | ND | 0.1 | 3.9 | 0.3 | 2.1 | 62.3 | 23.6 | 3.0 | 0.7 | 1.5 | 0.1 | 0.3 | 0.4 | 0.2 | 0.2 |
| 1995 | *B. napus* | Laurical | 36.2 | 4.4 | 3.3 | 0.3 | 1.3 | 31.6 | 13.4 | 7.0 | 0.5 | 1.0 | 0.1 | 0.3 | 0.2 | ND | 0.1 |
| 2000 | *B. napus* | Nexera 715 | ND | 0.1 | 3.3 | 0.2 | 1.7 | 73.8 | 15.3 | 2.6 | 0.6 | 1.5 | 0.1 | 0.3 | 0.1 | 0.2 | 0.1 |
| 2001 | *B. napus* | IMC 302 | 0.0 | 0.1 | 3.0 | 0.3 | 1.9 | 75.9 | 10.3 | 4.9 | 0.7 | 1.6 | 0.1 | 0.4 | 0.1 | 0.2 | 0.2 |
| 2002 | *B. napus* | IMC | ND | 0.1 | 2.8 | 0.3 | 1.8 | 80.5 | 7.4 | 3.4 | 0.7 | 1.7 | 0.1 | 0.4 | 0.1 | 0.3 | 0.2 |
|  |  |  |  |  |  |  |  |  | Mustard |  |  |  |  |  |  |  |
| 1965 | *B. juncea* | Common brown | ND | 0.1 | 2.9 | 0.2 | 1.5 | 21.4 | 22.0 | 12.3 | 0.9 | 12.3 | 1.1 | 0.5 | 22.0 | 0.4 | 0.3 |
| 2002 | *B. juncea* (canola) | Amulet | ND | 0.1 | 3.3 | 0.2 | 2.3 | 65.3 | 14.2 | 10.9 | 0.6 | 1.6 | 0.1 | 0.3 | 0.1 | 0.2 | 0.4 |

ND = not detected.

**Table 3.2** Range in fatty acid composition for rapeseed and mustardseed oils accepted by Codex Alimentarius Commission (2001)

| Fatty acid | 14:0 | 16:0 | 16:1 | 17:0 | 17:1 | 18:0 | 18:1 | 18:2 | 18:3 | 20:0 | 20:1 | 20:2 | 22:0 | 22:1 | 22:2 | 24:0 | 24:1 |
|---|---|---|---|---|---|---|---|---|---|---|---|---|---|---|---|---|---|
| Mustardseed oil | ND–1.0 | 0.5–4.5 | ND–0.5 | ND | ND | 0.5–2.0 | 8.0–23.0 | 10.0–24.0 | 6.0–18.0 | ND–1.5 | 5.0–13.0 | ND–1.0 | 0.2–2.5 | 22.0–50.0 | ND–1.0 | ND–0.5 | 0.5–2.5 |
| Rapeseed oil | ND–0.2 | 1.5–6.0 | ND–3.0 | ND–0.1 | ND–0.1 | 0.5–3.1 | 8.0–60.0 | 11.0–23.0 | 5.0–13.0 | ND–3.0 | 3.0–15.0 | ND–1.0 | ND–2.0 | >2.0–60.0 | ND–2.0 | ND–2.0 | ND–3.0 |
| Rapeseed oil (low erucic acid) | ND–0.2 | 2.5–7.0 | ND–0.6 | ND–0.3 | ND–0.3 | 0.8–3.0 | 51.0–70.0 | 15.0–30.0 | 5.0–14.0 | 0.2–1.2 | 0.1–4.3 | ND–0.1 | ND–0.6 | ND–2.0 | ND–0.1 | ND–0.3 | ND–0.4 |

ND = not determined.

## 3.2 Development of specialty types of rapeseed oils

Success in removal of erucic acid led many plant breeders to look at other possible modifications in rapeseed. One of the first objectives was to increase the level of erucic acid in order to increase the value of oil used for the industrial production of chemicals (Latta, 1990; Sonntag, 1991). While Canadian production of high erucic acid rapeseed (HEAR) oil originally centered on a *B. rapa* Sarson type *R-500*, the demand for seed with better agronomics resulted in the development of several lines of high erucic acid seed with improved erucic acid content. One of the challenges of the high erucic acid rapeseed program is overcoming the inability of *B. napus* to place long-chain fatty acids in the 2-position of triacylglycerol molecules (Brockerhoff and Yurkowski, 1966; Grynberg *et al.*, 1966). Several strategies have been developed (Taylor *et al.*, 1992) but there has been no report of cultivars with erucic acid levels greater than 55% to date.

The relatively high level of α-linolenic acid (18:3($n$-3)) in LEAR oil may have some nutritional advantages but it is not desirable for vegetable oils used in frying since it is rapidly oxidized and produces off flavors. Mutation experiments at the Institute for Plant Breeding, Göttingen, Germany, resulted in the isolation of a line with a block in the desaturation pathway between linoleic acid and α-linolenic acids (Rakow, 1973). Lines from this work have been used to develop low linolenic acid lines of canola (Table 3.3). While oil from these varieties gave improved oxidative stability (Eskin *et al.*, 1989), they have seen only limited acceptance in the general marketplace until recently due to expenses associated with agronomic problems and the need to segregate the seed.

The need to develop oils with high degrees of oxidative stability with low levels of *trans* fatty acids for use as frying oils led to the development of rapeseed oils with high levels of oleic acid (Table 3.1). These types can be subdivided into:

- types with levels of oleic acid (18:1($n$-9)) above 72% and levels of α-linolenic acid slightly reduced (4–5%) because of selection for the high oleic acid trait (e.g. IMC 302 in Table 3.1);
- types with levels of oleic acid above 72% and levels of linolenic acid less than 3% because of selection from low linolenic varieties such as Allons (e.g. Nexera 715 in Table 3.1);
- types with oleic acid above 80% from mutation selection (e.g. IMC in Table 3.1) types with levels of lauric acid (12:0) greater than 30% from insertion of recombinant DNA (Laurical in Table 3.1) produced in the mid-1990s but now no longer in production.

Several other fatty acid modification types have been reported in the literature but do not seem to be under commercial production – these include types with high or low levels of saturated fatty acids (Persson, 1985; Friedt and Lühhs, 1999;

**Table 3.3** Uncommon minor fatty acids and totals for saturated, mono and polyunsaturated fatty acids of some LEARs (*Brassica napus* or *campestris*) as a percentage of total fatty acids

| Fatty acid | Laboratory-extracted oil from *B. napus* variety tower seeds[a] | Laboratory-extracted oil from *B. campestris* variety candle seeds[b] Saskatoon | Laboratory-extracted oil from *B. campestris* variety candle seeds[b] Nipawin | Commercial oil from *B. campestris* variety candle at various stages of refining[b] Crude | Commercial oil from *B. campestris* variety candle at various stages of refining[b] Degummed | Commercial oil from *B. campestris* variety candle at various stages of refining[b] Refined |
|---|---|---|---|---|---|---|
| 13:0 | NR | NR | NR | NR | NR | NR |
| 14:0 | 0.02 | 0.01 | 0.02 | 0.01 | 0.01 | 0.02 |
| *Iso*-14:0 | NR | NR | NR | NR | NR | NR |
| anteiso-15:0 | NR | NR | NR | NR | NR | NR |
| Total saturated[c] | 6.46 | 5.16 | 6.48 | 6.7 | 6.17 | 5.75 |
| 14:1(*n*-9) | Trace | ND | Trace | Trace | Trace | Trace |
| 14:1(*n*-7) | 0.007 | ND | 0.01 | 0.01 | 0.01 | 0.01 |
| 14:1(*n*-5) | Trace | ND | Trace | Trace | Trace | Trace |
| *cis*-15:1(*n*-10) | 0.02 | 0.01 | 0.02 | 0.01 | 0.01 | 0.01 |
| *trans*-15:1(*n*-10) | 0.01 | 0.01 | 0.02 | 0.01 | 0.01 | 0.02 |
| 15:1(*n*-8) | Trace | 0.02 | 0.01 | Trace | Trace | Trace |
| 16:1(*n*-9) | 0.06 | NR | NR | NR | 0.038 | NR |
| 16:1(*n*-7) | 0.21 | NR | NR | NR | 0.202 | NR |
| 16:1(*n*-5) | 0.02 | NR | NR | NR | 0.01 | NR |
| 17:1(*n*-8) | 0.06 | 0.03 | 0.04 | 0.05 | 0.04 | 0.03 |
| 19:1 | 0.02 | Trace | 0.02 | 0.02 | 0.02 | 0.03 |

| | | | | | |
|---|---|---|---|---|---|
| 20:1(n-7) | 0.03 | NR | NR | NR | 0.05 | NR |
| 22:1(n-7) | 0.002 | NR | NR | NR | 0.01 | NR |
| 24:1(n-7) | 0.06 | NR | NR | NR | NR | NR |
| Total monoenes[c] | 65.8 | 62.11 | 54.75 | 55.69 | 55.69 | 56.46 |
| 16:2(n-6) | 0.09 | 0.03 | 0.03 | 0.03 | 0.03 | 0.02 |
| 16:2(n-4) | Trace | 0.01 | 0.01 | Trace | Trace | Trace |
| 16:3(n-3) | 0.08 | 0.11 | 0.14 | 0.13 | 0.13 | 0.15 |
| 20:2(n-6) | 0.05 | 0.08 | 0.1 | 0.14 | 0.11 | 0.11 |
| 20:3(n-3) | 0.01 | 0.01 | 0.03 | ND | ND | ND |
| c9, c12, t15–18:3[d] | ND | 0.02 | 0.04 | 0.18 | 0.25 | 0.15 |
| t9, c12, c15–18:3[d] | ND | Trace | Trace | 0.15 | 0.09 | 0.02 |
| Total polyenes[c] | 27.61 | 32.77 | 32.5 | 38.68 | 38.19 | 37.7 |

[a] Data adapted from Sebedio and Ackman (1979, 1981).
[b] Data adapted from Sebedio and Ackman (1981).
[c] Total includes all the other fatty acids.
[d] c = *cis*, t = *trans*.
NR = values not reported.
ND = not detected.

Thelen and Ohlrogge, 2002), as well as alcohols and waxes (Anon., 2003). Some of the delay in producing commercialized varieties of these types has been the general resistance towards the acceptance of genetically modified oils. There has also been a problem with developing new types with sufficient agronomic properties, especially yield potential, to allow economic, commercial production in systems where the production includes the additional cost of identity preservation.

Despite these problems, the utilization of modified oils, especially those modified to give greater stability or replace hydrogenated oils due to the *trans* issue, has increased in recent years. It has been estimated that by 2010, canola with specialty traits will compose about one third of Canadian production (Anon., 2003).

## 3.3 Minor fatty acids

Rapeseed oils, in addition to the usual, common fatty acids listed in Table 3.1, contain a number of minor fatty acids. Table 3.3 shows the minor fatty acids identified by Sebedio and Ackman (1979, 1981) in some laboratory-extracted and commercial samples of low-erucic rapeseed oil. Figure 3.1 shows the gas–liquid chromatogram of fatty acid methyl esters from a retail canola oil sample, which

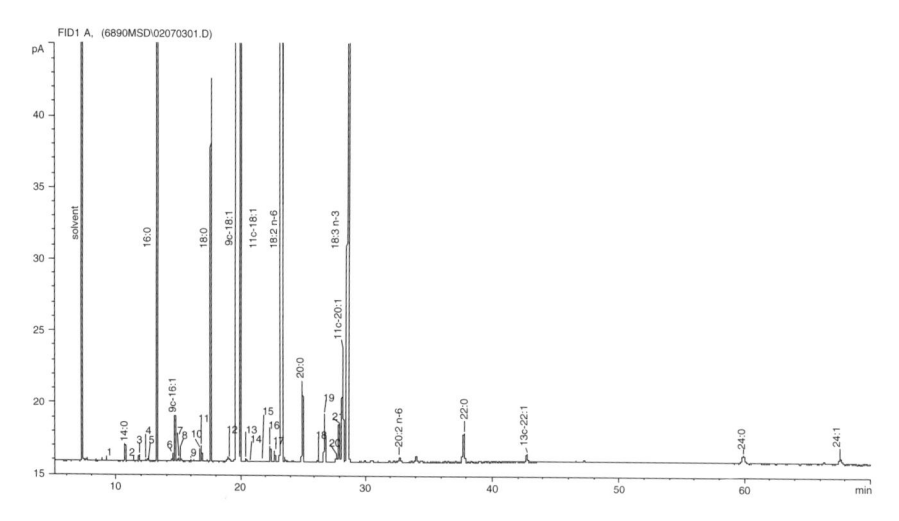

**Figure 3.1** Gas chromatogram of fatty acid methyl esters from a retail canola oil sample. Analysis on an SP-2560 capillary column (100 m × 0.25 mm i.d. × 20 μm film). Operated isothermally at 180°C. Peak identification: 1,12:0; 2,14:2 *n*-6; 3,15:0; 4, 5c-15:1; 5, 7c-15:1; 6, 7c-16:1; 7, 11c-16:1; 8, 17:0; 9, 16:2 *n*-6; 10, 17:1; 11, 16:3 *n*-3; 12, t-18:1; 13, 13c-18:1; 14, 15c-18:1; 15, 9t,12t-18:2; 16, 9c,12t-18:2; 17, 9t,12c-18:2; 18, 9t,12c,15t-18:3; 19, 9c,12c,15t-18:3; 20, 9c,12t,15c-18:3 and 21, 9t,12c,15c-18:3.

illustrates the gas chromatography (GC) position of some of the minor fatty acids relative to the major and more common fatty acids of vegetable oils. Most of the minor fatty acids present in rapeseed oil differ from the usual common fatty acids in the location of the double bond and a few other minor fatty acids may have unusual combination of the common structural features. A majority of the minor fatty acids in rapeseed oils belong to the n-7 series of monoethylenic fatty acids. Rapeseed oils are somewhat unusual in containing significant quantities of n-7 fatty acids. These fatty acids are formed by elongation of palmitoleic acid formed by the action of a Δ9-desaturase on palmitoyl-acyl carrier protein (Hu et al., 1994). While many oils contain traces of these fatty acids, canola oil contains 2–3% (Table 3.4). The elongation of the n-7 fatty acids follows approximately the same pattern as the elongation of n-9 fatty acids, although the relative proportion of n-7 to n-9 for the 20:1 fatty acids (1.5% n-7 to 5.2% n-9) is higher than the ratio for 18:1 or 22:1. The internal seed coat lipids of B. napus and B. rapa seem to have a high proportion (about 20% of the total fatty acids) of n-7 fatty acids (van de Loo et al., 1993).

Several other minor monoethylenic fatty acid isomers with the double bond in unusual positions are also present in rapeseed oils. Among them are 14:1(n-5), 16:1(n-5), 17:1(n-8) and 15:1(n-10), the latter accompanied by half as much trans-15:1(n-10). Sebedio and Ackman (1981) have hypothesized that the Δ9-desaturase enzyme acting on 14:0 in lieu of 16:0 would yield 14:1(n-5) and then 16:1(n-5) by elongation, but this would not extend further to 18:1(n-5). The 17:1(n-8), although an unusual fatty acid, was found in various fats and oils (Pohl and Wagner, 1972). The cis- and trans-15:1(n-10) fatty acids have not been found in other natural sources, except for a 15:1(n-10) identified in Norway spruce (Picea abies) (Ekman and Pesner, 1973).

Several minor all-cis-polyunsaturated fatty acids were also present in rapeseed oils (Table 3.3). Among them were four shorter chain structures, namely 14:2(n-6), 16:2(n-6), 16:2(n-4) and 16:3(n-3) and two long-chain structures, 20:2(n-6) and 20:3(n-3) (Sebedio and Ackman, 1979, 1981). The exact

Table 3.4 (n-7) Fatty acids in oils from different rapeseed and canola types[a]

| Seed type | Weight % of total fatty acids | | | |
|---|---|---|---|---|
| | 16:1(n-7) | 18:1(n-7) | 20:1(n-7) | 22:1(n-7) |
| B. rapa canola | 0.2 | 2.4 | ND | ND |
| B. napus canola | 0.2 | 2.5 | ND | ND |
| B. juncea canola | 0.2 | 1.5 | ND | ND |
| B. napus low linolenic acid canola | 0.2 | 2.7 | ND | ND |
| B. napus high erucic acid canola | 0.2 | 0.7 | 1.5 | 1.5 |
| B. napus high oleic acid canola | 0.3 | 2.3 | ND | ND |
| B. napus high lauric canola | 0.2 | 2.2 | ND | ND |

[a] Based on analysis of samples in an author's (Daun) laboratory.
ND = not detected.

biosynthetic routes of these minor polyunsaturated fatty acids are not known, but Sebedio and Ackman (1981) have suggested that 14:2(*n*-6) and 16:2(*n*-6) are most likely derived by chain shortening of 18:2(*n*-6) and 16:3(*n*-3) from 18:3(*n*-3). The two $C_{20}$ polyunsaturated fatty acids, 20:2(*n*-6) and 20:3(*n*-3), are plausibly derived from two carbon chain extension of 18:2(*n*-6) and 18:3(*n*-3) respectively. These minor polyunsaturated fatty acids found in rapeseed oil are also present in other edible seed oils. They are especially very common in seaweeds (Ratnayake, 1980), fresh water and marine fishes (Yukowski, 1989) and other marine organisms (Ackman *et al.*, 1970).

Minor amounts of geometrical isomers of linoleic (9 *trans*-, 12 *trans*-18:2; 9 *cis*-, 12 *trans*-18:2 and 9 *trans*-, 12 *cis*-18:2) and α-linolenic acids (9 *trans*-, 12 *cis*-, 15 *trans*-18:3; 9 *cis*-, 12 *cis*-, 15 *trans*-18:3; 9 *cis*-, 12 *trans*-, 15 *cis*-18:3 and 9 *trans*-, 12 *cis*-, 15 *cis*-18:3) are very often present in rapeseed oils (Ackman *et al.*, 1974). These are not natural products, but are artifacts formed by the isomerization of one or more double bonds of linoleic or α-linolenic acids during refining of oils. These *trans* isomers can be found in any refined oil containing linoleic and α-linolenic acids and their levels can reach up to 1% of total fatty acids (Ackman *et al.*, 1974).

Wijesundera and Ackman (1988) have indicated the possible presence of trace amounts (<0.01% of total fatty acids) of sulfur containing fatty acids in unprocessed canola oil. Their structures have tentatively been identified as isomeric 9,12-; 8,11-; and 7,10-epithiostearic acids, with a methyl substitution in the ring (Fig. 3.2). Canola is the only known edible oil that contains sulfur fatty acids. The structures of these sulfur fatty acids resemble the furanoid fatty acids (oxygen in the ring in lieu of sulfur) which are known as natural

| Structure | x | y |
|---|---|---|
| 1 | 5 | 7 |
| 2 | 6 | 6 |
| 3 | 7 | 5 |

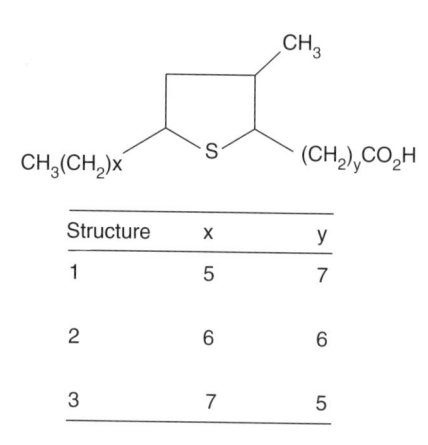

**Figure 3.2** Structures proposed for three sulfur-containing fatty acids found in canola oil (Wijesundera and Ackman, 1988).

components of plants but limited to the seeds of *Exocarpus cupressiformis* (Morris *et al.*, 1966) and the latex of the rubber tree *Hevea brasiliensis* (Hasma and Subramaniam, 1978). A series of furanoid fatty acids with one or two methyl substituents have been found in certain fish lipids (Gunstone *et al.*, 1974, 1978).

A study by Przybylski *et al.* (1993) of the composition of sediments from retail and industrial winterization of canola oil samples found a series of novel, long-chain saturated fatty acids and alcohols with 26–32 carbons (Table 3.5). Most of the long-chain fatty acids are present as esters of long-chain alcohols; however, a small portion might possibly be a part of triacylglycerols. These compounds probably originate from the waxes on the seed coat and hull and are thought to be the cause for occasional haze formation seen in some canola oil batches (Przybylski *et al.*, 1993).

**Table 3.5** Composition of sediments isolated from bottled canola oil and filter cake from commercial winterization[a]

| | Canola oil | | Winterization | |
|---|---|---|---|---|
| Compound | Fatty acid | Alcohol | Fatty acid | Alcohol |
| 15:0 | 0.31 | ND | ND | ND |
| 16:0 | 2.38 | 0.18 | 1.83 | 0.56 |
| 16:0-branched chain | 0.73 | ND | 1.27 | ND |
| 17:0 | 0.24 | ND | 0.23 | 0.35 |
| 18:0 | 3.36 | 0.43 | 1.81 | 0.43 |
| 18:1 | 2.81 | ND | ND | ND |
| 18:2 | 0.44 | ND | ND | ND |
| 19:0 | 0.27 | ND | ND | 3.96 |
| 20:0 | 54.45 | 2.06 | 35.46 | 0.46 |
| 21:0 | 1.21 | ND | 0.83 | 0.46 |
| 22:0 | 21.52 | 11.45 | 20.47 | 1.45 |
| 23:0 | 0.57 | 0.93 | 0.72 | 4.97 |
| 24:0 | 3.84 | 31.01 | 8.03 | 5.03 |
| 25:0 | 0.23 | 2.45 | 0.65 | 5.03 |
| 26:0 | 1.25 | 26.51 | 7.17 | 23.64 |
| 27:0 | ND | 0.73 | 0.51 | 5.74 |
| 28:0 | 0.63 | 10.47 | 8.97 | 12.22 |
| 29:0 | ND | 0.41 | 0.43 | 1.76 |
| 30:0 | 0.12 | 4.43 | 3.25 | 6.75 |
| 31:0 | ND | 0.16 | 0.13 | 0.23 |
| 32:0 | 0.03 | 3.23 | 0.42 | 6.27 |
| Others[b] | 5.61 | 5.55 | 7.82 | 5.13 |

Composition (% total fatty acids or alcohols)

[a] Table adapted from Przybylski *et al.* (1993).
[b] Total of unidentified compounds.
ND = not detected.

## 3.4 Triacylglycerols of rapeseed and canola oils

In common with other vegetable oils, approximately 95–99% of the fatty acids in rapeseed and canola oils are present as triacylglycerols (also named as triglycerides) (Unger, 1991). The structures of triacylglycerols (TAG) are characterized by three esterified fatty acids and their positions in the glycerol molecule. Consequently, complete characterization of TAG molecular species involves determination of the three fatty acids and their positions in the glycerol molecule. The TAG molecular structures govern the physical and chemical properties of fats and oils, and possibly some nutritional aspects. Theoretically, an oil with $n$ fatty acids could consist of $(n^3+3n^2+2n)/6$ TAG molecular species without consideration of potential stereoisomers (Litchfield, 1972). Rapeseed and canola oils are usually characterized by nine fatty acids with levels >0.5% of total fatty acids (Table 3.1) and therefore, without distinguishing isomers, these oils could be expected to be composed of at least 165 TAG molecular species. The identification of all the 165 TAG species has not been reported so far and it might be an impossible task even with the modern analytical techniques. Some aspects of TAG molecular structures of rapeseed and canola oil have been described by Persmark (1972), Ackman (1983), Eskin (1996) and Przybylski and Mag (2002).

In the earlier days, before the advent of GC and high performance liquid chromatography (HPLC), classical methods of separation (e.g. fractional crystallization primarily according to the degree of unsaturation) were used to identify TAG structures. Since the resolution of TAG mixtures by fractional crystallization is never complete, there is always some element of uncertainty in interpretation of the data. Among the earlier workers, Hilditch et al. (1947) were the very first group to attempt to elucidate the TAG structures of traditional rapeseed oil. They separated TAG by fractional crystallization from acetone and identified the main classes as $SU(C_{18})$ U(22:1) (18%), $U(C_{18})$ U(22:1) U(22:1) (54%) and $U(C_{18})$ $U(C_{18})$ U(22:1) (28%) (S=saturated, $U(C_{18})$=unsaturated $C_{18}$, U(22:1)=erucic acid). Subsequent workers fractionated rapeseed oil TAG initially by silver nitrate ($AgNO_3$) chromatography followed by reversed-phase thin-layer chromatography (Kaufmann and Wessels, 1964; Grynberg et al., 1966). Using this procedure, Kaufmann and Wessels (1964) detected 50 different TAG molecular species, whereas Grynberg et al. (1966) found only 11. However, none of the workers fully identified the separated fractions and no quantitative data were reported.

The introduction of GC provided a novel tool for lipid chemists for examination of the fatty acids as well as the TAG molecular species of fats and oils. Subbaram and Young (1967) analysed traditional rapeseed oil TAG by packed-column GC after oxidative fission of the double bonds. The chief TAG classes reported were tri-$C_{18}$-unsaturated (20%), mono-20:1-di-$C_{18}$-unsaturated (15%) and mono-22:1-di-$C_{18}$-unsaturated (35%). These data, however, were not in agreement with those of Hilditch *et al.* (1947), probably because the erucic acid content of the rapeseed oil used by Subbaram and Young (1967) was 22% suggesting that the oil originated from *B. rapa*. This is somewhat lower than the usual 40–45% reported for traditional *B. napus* oils. The proportions of eicosenoic, linolenic and linoleic acids were accordingly somewhat higher than is usual for traditional rapeseed oil. Subbaram and Young (1967) analysed this same oil by pancreatic-lipase hydrolysis, which selectively cleaves the fatty acids in the 1,3-positions. The TAG composition was subsequently calculated according to the 1,3-random-2-random distribution theory of Coleman (1960). One hundred and fourteen TAG were calculated and their proportions are presented in Table 3.6. Compared to TAG composition by GLC analysis of oxidized glycerides, the corresponding data calculated by the pancreatic-lipase hydrolysis showed a higher percentage of the tri-$C_{18}$-unsaturated TAG (20% vs 8.5% by GLC) and lower percentage of mono-erucic-di-$C_{18}$-unsaturated TAG (35% vs 26.2%). This discrepancy might result from the possibility that oils rich in erucic acid tend to be poor substrates for pancreatic-lipase hydrolysis (Subbaram and Young, 1967).

The direct analysis of TAG by GLC, using short non-polar columns, is a standard procedure for separation of TAG molecular species according to their total acyl carbon number. Although this technique alone is not sufficient to identify the complete TAG structures, it is very useful in comparing different oils (Ackman, 1983). Most oils including canola oil (low erucic acid and low glucosinolate rapeseed oil) have a maximum at a carbon number of 54, which primarily represents TAG molecular species with three $C_{18}$ fatty acids (Table 3.7). The high erucic acid rapeseed oils have a more complex TAG profile with a maximum at a carbon number of 62 probably reflecting the fatty acid combination $2 \times C_{22} + 1 \times C_{18}$ (Ackman, 1983). As the proportion of erucic acid reduces from the usual value of approximately 48%, this shifts the emphasis from carbon number 62 to lesser carbon numbers as the probability of two erucic acid moieties being in one TAG molecular species is low.

Nowadays, much better data on TAG molecular species can be obtained by a combination of reversed-phase HPLC with flame ionization detection (FID), fatty acid analysis by GC, mass spectrometry detection of TAG species and stereospecific TAG structure analysis by lipolysis and GC (Neff *et al.*, 1994a,b, 1997; Byrdwell and Neff, 1996; Andrikopoulos, 2002). This procedure might be complex and time-consuming, but it has the capacity to provide complete identification and quantification of all of the major and most of the minor TAG molecular species. Neff and colleagues (Neff *et al.*, 1994a, 1997; Byrdwell and Neff, 1996) using the combined technique examined the TAG composition of a number of genetically modified low α-linolenic, high stearic and high lauric acid canola lines. These lines were developed with the aim of providing oils with improved storage and frying stabilities, by decreasing

**Table 3.6** Triacylglycerol composition of high erucic rapeseed oil[1] (mole %)

| Molecular species[2] | From lipase data[3] | GLC det.[4] | Molecular species[2] | From lipase data[3] | GLC det.[4] |
|---|---|---|---|---|---|
| OOO | 3.93 | | POP | 1.11 | |
| OOL | 3.21 | | POEr | 1.22 | |
| OOLn | 1.04 | | SOE | 0.3 | |
| LOL | 0.6 | | ErOEr | 3.48 | |
| LOLn | 0.4 | | PLP | 0.09 | |
| LnOLn | 0.06 | | PLEr | 1.0 | |
| OLO | 3.36 | | SLE | 0.24 | |
| OLL | 2.6 | | ErLEr | 2.9 | |
| OLLn | 0.85 | | PLnP | 0.04 | |
| LLL | 0.5 | | PLnEr | 0.42 | |
| LLLn | 0.34 | | SLnE | 0.1 | |
| LLLn | 0.06 | | ErLnEr | 1.16 | |
| LnLLn | 1.3 | | EEEr | 0.2 | |
| OLnO | 1.1 | | PErO | 0.22 | |
| OLnL | 0.34 | | PErL | 0.08 | |
| OLnLn | 0.2 | | OErEr | 1.22 | |
| LLnL | 0.2 | | LErEr | 0.52 | |
| LLnLn | 0.14 | | LnErEr | 0.18 | |
| Subtotal | 20.03 | 8.5 | EErE | 0.2 | |
| OOE | 4.5 | | Subtotal | 13.68 | 13.2 |
| LOE | 1.71 | | POS | 0.08 | |
| LnOE | 0.56 | | SOEr | 0.48 | |
| OLE | 3.6 | | PLS | 0.06 | |
| LLE | 1.42 | | SLEr | 0.52 | |
| LnLE | 0.48 | | PLnS | 0.02 | |
| OLnE | 1.48 | | SLnEr | 0.16 | |
| LLnE | 0.56 | | PEEr | 0.06 | |
| LnLnE | 0.2 | | ErEEr | 0.2 | |
| OEO | 0.27 | | PErE | 0.12 | |
| OEL | 0.18 | | SErO | 0.1 | |
| LEL | 0.03 | | SErEL | 0.04 | |
| Subtotal | 14.99 | 12.6 | EErEr | 0.7 | |
| POO | 1.3 | | Subtotal | 2.54 | 2.8 |
| POL | 0.5 | | PErP | 0.02 | |
| POLn | 0.16 | | PErEr | 0.22 | |
| OOEr | 7.39 | | SErE | 0.06 | |
| LOEr | 2.99 | | ErErEr | 0.58 | |
| LnOEr | 0.98 | | Subtotal | 0.88 | 0.8 |
| EOE | 1.22 | | | | |
| PLO | 1.08 | | SErEr | 0.08 | |
| PLL | 0.42 | | ELE | 1.02 | |
| PLLn | 0.14 | | PLnO | 0.44 | |
| PLL | 0.14 | | PLnL | 0.16 | |
| OLEr | 6.3 | | PLnLn | 0.04 | |
| LLEr | 2.44 | | OLnEr | 2.52 | |
| LnLEr | 0.8 | | LLnEr | 0.98 | |
| LnLnEr | 0.32 | | SLO | 0.44 | |

| | | | | | |
|---|---|---|---|---|---|
| ElnE | 0.4 | | SLL | 0.18 | |
| OEE | 0.23 | | SLLn | 0.08 | |
| LEE | 0.08 | | ELEr | 3.5 | |
| LnEE | 0.04 | | PLnE | 0.24 | |
| OErO | 0.65 | | SliO | 0.2 | |
| OErL | 0.6 | | SLnL | 0.08 | |
| OErLn | 0.16 | | ELnEr | 1.4 | |
| LEL | 0.11 | | PEO | 0.06 | |
| Subtotal | 33.47 | 43.9 | OEEr | 0.45 | |
| | | | LEEr | 0.14 | |
| POE | 0.72 | | EEE | 0.06 | |
| SOO | 0.52 | | OErE | 0.74 | |
| SOL | 0.21 | | LErE | 0.3 | |
| SOLn | 0.08 | | LnErE | 0.1 | |
| EOEr | 4.25 | | | | |
| PLE | 0.58 | | Total | 14.33 | 18.2 |

[1] Data from Subbaram and Young (1967). Fatty acid abbreviations: P = palmitic; S = stearic; O = oleic; E = eicosenic; Er = erucic; L = linoleic; Ln = linolenic.
[2] The molecular species are represented (left to right) by fatty acids in the *sn*-1, *sn*-2 and *sn*-3 positions.
[3] Values derived from 1,3-random, 2-random calculation using the fatty acid compositions of the appropriate *sn* positions.
[4] Values derived by GLC analysis of oxidized glycerides.

**Table 3.7** Distribution of triacylglycerol molecular species according to total carbon number of some vegetable oils as determined by gas–liquid chromatography[a]

| Oil | Mole % erucic acid | Gas–liquid chromatography carbon number (GLC area %) | | | | | | | | |
|---|---|---|---|---|---|---|---|---|---|---|
| | | $C_{48}$ | $C_{50}$ | $C_{52}$ | $C_{54}$ | $C_{56}$ | $C_{58}$ | $C_{60}$ | $C_{62}$ | $C_{64}$ |
| HEAR | 48 | – | – | 1 | 2 | 6 | 8 | 18 | 61 | 5 |
| HEAR | 38 | – | <1 | 2 | 4 | 10 | 20 | 25 | 39 | <1 |
| HEAR | 45 | 1 | 2 | 6 | 13 | 17 | 20 | 20 | 20 | <1 |
| HEAR | 40 | <1 | 2 | 6 | 9 | 11 | 18 | 19 | 32 | <1 |
| HEAR | 25 | – | – | 11 | 21 | 46 | 11 | 5 | 2 | 3 |
| Mustard | 32 | – | <1 | 3 | 8 | 14 | 29 | 20 | 21 | – |
| Canola | 0.7 | 3 | 4 | 18 | 70 | 5 | 1 | <1 | – | – |
| Canola | 0.4 | 4 | 5 | 24 | 63 | 3 | – | – | – | – |
| Sunflower | – | – | 2 | 21 | 75 | 1 | – | – | – | – |
| Sunflower | – | – | 4 | 29 | 67 | – | – | – | – | – |
| Soybean | – | – | 5 | 42 | 52 | – | – | – | – | – |
| Peanut | – | – | 4 | 26 | 60 | 7 | 4 | – | – | – |

[a] Table adapted from Ackman (1983).
HEAR = high erucic acid rapeseed oil.

the contents of linoleic and α-linoleic acids and increasing the content of oleic and saturated fatty acids.

Table 3.8 lists the fatty acid composition of 11 genetically modified low α-linolenic acid varieties. These included some commercially grown cultivars as well as some experimental varieties (Neff *et al.*, 1994a). The α-linolenic acid

**Table 3.8** Fatty acid composition (GC area % of total fatty acids) of oils from some genetically modified low α-linolenic canola lines[a]

| IMC variety[b] | 4:0 | 16:0 | 18:0 | 20:0 | 22:0 | 24:0 | 16:1 | 18:1 | 20:1 | 22:1 | 18:2 | 18:3 |
|---|---|---|---|---|---|---|---|---|---|---|---|---|
| 600 | 0.2 | 3.9 | 2.2 | 0.7 | 0.3 | 0.3 | 0.1 | 60.0 | 1.2 | 0.0 | 22.4 | 8.8 |
| 500 | 0.0 | 3.5 | 2.0 | 0.8 | 0.4 | 0.3 | 0.0 | 65.2 | 1.4 | 0.0 | 18.7 | 7.6 |
| 1000 | 0.1 | 3.8 | 2.0 | 0.6 | 0.3 | 0.2 | 0.2 | 66.4 | 1.4 | 0.2 | 18.0 | 6.9 |
| 400 | 0.0 | 3.7 | 1.8 | 0.6 | 0.3 | 0.2 | 0.2 | 67.2 | 1.4 | 0.2 | 18.2 | 6.6 |
| 300 | 0.0 | 2.7 | 2.5 | 1.0 | 0.4 | 0.4 | 0.0 | 68.1 | 1.9 | 0.0 | 17.7 | 5.4 |
| 100 | 0.0 | 3.9 | 2.5 | 0.8 | 0.8 | 0.5 | 0.0 | 66.3 | 1.1 | 0.0 | 20.4 | 3.6 |
| 700 | 0.0 | 3.7 | 2.3 | 0.7 | 0.3 | 0.2 | 0.2 | 66.3 | 1.1 | 0.0 | 23.1 | 2.1 |
| 800 | 0.0 | 3.3 | 2.2 | 0.8 | 0.4 | 0.3 | 0.2 | 69.7 | 1.2 | 0.0 | 18.7 | 3.3 |
| 900 | 0.1 | 3.8 | 2.3 | 0.7 | 0.3 | 0.2 | 0.2 | 68.5 | 1.2 | 0.0 | 20.4 | 2.4 |
| 1100 | 0.0 | 3.4 | 2.5 | 0.7 | 0.4 | 0.2 | 0.2 | 77.7 | 1.4 | 0.0 | 8.1 | 6.4 |
| 200 | 0.1 | 3.2 | 2.5 | 0.7 | 0.2 | 0.0 | 0.1 | 81.3 | 1.5 | 0.0 | 6.5 | 4.1 |

[a] Table adapted from Neff *et al.* (1994a).
[b] InterMountain Canola Company (IMC) (Cinnaminson, NJ) variety. Variety identification with IMC numbers are: 600 Hyola, commercially grown generic spring hybrid cultivar, developed by Zeneca Seeds, Wilmington, Delaware; 500 Legend, commercially grown generic spring canola cultivar, developed by Svalof, Sweden; 1000 Westar, commercially grown generic spring canola cultivar, developed by Agriculture Canada, Saskatchewan; 400 IMC Westar, commercially grown generic spring canola cultivar, developed by Agriculture Canada, Saskatchewan, Canada; 300 IMC, commercially grown low saturates canola; 100 IMC, commercially grown low α-linolenic acid (ALA) spring canola variety; 700 Stellar, commercially grown low α-linolenic spring canola cultivar, developed by University of Manitoba, Winnipeg, Manitoba; 800, experimental IMC low α-linolenic breeding germplasm; 900, experimental IMC low α-linolenic breeding germplasm, Canada; 1100, experimental IMC low α-linolenic breeding germplasm; 200 IMC, commercially grown high oleic acid variety.

content showed a considerable variation ranging from 9 (IMC 600) to 2% (IMC 700). This modification was accompanied by changes in the proportions of linoleic and oleic acids. Other fatty acids (saturates, 16:1, 20:1 and 22:1) showed very little variations. In these low α-linolenic varieties, two TAG molecular species, OOO (22–50%) and LOO (13–28%), predominated (Table 3.9). Depending on the variety, these two TAGs comprised 45 (IMC 600) to 65% (IMC 1100) of the total TAG. In variety 600, the contents of LOO and OOO were approximately the same, whereas in the other canola varieties, the content of OOO was considerably higher. Other important triglycerides were LnOO (3–10%), LLO (1–11%), LnLO (1–8%), LOP (2–6%), POO (4–8%) and SOO (2–5%). Varieties 200 and 1100 contained the highest contents of OOO, with lower contents of α-linolenic and linoleic acids containing TAGs such as LnLO, LLO and LOO. Oils with a high OOO content have high oxidative stability (Neff *et al.*, 1994a).

The fatty acid composition of oils from six genetically modified normal and high stearic and high lauric acid canola varieties reported by Neff *et al.* (1997)

**Table 3.9** Triacylglycerol composition (HPLC-FID area %) of oils from genetically modified low α-linolenic acid canola lines[a]

| Triacylglycerol[b] | \multicolumn{11}{c}{InterMountain Canola Company (IMC) variety[c]} |
|---|---|---|---|---|---|---|---|---|---|---|---|
|  | 600 | 500 | 1000 | 400 | 300 | 100 | 700 | 800 | 900 | 1100 | 200 |
| LnLnLn | 0.2 | 0.1 | 0.1 | 0.0 | 0.1 | 0.1 | 0.0 | 0.0 | 0.0 | 0.0 | 0.0 |
| LnLnL | 0.6 | 0.4 | 0.2 | 0.3 | 0.3 | 0.5 | 0.0 | 0.1 | 0.2 | 0.1 | 0.5 |
| LnLL | 1.4 | 0.9 | 0.5 | 0.6 | 0.7 | 0.3 | 0.0 | 0.5 | 0.0 | 0.1 | 0.3 |
| LnLnO | 1.7 | 1.3 | 0.9 | 1.1 | 1.0 | 1.1 | 0.4 | 0.4 | 0.3 | 0.1 | 0.1 |
| LnLnP | 0.2 | 0.1 | 0.1 | 0.1 | 0.1 | 0.0 | 0.2 | 0.1 | 0.2 | 0.0 | 0.0 |
| LLL | 1.3 | 0.7 | 0.5 | 0.6 | 0.8 | 1.1 | 1.6 | 0.9 | 1.0 | 0.1 | 0.2 |
| LnLO | 7.6 | 5.3 | 4.4 | 4.9 | 4.2 | 1.8 | 1.7 | 3.2 | 1.1 | 1.6 | 1.5 |
| LnLP | 0.9 | 0.6 | 0.5 | 0.9 | 0.6 | 0.4 | 0.3 | 0.6 | 0.7 | 0.2 | 0.5 |
| LLO | 8.6 | 6.3 | 5.9 | 6.5 | 6.5 | 8.3 | 11.0 | 7.4 | 7.8 | 1.2 | 1.1 |
| LnOO | 10.4 | 10.2 | 8.9 | 8.7 | 8.1 | 4.9 | 2.6 | 5.5 | 2.7 | 10.1 | 8.6 |
| LLP | 1.4 | 0.9 | 0.9 | 1.2 | 1.0 | 1.2 | 1.1 | 1.2 | 0.9 | 0.3 | 0.8 |
| LnOP | 2.1 | 1.5 | 1.1 | 1.6 | 1.3 | 0.8 | 0.5 | 0.8 | 0.3 | 0.8 | 1.1 |
| LOO | 22.5 | 21.8 | 23.8 | 20.8 | 22.4 | 24.9 | 28.4 | 21.8 | 26.6 | 14.6 | 12.7 |
| LLS | 0.3 | 0.3 | 0.3 | 0.6 | 0.7 | 0.7 | 0.6 | 0.4 | 0.6 | 0.3 | 0.4 |
| LOP | 5.7 | 4.6 | 3.4 | 5.4 | 3.6 | 4.6 | 4.2 | 4.8 | 3.7 | 2.2 | 2.2 |
| PLP | 0.3 | 0.0 | 0.2 | 0.3 | 0.3 | 0.3 | 0.3 | 0.2 | 0.5 | 0.2 | 0.2 |
| OOO | 22.4 | 30.1 | 36.2 | 29.1 | 31.5 | 31.7 | 32.8 | 31.9 | 37.9 | 50.4 | 49.5 |
| LOS | 1.6 | 1.7 | 1.3 | 2.7 | 2.1 | 2.3 | 1.9 | 1.8 | 2.2 | 1.3 | 1.0 |
| POO | 4.6 | 5.3 | 4.2 | 6.4 | 4.3 | 6.0 | 4.8 | 6.8 | 5.3 | 6.9 | 7.7 |
| SLP | 0.2 | 0.2 | 0.2 | 0.4 | 0.2 | 0.0 | 0.2 | 0.3 | 0.0 | 0.0 | 0.4 |
| POP | 0.2 | 0.2 | 0.1 | 0.2 | 0.0 | 0.0 | 0.2 | 0.3 | 0.0 | 2.1 | 0.3 |
| PPP | 0.1 | 1.7 | 0.1 | 1.5 | 2.5 | 1.5 | 1.4 | 2.3 | 1.5 | 0.4 | 2.8 |
| SOO | 2.6 | 2.6 | 1.9 | 2.8 | 2.9 | 3.1 | 2.4 | 4.0 | 2.7 | 3.9 | 5.0 |
| SLS | 0.4 | 0.1 | 0.2 | 0.2 | 0.1 | 0.0 | 0.2 | 0.2 | 0.0 | 0.0 | 0.4 |
| SOP | 0.1 | 0.3 | 0.2 | 0.3 | 0.2 | 0.3 | 0.2 | 0.2 | 0.1 | 0.2 | 0.3 |
| PPS | 0.5 | 0.6 | 0.6 | 0.1 | 0.9 | 0.8 | 0.6 | 1.3 | 0.2 | 0.9 | 0.6 |
| SPS | 0.2 | 0.3 | 0.0 | 0.1 | 0.5 | 0.5 | 0.3 | 0.6 | 0.9 | 0.2 | 0.8 |
| SSS | 0.1 | 0.1 | 0.1 | 0.2 | 0.3 | 0.3 | 0.2 | 0.4 | 0.3 | 0.0 | 0.1 |
| Unidentified | 2.2 | 2.9 | 1.9 | 3.0 | 3.5 | 3.4 | 2.2 | 2.6 | 2.3 | 0.4 | 1.1 |

[a] Table adapted from Neff et al. (1994a). Triacylglycerol composition determined by reversed-phase high performance liquid chromatography with flame ionization detection (HPLC-FID).
[b] Fatty acid abbreviations as in Table 3.6.
[c] InterMountain Canola Company (IMC) (Cinnaminson, NJ) variety. Variety identification as in Table 3.8.

is shown in Table 3.10. These are experimental varieties produced by Calgene, Inc. (Davis, CA). Two of the canola varieties (samples 5 and 6) contained lauric acid, not previously a canola fatty acid. Their lauric acid content ranged from 9 to 31%. The other modified oils (samples 2, 3 and 4) were high in stearic acid, whose levels ranged from 17 to 28%. These values for stearic acid were substantially higher than the values of 1–2% normally seen for regular canola oil. The triacylglycerol composition showed considerable variation

**Table 3.10** Fatty acid composition (% of total fatty acids) for oils from genetically modified normal, high stearic acid and high lauric acid canola varieties[a]

| | Canola varieties[b] | | | | | |
|---|---|---|---|---|---|---|
| | Normal fatty acid composition | High stearic | | | High lauric | |
| Fatty acid | Sample 1 | Sample 2 | Sample 3 | Sample 4 | Sample 5 | Sample 6 |
| 12:0 | 0.0 | 0.0 | 0.0 | 0.0 | 8.7 | 31.3 |
| 14:0 | 0.1 | 0.1 | 0.1 | 0.1 | 1.2 | 4.2 |
| 16:0 | 4.6 | 3.6 | 3.6 | 3.6 | 4.3 | 3.3 |
| 18:0 | 1.6 | 28.0 | 27.5 | 16.6 | 1.4 | 1.1 |
| 20:0 | 0.6 | 2.2 | 1.2 | 3.3 | 0.4 | 0.3 |
| 22:0 | 0.3 | 0.4 | 0.4 | 1.0 | 0.2 | 0.2 |
| 24:0 | 0.2 | 0.2 | 0.2 | 0.3 | 0.0 | 0.0 |
| 16:1 | 0.3 | 0.2 | 0.2 | 0.2 | 0.2 | 0.2 |
| 18:1 | 57.5 | 28.8 | 33.5 | 45.4 | 56.5 | 35.1 |
| 20:1 | 1.4 | 0.4 | 0.3 | 0.7 | 1.0 | 0.7 |
| 24:1 | 0.2 | 0.2 | 0.3 | 0.3 | 0.1 | 0.0 |
| 18:2 | 19.4 | 18.4 | 18.4 | 24.1 | 15.5 | 14.6 |
| 18:3 | 13.8 | 17.4 | 13.7 | 4.3 | 10.6 | 8.8 |

[a] Table adapted from Neff *et al.* (1997).
[b] Canola varieties were from Calgene, Inc. (Davis, CA).

(Table 3.11). For example, for the control oil (sample 1), triacylglycerols LnLO, LnOO, LOO, LOP and OOO occurred in concentrations, each greater than 5%. Whereas in the high stearic oils, these triacylglycerols varied as follows: LnLO (1–9%), LnOO (2–16%), LOO (5–19%), LOP (1–5%) and OOO (4–23%). In addition, compared to the control oil (sample 1), LLS, LnOS, LOS, LnSS, SLS and SOS became abundant or appeared as the stearic acid content in the oils is increased. The triacylglycerol composition of lauric acid containing canola oils is presented in Table 3.12. These oils have new triacylglycerol species such as LaLaO, LaLaL, LaOO, LaOL and LaMO, not previously reported in canola varieties. These and other saturated TAGs such as SOO, POO and LOP generally improve vegetable-oil oxidative stability (Neff *et al.*, 1992, 1994b).

In addition to the triacylglycerol composition, the positional distribution of fatty acids in the glycerol molecule also influences the chemical characteristics, in particular the oxidative stability of oils. It is generally assumed that unsaturated fatty acids attached to carbon 2-position of the glycerol molecule are less prone to oxidation than those at carbon 1 or 3-position (Wada and Koizumi, 1983; Neff *et al.*, 1992, 1994b). The distribution of fatty acids is also of some interest to nutritionists, because of the preferential cleavage of fatty acids in the intestine by pancrease from positions 1 and 3 to yield 2-monoacylglycerols and subsequent absorption, and transportation of these hydrolyzed products via different routes (Tso, 1985). It is well established that in most common vegetable oils, including the various types of rapeseed oils, saturated fatty acids

**Table 3.11** Triacylglycerol composition (HPLC peak area %) of oils from genetically modified normal and high stearic canola lines[a]

| Triacylglycerol[b] | Normal fatty acid composition Sample 1 | High stearic acid composition Sample 2 | Sample 3 | Sample 4 |
|---|---|---|---|---|
| LnLnLn | 0.2 | 0.1 | 0.0 | 0.0 |
| LnLnL | 0.9 | 0.3 | 0.3 | 0.2 |
| LnLL | 0.9 | 0.3 | 0.2 | 0.3 |
| LnLnO | 3.7 | 0.5 | 1.0 | 0.6 |
| LnLnP | 0.4 | 0.3 | 0.2 | 0.1 |
| LLL | 0.7 | 0.7 | 1.0 | 0.3 |
| LnLO | 8.9 | 1.4 | 1.5 | 1.3 |
| LnLP | 1.3 | 0.6 | 2.6 | 1.4 |
| LnLnS | 0.0 | 2.6 | 0.8 | 0.7 |
| LnLS | 0.0 | 2.2 | 2.3 | 0.0 |
| LLO | 4.5 | 2 | 2.2 | 6.6 |
| LnOO | 16.1 | 6.2 | 5.7 | 2.0 |
| LLP | 1.4 | 0.0 | 1.0 | 2.5 |
| LnOP | 3.1 | 1.5 | 0.2 | 0.7 |
| LnPP | 0.0 | 0.7 | 0.1 | 0.1 |
| LOO | 19.1 | 4.7 | 5.9 | 13.3 |
| LLS | 0.0 | 3.4 | 2.9 | 4.5 |
| LnOS | 0.0 | 11.1 | 11.1 | 0.1 |
| LOP | 5.2 | 0.8 | 1.0 | 5.3 |
| PLP | 0.2 | 1.4 | 1.0 | 0.6 |
| OOO | 23.1 | 4.3 | 6.2 | 13.3 |
| LOS | 1.5 | 11.1 | 11.9 | 12.1 |
| POO | 4.8 | 2.2 | 1.9 | 2.6 |
| LnSS | 0.0 | 11.8 | 9.1 | 2.1 |
| SLP | 0.2 | 0.0 | 0.0 | 0.0 |
| POP | 0.0 | 0.3 | 0.4 | 0.3 |
| PPP | 1.0 | 0.6 | 0.1 | 0.1 |
| SOO | 1.1 | 7.6 | 9.7 | 10.6 |
| SLS | 0.0 | 10.7 | 9.3 | 5.2 |
| SOP | 0.1 | 1.5 | 1.4 | 1.1 |
| PPS | 0.3 | 1.0 | 0.7 | 1.2 |
| SOS | 0.1 | 5.0 | 6.1 | 3.2 |
| PSS | 0.2 | 0.7 | 0.6 | 1.0 |
| SSS | 0.1 | 0.3 | 0.2 | 0.4 |
| Unidentified | 0.9 | 2.6 | 1.4 | 6.2 |

[a] Table adapted from Neff *et al.* (1997). Canola varieties were from Calgene, Inc. (Davis, CA). Triacylglycerol composition determined by reversed-phase high performance liquid chromatography with flame ionization detection (HPLC-FID).
[b] Fatty acid abbreviations as in Table 3.6.

are preferentially concentrated in positions 1 and 3, whilst unsaturated fatty acids, especially linoleic and α-linolenic acids, tend to prefer the 2-position (Appelqvist, 1972; Litchfield, 1972; Ohlson *et al.*, 1975; Zadernowski and Sosulski, 1979; Jàky and Kurnik, 1981). This was recently confirmed by Richards *et al.* (2002) who examined the regiospecific triacylglycerol composition of

**Table 3.12** Triacylglycerol composition (HPLC peak area %) of oils from two genetically modified high lauric canola lines[a]

| Triacylglycerol[b] | Sample 5 | Sample 6 |
|---|---|---|
| LaLnLn | 0.3 | 0.1 |
| LaLaLn | 0.7 | 7.4 |
| LaLLn | 0.6 | 2.3 |
| LaLaL | 1.0 | 14.2 |
| LaOLn | 3.4 | 4.0 |
| LaML + LaPLn | 0.0 | 2.0 |
| LaLaO | 2.7 | 27.6 |
| LnLO | 2.6 | 0.0 |
| LaOL | 6.8 | 6.6 |
| LaMO | 2.8 | 5.6 |
| LLO | 5.8 | 0.0 |
| POLn | 2.0 | 0.0 |
| LaOO | 9.1 | 11.2 |
| LaPO | 0.0 | 3.6 |
| LOO | 11.0 | 2.1 |
| MOO | 0.0 | 1.3 |
| LaSO | 5.6 | 1.4 |
| PLP | 1.0 | 0.0 |
| OOO | 17.0 | 3.3 |
| LOS | 4.5 | 0.0 |
| POO | 3.9 | 0.7 |
| SLP | 3.0 | 0.0 |
| POP | 0.4 | 0.0 |
| PPP | 0.7 | 0.0 |
| SOO | 2.7 | 0.0 |
| SLS | 1.9 | 0.0 |
| SOP | 0.2 | 0.0 |
| PPS | 1.1 | 0.0 |
| SOS | 0.2 | 0.0 |
| PSS | 0.2 | 0.5 |
| SSS | 0.1 | 0.0 |
| Unidentified | 8.3 | 6.1 |

[a] Table adapted from Neff *et al.* (1997). Canola varieties were from Calgene, Inc. (Davis, CA). Triacylglycerol composition determined by reversed-phase high performance liquid chromatography with flame ionization detection (HPLC-FID).
[b] La = lauric, M = myristic. Other fatty acid abbreviations as in Table 3.6.

some cooking oils consumed in Australia (Table 3.13). There are exceptions however, most notably for highly saturated oils such as coconut and palm-kernel oils, where appreciable amounts of saturated fatty acids are distributed in the 2-position. Jàky and Kurnick (1981), more than two decades ago, after examining several types of rapeseed oils, suggested that in HEAR oils with 14 and 13% of linoleic acid, at least 95% was concentrated in the 2-position, whereas in an LEAR oil with 19% of linoleic acid, the percentage of linoleic acid in the

Table 3.13 Total fatty acid profile and regiospecific fatty acid composition at glycerol carbons 1,3 and 2 of some cooking oils consumed in Australia[a]

| Cooking oil | 8:0 | 10:0 | 12:0 | 14:0 | 16:0 | 16:1 | 18:0 | 18:1 | 18:2 | 18:3 |
|---|---|---|---|---|---|---|---|---|---|---|
| **Canola** | | | | | | | | | | |
| Total profile | – | – | – | – | 4.7 | 0.3 | 2.1 | 58.1 | 20.8 | 10.2 |
| 1,3-position | – | – | – | – | 6.6 | 0.3 | 2.8 | 60.9 | 16.4 | 6.9 |
| 2-position | – | – | – | – | 1.4 | 0.1 | 0.7 | 52.7 | 29.5 | 13.7 |
| **Cold-pressed canola** | | | | | | | | | | |
| Total profile | – | – | – | – | 5.4 | 0.2 | 1.8 | 61.1 | 20.2 | 8.0 |
| 1,3-position | – | – | – | – | 6.3 | 0.3 | 2.1 | 62.5 | 17.6 | 7.5 |
| 2-position | – | – | – | – | 3.5 | 0.1 | 1.2 | 58.2 | 25.5 | 9.1 |
| **Olive (early harvest)** | | | | | | | | | | |
| Total profile | – | – | – | – | 12.9 | 0.9 | 2.7 | 71.5 | 8.8 | 0.7 |
| 1,3-position | – | – | – | – | 17.7 | 1.2 | 3.6 | 67.1 | 7.6 | 0.1 |
| 2-position | – | – | – | – | 3.5 | 0.5 | 1.0 | 81.1 | 11.2 | 1.9 |
| **Olive (late harvest)** | | | | | | | | | | |
| Total profile | – | – | – | – | 13.5 | 1.1 | 2.8 | 71.3 | 7.9 | 0.7 |
| 1,3-position | – | – | – | – | 20.0 | 1.3 | 3.9 | 64.9 | 6.5 | 0.6 |
| 2-position | – | – | – | – | 0.5 | 0.8 | 0.6 | 83.9 | 10.7 | 0.9 |
| **Sunola** | | | | | | | | | | |
| Total profile | – | – | – | – | 5.0 | 0.1 | 4.5 | 77.2 | 12.1 | 0.4 |
| 1,3-position | – | – | – | – | 5.8 | 0.4 | 5.2 | 75.4 | 11.6 | 0.8 |
| 2-position | – | – | – | – | 3.3 | Tr | 3.1 | 80.9 | 13.1 | Tr |
| **Sunflower** | | | | | | | | | | |
| Total profile | – | – | – | – | 6.7 | – | 3.9 | 28.2 | 60.3 | 0.4 |
| 1,3-position | – | – | – | – | 8.0 | – | 4.7 | 28.3 | 57.9 | 0.7 |
| 2-position | – | – | – | – | 4.1 | – | 2.5 | 27.9 | 65.0 | Tr |
| **Cottonseed** | | | | | | | | | | |
| Total profile | – | – | – | 1.0 | 24.9 | 0.5 | 2.3 | 14.8 | 56.8 | 0.2 |
| 1,3-position | – | – | – | 1.0 | 37.8 | 0.1 | 3.5 | 13.9 | 43.4 | 0.3 |
| 2-position | – | – | – | 0.9 | Tr | 1.3 | Tr | 16.4 | 83.5 | Tr |
| **Palm** | | | | | | | | | | |
| Total profile | – | – | – | – | 48.8 | 0.2 | 4.8 | 36 | 9.1 | 0.1 |
| 1,3-position | – | – | – | – | 61.0 | 0.1 | 6.0 | 25.7 | 5.8 | 0.2 |
| 2-position | – | – | – | – | 24.5 | 0.2 | 2.5 | 56.5 | 15.6 | Tr |
| **Flaxseed** | | | | | | | | | | |
| Total profile | – | – | – | – | 5.9 | – | 3.9 | 14.9 | 15.6 | 59.1 |
| 1,3-position | – | – | – | – | 9.3 | – | 6.0 | 13.9 | 13.0 | 57.0 |
| 2-position | – | – | – | – | Tr | – | Tr | 16.9 | 20.9 | 63.3 |
| **Sesame** | | | | | | | | | | |
| Total profile | – | – | – | – | 9.4 | – | 6.6 | 42.6 | 40.4 | 0.3 |
| 1,3-position | – | – | – | – | 16.9 | – | 9.4 | 38.1 | 34.1 | 0.3 |
| 2-position | – | – | – | – | Tr | – | 0.9 | 49.2 | 48.3 | 0.3 |
| **Coconut** | | | | | | | | | | |
| Total profile | 7.0 | 5.1 | 48 | 21.3 | 8.6 | – | 2.5 | 5.9 | 1.6 | Tr |
| 1,3-position | 14.0 | 7.0 | 34.7 | 24.2 | 11.0 | – | 3.1 | 5.0 | 0.9 | Tr |
| 2-position | – | 1.2 | 74.8 | 15.5 | 3.6 | – | 1.2 | 7.7 | 3.0 | Tr |

[a] Table adopted from Richards et al. (2002).
Tr = zero to trace amounts.

2-position was only 54%. The increased level of linoleic acid appears to be distributed into the 1- and 3-positions. In a much earlier study by Ohlson et al. (1975) of rapeseed oils containing different levels of erucic acid (22:1), it was suggested that as 22:1 is reduced the increased amount of linoleic acid is preferentially placed in the 1-position, while reduced levels of 20:1 (gadolic) and 22:1 are preferentially placed in the 3-position.

As shown in Tables 3.14–3.17, in the novel genetically modified canola varieties, α-linolenic and linoleic acids are also preferentially distributed in the 2-position, while saturated fatty acids are found at positions 1 or 3 (Neff et al.,

**Table 3.14** Fatty acid composition (GC area % of total fatty acids) at glycerol carbon 2 for oils from genetically modified α-linolenic acid canola lines[a]

| IMC variety[b] | 14:0 | 16:0 | 18:0 | 20:0 | 22:0 | 24:0 | 16:1 | 18:1 | 20:1 | 22:1 | 18:2 | 18:3 |
|---|---|---|---|---|---|---|---|---|---|---|---|---|
| 600 | 0.3 | 5.8 | 3.4 | 1.1 | 0.4 | 0.5 | 0.2 | 64.9 | 1.8 | 0.0 | 15.5 | 6.5 |
| 500 | 0.0 | 5.3 | 3.0 | 1.2 | 0.7 | 0.5 | 0.0 | 67.1 | 2.1 | 0.0 | 13.8 | 6.3 |
| 1000 | 0.1 | 5.5 | 2.8 | 0.9 | 0.4 | 0.3 | 0.2 | 69.2 | 2.1 | 0.3 | 12.9 | 5.5 |
| 400 | 0.0 | 5.5 | 2.6 | 0.9 | 0.5 | 0.3 | 0.3 | 70.4 | 2.1 | 0.3 | 12.9 | 5.0 |
| 300 | 0.0 | 4.0 | 3.7 | 1.5 | 0.6 | 0.6 | 0.0 | 68.3 | 2.9 | 0.0 | 13.7 | 4.8 |
| 100 | 0.0 | 5.8 | 3.8 | 1.4 | 1.3 | 0.8 | 0.0 | 68.2 | 1.7 | 0.0 | 15.2 | 2.9 |
| 700 | 0.0 | 5.5 | 3.1 | 0.9 | 0.5 | 0.3 | 0.2 | 68.7 | 1.5 | 0.0 | 17.5 | 1.8 |
| 800 | 0.0 | 4.9 | 3.1 | 1.1 | 0.6 | 0.5 | 0.2 | 71.7 | 1.8 | 0.0 | 13.6 | 2.5 |
| 900 | 0.1 | 5.5 | 3.3 | 0.9 | 0.5 | 0.3 | 0.3 | 70.3 | 1.7 | 0.0 | 15.2 | 2.1 |
| 1100 | 0.0 | 5.0 | 3.3 | 1.1 | 0.5 | 0.2 | 0.2 | 78.2 | 2.2 | 0.0 | 5.2 | 4.2 |
| 200 | 0.1 | 4.6 | 3.3 | 1.1 | 0.3 | 0.0 | 0.1 | 81.4 | 2.3 | 0.0 | 4.4 | 2.9 |

[a] Table adapted from Neff et al. (1994b).
[b] InterMountain Canola Company (IMC) (Cinnaminson, NJ) variety. Variety identification as in Table 3.8.

**Table 3.15** Fatty acid composition at glycerol carbon 1(3) for oils from genetically modified low α-linolenic acid canola lines[a]

| IMC variety[b] | 14:0 | 16:0 | 18:0 | 20:0 | 22:0 | 24:0 | 16:1 | 18:1 | 20:1 | 22:1 | 18:2 | 18:3 |
|---|---|---|---|---|---|---|---|---|---|---|---|---|
| 600 | 0.3 | 5.8 | 3.4 | 1.1 | 0.4 | 0.5 | 0.2 | 64.9 | 1.8 | 0.0 | 15.5 | 6.5 |
| 500 | 0.0 | 5.3 | 3.0 | 1.2 | 0.7 | 0.5 | 0.0 | 67.1 | 2.1 | 0.0 | 13.8 | 6.3 |
| 1000 | 0.1 | 5.5 | 2.8 | 0.9 | 0.4 | 0.3 | 0.2 | 69.2 | 2.1 | 0.3 | 12.9 | 5.5 |
| 400 | 0.0 | 5.5 | 2.6 | 0.9 | 0.5 | 0.3 | 0.3 | 70.4 | 2.1 | 0.3 | 12.9 | 5.0 |
| 300 | 0.0 | 4.0 | 3.7 | 1.5 | 0.6 | 0.6 | 0.0 | 68.3 | 2.9 | 0.0 | 13.7 | 4.8 |
| 100 | 0.0 | 5.8 | 3.8 | 1.4 | 1.3 | 0.8 | 0.0 | 68.2 | 1.7 | 0.0 | 15.2 | 2.9 |
| 700 | 0.0 | 5.5 | 3.1 | 0.9 | 0.5 | 0.3 | 0.2 | 68.7 | 1.5 | 0.0 | 17.5 | 1.8 |
| 800 | 0.0 | 4.9 | 3.1 | 1.1 | 0.6 | 0.5 | 0.2 | 71.7 | 1.8 | 0.0 | 13.6 | 2.5 |
| 900 | 0.1 | 5.5 | 3.3 | 0.9 | 0.5 | 0.3 | 0.3 | 70.3 | 1.7 | 0.0 | 15.2 | 2.1 |
| 1100 | 0.0 | 5.0 | 3.3 | 1.1 | 0.5 | 0.2 | 0.2 | 78.2 | 2.2 | 0.0 | 5.2 | 4.2 |
| 200 | 0.1 | 4.6 | 3.3 | 1.1 | 0.3 | 0.0 | 0.1 | 81.4 | 2.3 | 0.0 | 4.4 | 2.9 |

[a] Table adapted from Neff et al. (1994b).
[b] InterMountain Canola Company (IMC) (Cinnaminson, NJ) variety. Variety identification as in Table 3.8.

**Table 3.16** Fatty acid composition (% total fatty acids) at glycerol carbon 2 of oils from genetically modified normal, high stearic, high lauric acid canola lines[a]

| Canola | 14:0 | 16:0 | 18:0 | 20:0 | 22:0 | 24:0 | 16:1 | 18:1 | 20:1 |
|---|---|---|---|---|---|---|---|---|---|
| Sample 1 (normal fatty acid composition) | 0.0 | 0.0 | 0.0 | 0.0 | 0.0 | 0.0 | 49.2 | 30.5 | 20.3 |
| Sample 2 (high stearic) | 0.0 | 0.0 | 0.3 | 0.7 | 0.0 | 0.2 | 29 | 36.2 | 33.6 |
| Sample 3 (high stearic) | 0.0 | 0.0 | 0.4 | 1.0 | 0.0 | 0.2 | 42.8 | 34.7 | 21.0 |
| Sample 4 (high stearic) | 0.0 | 0.0 | 0.3 | 0.5 | 0.0 | 0.1 | 48.4 | 43.3 | 7.3 |
| Sample 5 (high lauric) | 0.0 | 0.0 | 0.3 | 0.3 | 0.1 | 0.1 | 55.1 | 27.4 | 16.7 |
| Sample 6 (high lauric) | 2.0 | 0.2 | 0.6 | 0.9 | 0.0 | 0.1 | 54.5 | 27.7 | 14.0 |

[a] Table adapted from Neff *et al.* (1997). Canola varieties were from Calgene, Inc. (Davis, CA).

**Table 3.17** Fatty acid composition (% total fatty acids) at glycerol carbon 1(3) of oils from genetically modified normal, high stearic, high lauric acid canola lines[a]

| Canola | 12:0 | 14:0 | 16:0 | 18:0 | 20:0 | 22:0 | 24:0 | 16:1 | 18:1 | 20:1 | 18:2 | 18:3 | 24:1 |
|---|---|---|---|---|---|---|---|---|---|---|---|---|---|
| normal (1) | 0.0 | 0.2 | 6.9 | 2.4 | 0.9 | 0.5 | 0.3 | 0.5 | 62.0 | 2.1 | 14.0 | 10.6 | 0.3 |
| high stearic (2) | 0.0 | 0.2 | 5.3 | 41.7 | 0.6 | 0.6 | 0.3 | 0.2 | 29.0 | 0.6 | 9.5 | 9.3 | 0.3 |
| high stearic (3) | 0.0 | 0.1 | 5.2 | 40.7 | 3.0 | 0.6 | 0.2 | 0.2 | 29.0 | 0.5 | 10.0 | 10.1 | 0.4 |
| high stearic (4) | 0.0 | 0.1 | 5.2 | 24.6 | 5.0 | 1.5 | 0.5 | 0.2 | 44.0 | 1.1 | 15.0 | 2.8 | 0.4 |
| high lauric (5) | 12.9 | 1.8 | 6.3 | 2.0 | 0.6 | 0.3 | 0.0 | 0.3 | 44.0 | 1.5 | 9.6 | 7.7 | 0.2 |
| high lauric (6) | 46.0 | 6.2 | 4.7 | 1.2 | 0.5 | 0.3 | 0.0 | 0.3 | 25.0 | 1.1 | 8.1 | 6.2 | 0.0 |

[a] Table adapted from Neff *et al.* (1997). Canola varieties were from Calgene, Inc. (Davis, CA).

1994b, 1997). Oleic acid and other monounsaturated fatty acids, except 20:1, show no preference for either position 2 or position 1(3). The minor unsaturated fatty acid 20:1 was predominantly concentrated at carbon 1(3). Genetically modified oils with saturated fatty acids predominantly at carbons 1,3 have good storage and oxidative stabilities (Neff *et al.*, 1994b, 1997).

## 3.5 Minor lipid components

### 3.5.1 Sterols

In recent years, there has been a surge of interest in sterols derived from seed oils because they represent an important base for the health and nutrition industry. Plant-derived sterols (known as plant sterols or phytosterols) and their fully saturated derivatives, stanols, have been shown to significantly lower blood-serum cholesterol in people with mildly to moderately elevated cholesterol levels (Ling and Jones, 1995; Law, 2000; Plat and Mensink, 2001). Phytosterols have also been used in the cosmetic industry as emulsifiers and precursors to hormonal sterols (Clark, 1996). These natural-source phytosterols are recovered as a co-product from the deodorization step during processing of crude soybean, canola and other vegetable oils.

Sterols and sterol esters of fatty acids predominate among the non-acylglycerol lipids of vegetable oils. The total sterols (sum of esterified and non-esterified sterols) generally account for 0.2–1.0% of total lipids for most vegetable oils. Among the common vegetable oils, rapeseed and canola oils are considered as good convenient sources of phytosterols. According to Codex Committee on Fats and Oils (Codex Alimentarius Commission, 2001) that sets specifications for common dietary fats and oils from data compiled from various regions of the world, the total amount of sterols in low erucic rapeseed canola oil ranges from 0.45 to 1.13% (Table 3.18). This is about 50% higher than in soybean oil. Corn oil, produced from the corn seed embryo, contains the highest amount of sterols (0.7–2.21%) or approximately two times that found in canola oil (Table 3.18). The total sterol content of refined oils is generally lower than those of the crude oils. This is because significant portions, up to 40% of sterols, are removed from the oil during refining process, in particular at the deodorization step (Kochhar, 1983; Morchio *et al.*, 1987).

All the different varieties of rapeseed, including high erucic rapeseed, canola and specialty canola, contain moderate amounts of brassicasterol, a $C_{28}$ sterol, characteristic of *Brassica* oils (Table 3.19). However, it also occurs in other common edible vegetable oils but at extremely low levels (Table 3.18). Brassicasterol is, thus, a biological marker to identify *Brassica* oils and to detect adulterations of other oils with rapeseed and canola oils (Wolff, 1980; Strocchi, 1981). The other major sterols in rapeseed and canola oils include β-sitosterol and campesterol. These two sterols together account for approximately 80–88% of total sterols. Cholesterol, stigmasterol, Δ5-avenasterol, Δ7-stigmastenol, 24-methylene cholesterol, campestanol, Δ5,23-stigmastadienol, sitostanol, Δ5,24-stigmastenol and Δ7-avenasterol are also present in rapeseed and canola oils but at lower levels (Tables 3.18 and 3.19). Stigmasterol is a major sterol in most common vegetable oils (Table 3.18); however, it is a minor component in rapeseed and canola oils; usually its proportion does not exceed 1% of total sterols.

The sterol data presented in Table 3.19 suggest that a slight variation of the total sterol content may occur among the different species and cultivars of canola; however, the sterol composition is virtually unaffected by genetic variations. A study by Vlahakis and Hazebroek (2000) also found a limited variation of the total sterol content in laboratory-extracted oils from canola as well as sunflower and soybean seeds, produced from commercial lines developed through mutation and selection breeding. This suggests that a traditional breeding approach would not lead to dramatic alterations in phytosterol content or composition. In contrast, Abidi *et al.* (1999) found that the phytosterol content and composition of canola oils were markedly influenced by genetic modification of oilseeds. Brassicasterol, campesterol and β-sitosterol levels were consistently lowered in one genotype, whereas increased brassicasterol content was observed in another variety (Abidi *et al.*, 1999). The high stearic canola oil appeared to contain higher levels of phytosterols than the other

Table 3.18 Levels of sterols in some common crude vegetable oils as a percentage of total sterols[a]

| Sterol | Peanut | Coconut | Cotton seed | Corn | Palm | Canola | Safflower seed | Soybean | Sunflower seed |
|---|---|---|---|---|---|---|---|---|---|
| Cholesterol | 0–3.8 | 0–3.0 | 0.7–2.3 | 0.2–0.6 | 2.6–6.7 | 0–1.3 | 0–0.7 | 0.2–1.4 | 0–0.7 |
| Brassicasterol | 0–0.2 | 0–0.3 | 0.1–0.3 | 0–0.2 | 0 | 5.0–13.0 | 0–0.4 | 0–0.3 | 0–0.2 |
| Campesterol | 12.0–19.8 | 6.0–11.2 | 6.4–14.5 | 16.0–24.1 | 18.7–27.5 | 24.7–38.6 | 9.2–13.3 | 15.8–24.2 | 6.5–13.0 |
| Stigmasterol | 5.4–13.2 | 11.4–15.6 | 2.1–6.8 | 4.3–8.0 | 8.5–13.9 | 0.2–1.0 | 4.5–9.6 | 14.9–19.1 | 6.0–13.0 |
| β-sitosterol | 47.4–69.0 | 32.6–50.7 | 76.0–87.1 | 54.8–66.6 | 50.2–62.1 | 45.1–57.9 | 40.2–50.6 | 47.0–60 | 50–70 |
| Δ5-avenasterol | 5.0–18.8 | 20.0–40.7 | 1.8–7.3 | 1.5–8.2 | 0–2.8 | 2.5–6.6 | 0.8–4.8 | 1.5–3.7 | 0–6.9 |
| Δ7-stigmastenol | 0–5.1 | 0–3.0 | 0–1.4 | 0.2–4.2 | 0.2–2.4 | 0–1.3 | 13.7–24.6 | 1.4–5.2 | 6.5–24.0 |
| Δ7-avenasterol | 0–5.5 | 0–3.0 | 0.8–3.3 | 0.3–2.7 | 0–5.1 | 0–0.8 | 2.2–6.3 | 1.0–4.6 | 3.0–7.5 |
| Others | 0–1.4 | 0–3.6 | 0–1.5 | 0–2.4 | 0 | 0–4.2 | 0.5–6.4 | 0–1.8 | 0–5.3 |
| Total sterols (mg/kg) | 900–2900 | 400–1200 | 2700–6400 | 7000–22 100 | 300–700 | 4500–11 300 | 2100–4600 | 1800–4500 | 2400–5000 |

[a] Data adapted from Codex Alimentarius Commission (2001).

**Table 3.19** Composition (% total sterols) and total content (mg/kg oil) of sterols from oils extracted from high erucic rapeseed, regular canola and genetically modified canola seeds[a]

| Sterol | A | B | C | D | E | F | G | H | I |
|---|---|---|---|---|---|---|---|---|---|
| Cholesterol | 0.6 | 0.3 | 0.2 | 0.6 | 0.4 | 0.4 | 0.4 | 0.4 | 0.3 |
| Brassicasterol | 11.7 | 11.8 | 16.2 | 9.5 | 16.7 | 14.6 | 11.9 | 12.1 | 10.1 |
| 24-Methylene cholesterol | 1.7 | 1.0 | 1.2 | 1.5 | 1.2 | 1.8 | 1.7 | 1.6 | 1.3 |
| Campesterol | 30.2 | 31.2 | 27.1 | 35.2 | 30.1 | 32.6 | 30.3 | 31.4 | 30.7 |
| Stigmasterol | 0.3 | 0.4 | 0.5 | 0.3 | 0.5 | 0.4 | 0.4 | 0.4 | 0.5 |
| Campastanol | 0.5 | 0.5 | 0.5 | 0.6 | 0.6 | 0.6 | 0.4 | 0.3 | 0.6 |
| Δ5,23-Stigmastadienol | 0.6 | 0.7 | 0.5 | 0.5 | 0.6 | 0.5 | 0.5 | 0.6 | 0.6 |
| Δ5-Avenasterol | 0.8 | 0.9 | 1.2 | 0.8 | 0.6 | 0.7 | 0.8 | 0.7 | 1.6 |
| β-Sitosterol | 48.5 | 49.3 | 48 | 46.9 | 44.5 | 44.2 | 49.6 | 48.6 | 48.3 |
| Sitostanol | 0.6 | 0.4 | 0.5 | 1.0 | 1.5 | 1.0 | 0.7 | 1.0 | 0.5 |
| Δ7-Avenasterol | 3.1 | 1.7 | 2.1 | 2.4 | 1.9 | 2.6 | 2.4 | 2.4 | 2.5 |
| Δ5,24-Stigmastadienol | 1.5 | 1.8 | 2.2 | 0.8 | 1.4 | 0.9 | 0.9 | 0.6 | 3.0 |
| Total sterols (mg/kg oil) | 8523 | 8630 | 8009 | 7999 | 5119 | 8098 | 7946 | 8067 | 9601 |

[a] Based on analysis in an author's (Ratnayake) laboratory of oils extracted by the Canadian Grain Commission. Seed samples collected from plant breeders. Sample identification: A = regular canola, *B. rapa*, cultivar Parkland; B = regular canola, *B. rapa*, cultivar 46A65; C = regular canola, *B. juncea*, cultivar Arid; D = low linolenic acid canola, *B. napus*, cultivar IMC 105; E = high erucic rapeseed, *B. napus*, cultivar Millennium 03; F = high-oleic, regular linolenic acid canola, *B. napus*, cultivar Nex 500; G = high oleic acid, low linolenic acid canola, *B. napus*, cultivar Nexera; H = super high oleic acid canola, *B. napus*, unknown strain; I = high lauric canola, *B. napus*, cultivar LA161.

varieties. These contradictory reports (Abidi *et al.*, 1999; Vlahakis and Hazebroek, 2000) suggest that further research is required to fully evaluate the impact of genetic modifications of fatty acid composition of oilseeds on their sterol content and composition.

The growing temperature, however, has a considerable influence. In experiments performed by Vlahakis and Hazebroek (2000) on soybean, the total phytosterol levels increased with higher temperature. Composition also changed, with greater percentage of campesterol and lower percentages of stigmasterol and β-sitosterol at higher temperatures.

The sterols in canola oil are approximately equally distributed between the esterified and non-esterified (or free sterols) forms (Appelqvist *et al.*, 1981; Ackman, 1983; Evershed *et al.*, 1987). This is primarily due to equal amounts of the two major sterols, β-sitosterol and campesterol, in the esterified and free forms. Twice the amount of brassicasterol is found in the free sterols than in the esterified form. The compositions of sterols and fatty acids in the steryl ester fraction differ from their concentrations in the total sterols and fatty acids obtained by saponification of the oil (Tables 3.20 and 3.21) (Gordon *et al.*, 1997). In particular, the esterified sterols contain higher proportions of palmitic and stearic acids and lower proportion of oleic acid compared to canola oil.

**Table 3.20** Fatty acid compositions (% total fatty acids) of the steryl ester fraction and the whole oil from canola[a]

| Fatty acid | Steryl esters | Canola oil |
|---|---|---|
| 14:0 | 3.1 | 0.5 |
| 16:0 | 17.5 | 5.6 |
| 18:0 | 18.4 | 2.1 |
| 20:0 | 0.8 | 0.4 |
| 18:1 | 30.9 | 58.1 |
| 22:1 | 1.2 | 0.4 |
| 18:2 | 20.5 | 21.6 |
| 18:3 | 7.6 | 11.2 |

[a] Adapted from Gordon and Miller (1997).

**Table 3.21** Sterol composition (% total sterols) of the steryl fraction and whole oil from canola[a]

| Sterol | Steryl ester | Canola oil |
|---|---|---|
| Cholesterol | – | 0.6 |
| Brassicasterol | 5.1 | 12.7 |
| Campesterol | 32.0 | 31.1 |
| Stigmasterol | 0.0 | 0.0 |
| Unknown | 0.0 | 0.6 |
| β-Sitosterol | 55.2 | 52.2 |
| Avenasterol | 7.7 | 1.9 |

[a] Adapted from Gordon and Miller (1997).

## 3.5.2 Tocopherols

Tocopherols are natural antioxidants of physiological importance. These compounds are now considered to be essential dietary components for human beings (Institute of Medicine, 2000). They occur as a family of four derivatives, namely α-, β-, γ- and δ-tocopherols. The main biochemical function of the tocopherols is believed to be the protection of polyunsaturated fatty acids against oxidation. The tocopherols function primarily as chain-breaking antioxidants that prevent the propagation of lipid peroxidation. The different tocopherols have different antioxidant activity *in vitro* and *in vivo*. In food systems, the antioxidant activity of the tocopherol isomers decreases in the order: $\delta > \gamma > \beta > \alpha$ (Kamal-Eldin and Appelqvist, 1996) whereas in biological systems, the antioxidant activity appears to be limited to α-tocopherol (Institute of Medicine, 2000). Other naturally occurring forms of tocopherol (β-, γ- and δ-tocopherols) do not contribute toward meeting the vitamin E requirement of humans.

Vegetable oils, nuts and seeds are rich sources of tocopherols, but significant amounts are also found in green leafy vegetables and a variety of fish

(Institute of Medicine, 2000). The composition of tocopherol compiled by the Codex Committee on Fats and Oils (Codex Alimentarius Commission, 2001) for some common vegetable oils including canola oil is summarized in Table 3.22. These values reflect the data for crude oils from various regions of the world. Canola oil contains mostly α- and γ-tocopherols, with γ-tocopherol usually present in higher amounts. β- and δ-tocopherols are usually present in only trace amounts. Tocotrienols are completely absent in canola oil. The total content of tocopherols in canola oil is higher than in many other common vegetable oils such as sunflower, safflower, palm and coconut oils but lower than in soybean and corn oils. However, the higher content of tocopherols in soybean and corn oils is primarily due to the greater contribution from γ-tocopherol which has no biological activity. In contrast, canola oil, compared to soybean and corn oils, contains a higher amount of α-tocopherol, which is the only tocopherol with vitamin E activity in humans (Institute of Medicine, 2000).

Table 3.23 shows a comparison of data for high erucic rapeseed oil, regular canola oil and specialty canola oils (Przybylski and Mag, 2002). High oleic acid and low linolenic acid canola oils contained the lowest level of total tocopherols, whereas the high-oleic low-linolenic canola oil contained the highest amount of total tocopherols among the different varieties of canola. Abidi *et al.* (1999) examined oil derived from different lines of genetically modified canola varieties and found that the impact of oil modification on the content of tocopherols is variable and without a distinctive trend (Table 3.24). Greater variation was observed in the concentration of α- and γ-tocopherols than in that of δ-tocopherol. Dolde *et al.* (1999) and Goffman and Becker (2002) also examined different breeding lines of canola and found a slight variation in the total tocopherol content. Dolde *et al.* (1999) reported that tocopherol composition was consistent

**Table 3.22** Tocopherol content of some crude vegetable oils[a]

| Oil | α | β | γ | δ | Total[b] |
|---|---|---|---|---|---|
| Peanut | 49–373 | ND–41 | 88–389 | ND–22 | 170–1300 |
| Coconut | ND–17 | ND–11 | ND–14 | ND | ND–50 |
| Cotton seed | 136–674 | ND–29 | 138–746 | ND–21 | 380–1200 |
| Corn | 23–573 | ND–356 | 268–2468 | 23–75 | 330–3720 |
| Palm | 4–193 | ND–234 | ND–526 | ND–123 | 150–1500 |
| Canola | 100–386 | ND–140 | 189–753 | ND–22 | 430–2680 |
| Safflower | 234–660 | ND–17 | ND–12 | ND | 240–670 |
| High oleic safflower | 234–660 | ND–13 | ND–44 | ND–6 | 250–700 |
| Soybean | 9–352 | ND–36 | 89–2307 | 154–932 | 600–3370 |
| Sunflower | 403–935 | ND–45 | ND–34 | ND–7 | 440–1520 |
| High oleic sunflower | 400–1090 | 10–35 | 37709 | ND–17 | 450–1120 |

Tocopherol (mg/kg oil)

[a] Data adapted from Codex Alimentarius Commmission (2001).
[b] Total includes tocotrienols. Canola oil contains no tocotrienols.
ND = not detected.

**Table 3.23** Tocopherol content of crude oils from high erucic rapeseed, regular canola and genetically modified canola seeds[a]

| Oil | Tocopherol content (mg/kg oil) | | | | |
|---|---|---|---|---|---|
| | α | β | γ | δ | Total |
| High erucic rapeseed | 268 | – | 426 | – | 694 |
| Regular canola | 272 | – | 423 | – | 695 |
| Low linolenic canola | 150 | – | 423 | 7 | 580 |
| High oleic canola | 226 | – | 202 | 3 | 431 |
| High-oleic low-linolenic canola | 259 | – | 607 | 5 | 901 |

[a] Adapted from Przybylski and Mag (2002).

**Table 3.24** Tocopherol content of oils from different lines of genetically modified canola varieties[a]

| Canola variety[b] | Tocopherol content (mg/kg) | | | | |
|---|---|---|---|---|---|
| | α | β | γ | δ | Total |
| **IMC-varieties[b]** | | | | | |
| EX-1000 | 214.5 | ND | 445.4 | 5.4 | 665.3 |
| EX-100 | 179.9 | ND | 295.2 | 3.3 | 478.4 |
| EX-200 | 204.8 | ND | 405.0 | 6.8 | 616.6 |
| EX-600 | 231.5 | ND | 278.4 | 4.1 | 514.0 |
| EX-700 | 225.4 | ND | 446.6 | 4.7 | 676.7 |
| EX-900 | 269.5 | ND | 373.3 | 5.9 | 648.7 |
| EX-1100 | 188.0 | ND | 443.6 | 6.6 | 638.2 |
| **CG-varieties[c]** | | | | | |
| DS68494 | 169.5 | ND | 376.8 | 5.5 | 551.8 |
| DS68456 | 172.9 | ND | 472.8 | 5.8 | 651.3 |
| DS68482 | 231.5 | ND | 373.6 | 6.5 | 611.6 |
| DS68507 | 241.8 | ND | 381.3 | 9.5 | 632.6 |
| DS68519 | 245.5 | ND | 335.2 | 6.8 | 587.5 |

[a] Adapted from Abidi et al. (1999).
[b] Ex-series: InterMountain Canola Company (IMC) canola varieties (Cinnaminson, NJ); Ex-100 and EX-200, commercially grown low linolenic acid and high oleic acid spring canola respectively; EX-600, commercially grown generic spring canola developed by Zeneca Seeds (Wilmington, DE); EX-700, commercially grown low linolenic acid spring canola developed by University of Manitoba (Winnipeg, Manitoba, Canada); EX-900, EX-1000 and EX-1100, experimental IMC low linolenic acid breeding germ plasm.
[c] DS-varieties: Experimental transgenic canola from Calgene (Davis, California). DS68494, regular canola fatty acid composition; DS68456, high lauric variety; DS68482 and D568507, high stearic variety; D568519, high oleic variety.
ND = not detected.

with 60–74% γ-tocopherol and 26–35% α-tocopherol among the different canola varieties.

The total amount of tocopherols in seed oils is generally believed to be governed by the content of unsaturated fatty acids. In contrast, in experiments performed by Dolde et al. (1999) for canola oils and also sunflower oils with

**Table 3.25** Changes of tocopherols (mg/kg oil) during alkali and physical refining[a]

| Tocopherol | Alkali refining ||||  Physical refining |||
|---|---|---|---|---|---|---|---|
|  | Crude oil | Neutralized oil | Bleached oil | Deodorized oil | Crude oil | Bleached oil | Deodorized oil |
| α | 290 | 247 | 259 | 174 | 257 | 240 | 173 |
| γ | 382 | 333 | 360 | 220 | 361 | 340 | 212 |
| δ | 13.4 | 11.7 | 13.1 | 7 | 14.3 | 13.3 | 7.2 |
| Total | 685 | 522 | 632 | 402 | 632 | 595 | 390 |

[a] Adapted from Cmolik et al. (2000).

modified fatty acid profiles, the fatty acid compositions were not correlated with either total or individual tocopherol concentration. This might suggest that the tocopherol content in seed oils is influenced by a parameter other than the unsaturated fatty acid content. However, further studies are needed to confirm the lack of such correlations in genetically modified oils.

Similar to sterols, the content of tocopherols is also affected by oil extraction and subsequent refining methods (Willner, 1997; Cmolik et al., 2000). Solvent-extracted canola oils contain higher amounts of tocopherols whereas cold-pressed oils contain the lowest amount (Willner, 1997). When temperature of pressing was increased, the amount of tocopherols was doubled. A considerable amount of tocopherols is also removed during refining, especially during the deodorization step, where about 30–35% of the total tocopherols are distilled (Table 3.25) (Cmolik et al., 2000).

### 3.5.3 Carotenoids

Crude rapeseed and canola oils may contain as much as 95 mg/kg total carotenoids. These are predominantly xanthophylls (about 85–90%) including lutein (50%), neo-lutein A (15%) and neo-lutein B (20%). Approximately 7–10% of the carotenoids are present as carotene (Box and Boekenoogen, 1967; Hazuka and Drozdowski, 1987). Carotenoids are largely removed during the bleaching process (Boki et al., 1994; Chapman et al., 1994) and are further reduced to less than 1 mg/kg by oxidation and heating during deodorization (Warner et al., 1989).

### 3.5.4 Waxes

While rapeseed or canola oil was originally marketed as an oil that did not require winterization (wax removal), occasional occurrences of sedimentation in fully refined salad oils do occur, especially where oils have been cycled between room temperature and cool conditions. The onset of sedimentation problems is sporadic and may be associated with drought or high temperature conditions during crop development, but this connection has not been verified.

The general nature of this sediment has been reviewed recently (Hermann et al., 1999) and a number of quick tests for evaluating oils have been compared (Botha and Mailer, 2001).

The major component (up to 75%) of the sediment is wax esters (Daun and Jeffrey, 1991; Liu et al., 1996) with chain lengths from $C_{36}$ to $C_{56}$ made up of fatty acids ranging from $C_{16}$ to $C_{32}$ and fatty alcohols ranging from $C_{16}$ to $C_{30}$. Other components include triacylglycerols rich in long-chain fatty acids ($>C_{24}$) and also highly enriched with saturated fatty acids. Hydrocarbons (about 80% $C_{29}$ and $C_{31}$), non-esterified fatty acids and alcohols were also present. One report noted the inclusion of steryl esters (Gao and Ackman, 1995).

The relative composition of sediments varied with the techniques used to store the oil or isolate the sediment. Sediments from oils stored and isolated at room temperature were richer in long-chain fatty acids relative to those stored and collected at cool temperatures.

### 3.5.5 Polar lipids

Canola and rapeseed lipids contain about 3% phospholipid and 1% glycolipid (Sosulski et al., 1981). Most of these are removed in processing, especially by citric acid degumming (Table 3.26). Phospholipids in canola and rapeseed oils are generally divided into two classes, non-hydratable phospholipids (phosphatidic acid and phosphatidylinositol) and hydratable phospholipids. The cooking and flaking process is where the phospholipids are released from the seed to be extracted along with the oil, possibly through the breakdown of lipoproteins (Zajic et al., 1986).

Table 3.26 Polar lipids in canola oil at different stages of refining[a]

| Component | Expeller | Solvent | Crude | Degummed |
|---|---|---|---|---|
| Total phosphorus (mg/kg) | 217.0 | 468.0 | 293.0 | 45.0 |
| Total polar lipids (% total oil) | 0.57 | 1.2 | 0.76 | 0.12 |
| Phosphatidylcholine (% total PL[b]) | 15.2 | 24.1 | 17.9 | 2.6 |
| Phosphatidylethanolamine (% total PL) | 15.8 | 18.4 | 16.6 | 13.9 |
| Phosphatidylinositol (% total PL) | 10.1 | 16.4 | 12.0 | 18.7 |
| Phosphatidic acid (% total PL) | 50.9 | 31.2 | 45.0 | 49.0 |
| Phosphatidylserine (% total PL) | 4.3 | 3.7 | 4.2 | 12.6 |
| Lysophosphatidylethanolamine (% total PL) | 0.3 | 0.3 | 0.3 | 0.1 |
| Lysophosphatidylcholine (% total PL) | 0.2 | 0.3 | 0.3 | 0.1 |
| Monogalactosyldiacylglycerol (% total PL) | 0.2 | 0.2 | 0.2 | 0.2 |
| Digalactosyldiacylglycerol (% total PL) | 0.6 | 1.5 | 0.9 | 0.2 |
| Diacylglycerol (% total PL) | 2.1 | 3.0 | 2.4 | 2.4 |

[a] Data adapted from Przybylski and Eskin (1991). Polar lipids include phospholipids, galactolipids and diacylglycerols.
[b] PL = polar lipids.

**Table 3.27** Fatty acid composition of polar lipids in low erucic acid rapeseed[a]

| Polar lipid | Percentage of total fatty acids ||||||||| 
|---|---|---|---|---|---|---|---|---|---|
|  | 16:0 | 16:1 | 16:3 | 17:0 | 18:0 | 18:1 | 18:2 | 18:3 | 20:1 |
| PC | 8.7 | 0.8 | 1.2 | – | 55.8 | 30.9 | 1.9 | 0.2 | 0.5 |
| PI | 21.8 | 0.8 | 1.9 | – | 33.6 | 38.1 | 3.6 | 0.2 | – |
| PE | 17.7 | 1.8 | 2.0 | – | 47.7 | 27.3 | 2.7 | 0.3 | 0.5 |
| DGDG | 20.2 | 3.4 | Trace | 3.4 | 8.4 | 23.2 | 31.9 | 9.3 | – |
| MGDG | 19.1 | 9.9 | – | – | 6.7 | 43.3 | 15.9 | 9.5 | – |
| SG | 17.0 | 5.7 | – | 0.9 | 8.9 | 47.9 | 14.0 | 6.1 | – |

[a] Adapted from Sosulski *et al*. (1981).
PC = Phosphatidylcholine; PI = Phosphatidylinositol; PE = Phosphatidylethanolamine; DGDG = Digalactosyldiacylglycerol; MGDG = Monogalactosyldiacylglycerol; SG = Esterified sterol glycoside.

The fatty acid composition of phospholipids is somewhat different from that of the glyceride fraction, in that the phospholipids and polar lipids have larger amounts of saturated fatty acids (palmitic and stearic acid) (Table 3.27) while the glycolipids are more unsaturated than the phospholipids. Glycolipids contain significantly more palmitoleic acid than phospholipids.

Fatty acid data on polar lipids seem to be confined to canola and rapeseed data with normal fatty acid patterns. It would be interesting to find if modifications of fatty acids in the seed oils are reflected in the fatty acids in the phospholipids. Certainly, increased 22:1 fatty acids are reflected in the phospholipids' fatty acid composition and especially in the composition of the glycolipids.

## 3.6 Chlorophyll

Canola seed and rapeseed, and in particular *B. napus* varieties grown in short-season areas such as Canada and Scandinavia, contain significant amounts of chlorophyll. The chlorophyll is present in chloroplasts in the cotyledons and, if the seed does not fully mature, the chlorophyll pigments are extracted along with the oil. In general, top-grade crude canola oil is expected to have less than 30 mg/kg of chlorophyll pigments (Daun, 1987) but if the chlorophyll level of the seed is too high, it may not be possible to achieve this (Fig. 3.3). While chlorophyll *a* and chlorophyll *b* are the major chlorophyll pigments present in canola seed, they are rapidly broken down during processing (Table 3.28) into a mixture of pigments, including pheophytins (Phy), pheophorbides (Pho), methylpheophorbides (MePho) and pyropheophytens (Pyr) (Endo *et al*., 1992). While earlier processing stages have some reduction effect on chlorophyll, most is removed during the bleaching step.

Chlorophyll is usually measured analytically as the components contributing to the absorption near 670 nm, expressed as chlorophyll *a*. Since different chlorophyll-related pigments have different absorptivities, this leads to mass balance errors, but this technique has been a successful compromise up to now.

# CHEMICAL COMPOSITION

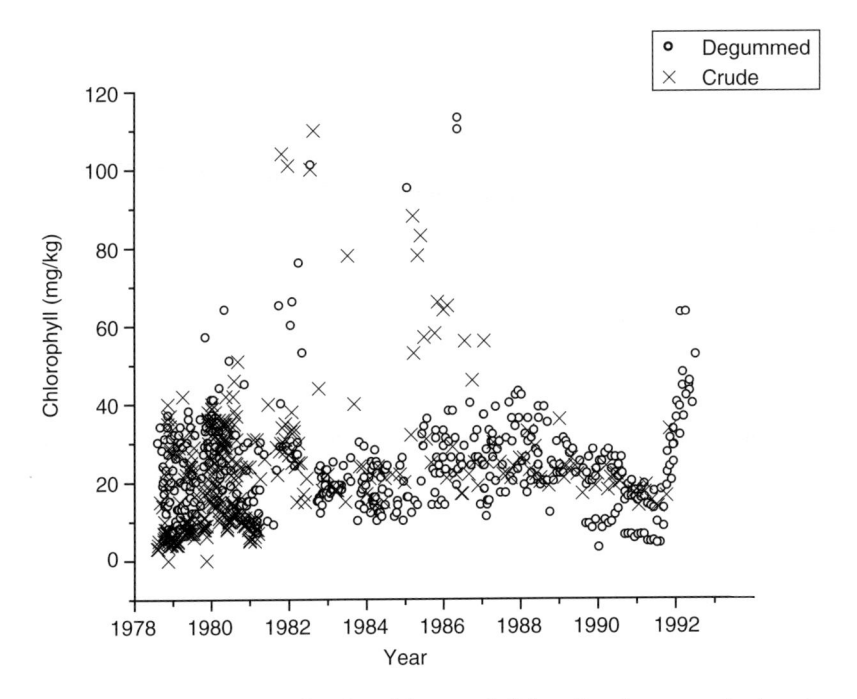

**Figure 3.3** Chlorophyll contents of crude and degummed oil from Canadian rapeseed and canola.

**Table 3.28** Chlorophyll pigments in canola during processing (ppm)[a]

| Process step | Chl $a$ | Chl $b$ | Phy $a$ and Pyr $a$ | Phy $b$ + Pyr $b$ | Pho $a$ | MePho $a$ | Pyr $a$ | Total |
|---|---|---|---|---|---|---|---|---|
| Seed | 8.4–23.7 | 2.4–8.3 | <0.1–1.0 | | 0–0.1 | 0–0.1 | | 11.3–33.2 |
| Meal | | | 0.7–1.3 | 0.1–0.5 | | 0–0.1 | 0.4–2.1 | 1.2–5.3 |
| Expelled oil | 6.3 | | Phy 4.5 | Phy 1.8 | <0.1–0.4 | | | 18.7 |
|  |  |  | Pyr 5.4 | Pyr 0.7 |  |  |  |  |
| Extracted oil | 1.9 | | Phy 3.3 | Phy 1.3 | | | | 24.5 |
|  |  |  | Pyr 16.6 | Pyr 0.7 |  |  |  |  |
| Crude oil | | | 13.4–35.2 | 2.0–5.8 | | 0.4–2.0 | 1.2–6.7 | 20.7–50.7 |
| Degummed oil | 0.3 | | 6.8–16.6 | 0.2–3.0 | 0.3–0.6 | 0.3–0.5 | 6.1–30.9 | 15.6–43.2 |
| Alkali refined | | | Phy 6.3 | Phy 1.1 | | | | 18.5 |
|  |  |  | Pyr 9.2 | Pyr 1.8 |  |  |  |  |
| Bleached | | | Phy 0.6 | Phy 0.3 | | | | 1.4 |
|  |  |  | Pyr 0.2 | Pyr 0.3 |  |  |  |  |

[a] Adapted from Endo *et al.* (1992) and Suzuki and Nishioka (1993).
Chl $a$ = Chlorophyll $a$; Chl $b$ = Chlorophyll $b$; Phy $a$ = pheophytin $a$; Phy $b$ = pheophytin $b$; Pho $a$ = pheophorbide $a$; Pho $b$ = pheophorbide $b$; MePho $a$ = methylpheophorbides; Pyr $a$ = pyropheophyten $a$; Pyr $b$ = pyropheophyten $b$.

Refined, bleached and deodorized oil usually has a specification of less than 25 parts per billion chlorophyll (Mag, 2001) although this measurement is based on an assumption that the component in question is chlorophyll. The actual identity of the components absorbing near 670 nm has not been confirmed. It is suspected that breakdown products of chlorophyll, even if the oil is within specification, may cause oxidation problems if the oil was originally high in chlorophyll (Tautorus and Low, 1993; Ramamurthi and Low, 1995).

## 3.7 Sulfur and sulfur-containing compounds

There is good evidence that many vegetable oils contain levels of sulfur in the order of a few mg/kg (Johansson, 1977; Wijesundera et al., 1988). The level of sulfur in rapeseed and canola oils (Table 3.29) has been of particular interest not only because the levels are higher than are found in many other oils but also because the sulfur has been implicated in problems related to hydrogenation of the oil (Drozdowski et al., 1973; de Man et al., 1983). The relatively high level of sulfur found in rapeseed oils is related to the presence of sulfur containing glucosinolates in the seed (Marcuse, 1977). When these compounds decompose during processing of the seed, a number of oil-soluble sulfur-containing compounds, including isothiocyanates and oxazolidinethiones, are formed and are transferred to the oil (Franzke et al., 1975; Daun and Hougen, 1977; Devinat et al., 1980; Abraham and de Man, 1987).

**Table 3.29** Sulfur content (mg/kg) of rapeseed and canola oils at different stages of processing. Determination by either reductive (Raney nickel) or oxidative (combustion) methods

| Reference[a] | A | B | C | D | E | F | G | H | I | J |
|---|---|---|---|---|---|---|---|---|---|---|
| Analytical method[b] | R | R | R | O | R | O | O | R | O | R | O | C | R | R | O |
| **Rapeseed oil** | | | | | | | | | | |
| Pressed | 9 | 22 | | | | | | | | |
| Extracted | 36 | 33 | | | | | 13 | 15 | | |
| Degummed | 11 | 16 | | 1.5 | 10 | | | | | |
| Crude | 10 | 25 | 3 | 24 | 2 | 30 | 33 | 17 | 28 | 47 | 8 | | 27 | | |
| Refined | 8 | 7 | 2 | 20 | | | 25 | 1 | 7 | | | | 14 | | |
| Bleached | 7 | 4 | | | | | 19 | | | | | | 12 | | |
| Deodorized | 1 | 1 | 1 | 16 | | | 10 | | | | | | 1 | | |
| **Canola oil** | | | | | | | | | | |
| Crude | | | | | | | | | | 21 | 4 | 25 | | 11 | 14 |
| Degummed | | | | | | | | | | | | | | 19 | 29 |
| Refined | | | | | | | | | | | | 10 | | 7 | 8 |

[a] A, Babuchowski and Zadernowski (1972); B, Daun and Hougen (1977); C, Abraham and de Man (1987); D, Cho and de Man (1987); E, Devinat et al. (1980); F, Drexler (1975); G, Krygier and Rutkowski (1991); H, Wijesundera et al. (1988); I, Franzke et al. (1975); J, de Clercq et al. (1991).
[b] R = reduction (Raney nickel), O = Oxidation, C = Combustion.

Different analytical methods used to determine sulfur content of oils have given different results. In general, methods relying on combustion gave larger values for sulfur than methods relying on reduction with Raney nickel. Results from the latter were well correlated with poisoning of the hydrogenation catalyst. A third method for sulfur determination, volatile sulfur, gave even lower values.

The nature of the sulfur compounds present in vegetable oils is not well understood. In general, the presence of volatile isothiocyanates and oxazolidinethiones has been reasonably well documented. In addition, as mentioned previously there has been some evidence to suggest that sulfur-containing fatty acids or other lipids are present, possibly after interaction with the sulfur-containing components formed from the breakdown of glucosinolates (Johansson, 1977; Wijesundera and Ackman, 1988).

## 3.8 Minerals

The Codex Alimentarius standard for vegetable oils specifies maximum levels for iron, copper, lead and arsenic (Codex Alimentarius Commission, 2001). Lead and arsenic are undesirable as they are toxic in nature. Iron and copper, as well as nickel, are undesirable as they may promote oxidation. The source of minerals in the oil is both the seed itself and the contact with processing equipment (Elson et al., 1979).

In studies of rapeseed and canola oils (Table 3.30), all oils were within the Codex specification except for two samples of degummed canola oil obtained

**Table 3.30** Content of minerals (mg/kg) in canola and rapeseed oils at various stages of refining[a]

| Mineral | Crude N | Mean (range) | CV | Degummed N | Mean (range) | CV | Bottled[b] N | Mean (range) | CV |
|---|---|---|---|---|---|---|---|---|---|
| Copper | 6 | 0.0200 (0.008–0.045) | 77 | 6 | 0.6990 (0.0124–0.078) | 24 | 20 | 0.06 (0.02–0.18) | 79 |
| Iron | 9 | 2.08 (1.31–4.17) | 42 | 6 | 0.78 (0.18–2.40) | 98 | | | |
| Lead[c] | 7 | 0.013 (0.008–0.019) | 32 | 6 | 2.074 (0.006–8.680) | 181 | | | |
| Nickel | 2 | 0.008 (0.005–0.010) | | 6 | 0.007 (0.005–0.009) | 40 | | | |
| Phosphorus | 9 | 419 (17–374) | | 6 | 43 (26–57) | 25 | | | |
| Zinc | 1 | | | | | | 20 | 0.19 (0.10–0.47) | 45 |
| Sodium | 1 | | | | | | 20 | 2.06 (1.21–4.85) | 40 |
| Potassium | 1 | | | | | | 20 | 0.45 (0.15–1.84) | 83 |
| Calcium | 1[d] | 118 | | | | | 20 | 4.05 (0.37–14.30) | 101 |
| Magnesium | 1 | | | | | | 20 | 3.04 (0.11–22.5) | 206 |

[a] Samples tested in the author's laboratory (Daun) using samples collected from Canadian Crushing Plants, 1991–1992.
[b] Bottle oils from Spanish supermarkets (Garrido et al., 1994).
[c] Two degummed oil samples from one crushing plant had lead levels of 1.67 mg/kg and 8.68 mg/kg respectively. Otherwise lead levels were well below the Codex maximum of 0.1%.
[d] From a single sample of commercially processed canola oil other samples prepared at a Pilot plant had significantly higher levels of calcium in the oil (Diosady et al., 1983).
N = number of samples, CV = Coefficient of variation.

from one plant in 1991. These oils had high levels of lead, probably due to contamination from lead piping within the system. It is noted that the oils from this plant are now back within specifications (0.1 mg/kg).

Phosphorus in oils is mostly present in the form of phospholipids. These are removed mostly through degumming procedures and also through the other refining steps. Fully refined oils usually contain less than 1 mg/kg phosphorus. Other metals such as iron and calcium may also be removed to a certain extent by processing.

## 3.9 Conclusions

Canola and rapeseed oils are characterized by a wide range of composition. While this composition is due partially to the different species making up the commodity, the main reasons for the variety are due to genetic selection to give improved oils for different purposes. While rapeseed oil has been used as a food staple for many thousands of years in China and India, its use in the Western Hemisphere has been more recent and the original rapeseeds grown in Canada were developed for industrial purposes for which their high levels of erucic acid suited them. Health concerns about erucic acid in the human diet resulted in the development of lines of rapeseed with very low levels of erucic acid (less than 1%). This type of oil has been approved by world health authorities and is the currently desired type of oil for edible purposes around the world.

At the same time as the erucic acid level was reduced the level of oleic acid was increased so that the new oil type, called canola if it is derived from seed with low levels of erucic acid and glucosinolates, was found to have a very favorable fatty acid composition for health, combining high levels of monounsaturated fatty acids with low levels of saturated fatty acids and a favorable ratio of dienoic and trienoic fatty acids. This led to canola oil's exploitation as a heart-healthy oil, especially in the US marketplace.

The malleability of the fatty acid composition of the *Brassica* species has been further exploited through a combination of genetic selection and predictable mutations to give commercial production of oils with levels of oleic acid as high as 82% and levels of linolenic acid lower than 3%. These oils show a remarkable stability in frying and are in demand for deep frying, especially in Japan, as there is a marked decrease in spatter in kitchen frying. These types of oils are also useful in reducing the amounts of *trans* fatty acids used for frying fats.

The use of recombinant DNA to incorporate a gene from the California bay tree resulted in the development of a canola-type oil with more than 30% lauric acid. While this type of oil was produced in commercial lots in the late 1990s, the lack of demand for the product has resulted in its withdrawal from the market.

The demand for high erucic acid oils for the chemical industry has also increased and breeding efforts have raised the level of erucic acid to more than

50%. In the *Brassica* species comprising rapeseed oil, erucic acid is not placed in the 2-position of the triglyceride molecule. In order to achieve levels of erucic acid greater than 66%, it will be necessary to find a means to incorporate genes that direct erucic acid into the 2-position. This may be achieved by recombinant DNA technology but may be more successfully achieved by interspecies crossing.

The future for canola and rapeseed oils appears to ride on a multiplicity of uses requiring a multiplicity of compositions. Genetic engineering has shown that the triacylglycerols normally produced by rapeseed plants can be replaced by wax esters. Fatty acid composition variants including high levels of stearic or palmitic acid can be produced. While the main uses of the oil will likely be as a dietary oil, both in its normal composition and in its high oleic and low linolenic acid variants, there will be an increased demand for other specialty type oils including, especially in Europe, oils with an optimum fatty acid composition for use as biodiesel.

While a great deal of information has been gathered on the fatty acid composition of the various rapeseed and canola oil types, there is less information available on the effect that variation of the fatty acid composition has on the composition of minor lipid (or non-lipid) components of the oils. While some information is slowly appearing in the literature, it is mostly in the area of seed composition, and little information is available on the composition of minor constituents in the processed oils from these new varieties. Changes in processing technologies leading to "green processing" or even cold-pressed oils mean that more study of minor constituents is needed. Without that information, it will be difficult to proceed with the full development of these new products, especially in the edible oil industry.

When Canada presented its case for GRAS (Generally Recognized As Safe) status of canola oil to the US-FDA (United States-Food and Drug Administration), a comment was made that canola oil was probably the most studied vegetable oil the world has ever known (Federal Register USA, 1985). The increasing complexity of the fatty acid variants of canola oil, being developed by both conventional and non-conventional genetic engineering coupled with radical changes occurring in the processing industry, requires that this level of study, at least on the components of the oil, be maintained.

## References

Abidi, S.L., List, G.R. and Rennick, K.A. (1999) Effect of genetic modification on the distribution of minor constituents in canola oil. *Journal of American Oil and Chemists' Society*, **76**, 463–467.

Abraham, V. and de Man, J.M. (1987) Determination of volatile sulfur compounds in canola oil. Winnipeg: Canola Council of Canada.

Ackman, R.G. (1983) Chemical composition of rapeseed oil. In: *High and Low Erucic Acid Rapeseed Oils* (eds J.K.G. Kramer, F.D. Sauer and W.J. Pigden). Academic Press, Toronto, pp. 85–129.

Ackman, R.G., Eaton, C.A., Sipos, J.C., Hooper, S.N. and Castell, J.D. (1970) Lipids and fatty acids of two species of north Atlantic krill (*Meganyctiphanes norvegica* and *Thysanoessa inermis*) and their role in the aquatic food web. *Journal of Fish Research Board*, **27**, 513–533.

Ackman, R.G., Hooper, S.N. and Hooper, D.L. (1974) Linoleic acid artifacts from the deodorization of oils. *Journal of American Oil and Chemists' Society*, **51**, 42–49.

Andrikopoulos, N.K. (2002) Chromatographic and spectrometric methods in the analysis of triacylglycerol species and regiospecific isomers of oils and fats. *Critical Review of Food Science Nutrition*, **42**, 473–505.

Anon. (2003) Improved canola holds the key to the future. *Canola Digest* 22.

Appelqvist, L. (1972) In: *Rapeseed: Cultivation, Composition, Processing and Utilization* (eds L.-. Appelqvist and R. Ohlson). Elsevier, Amsterdam, pp. 1–8.

Appelqvist, L.-. Kornfledt, A. and Wennerholm, J. (1981) Sterols and steryl esters in some *Brassica* and *Sinapis* seeds. *Phytochemistry*, **20**, 207–210.

Babuchowski, K. and Zadernowski, R. (1972) Zmiany w zawartosci zwiazkow siarki i fosforu iodczas przemyslowego przerobu nasion rzepaku ozimego czec I. Rafinacja. Tluszcze Jadalne, **16**(6), 298–305. (*Note*: Translation from Polish. Changes in contents of sulphur and phosphorus compounds during industrial processing of winter rape seeds. Part I. Refining).

Bengtsson, L., Von Hoften, A. and LööF, B. (1972) Botany of rapeseed. In: *Rapeseed Cultivation, Composition, Processing and Utilization* (eds L.-. Appelqvist and R. Ohlson), pp. 36–48. London: Elsevier Publishing Company.

Boki, K., Mori, H. and Kawasaki, N. (1994) Bleaching rapeseed and soybean oils with synthetic adsorbents and attapulgites. *Journal of American Oil and Chemists' Society*, **71**, 595–601.

Botha, I. and Mailer, R.J. (2001) Evaluation of cold-test methods for screening cloudy canola oils. *Journal of American Oil and Chemists' Society*, **78**, 395–399.

Box, J.A.G. and Boekenoogen, H.A. (1967) Vegetable oil pigments: carotenoids and pheophytines in soybean, rapeseed and linseed oils. *Fette Seifen Anstrich*, **69**, 724–729.

Brockerhoff, H. and Yurkowski, M. (1966) Stereospecific analyses of several vegetable fats. *Journal of Lipid Research*, **7**, 62–64.

Byrdwell, W.C. and Neff, W.E. (1996) Analysis of genetically modified canola varieties by atmospheric pressure chemical ionization, mass spectrometric and flame ionization, *Journal of Liquid Chromatography Related Technology*, **19**, 2203–2225.

Chapman, D.M., Pfannkoch, E.A. and Kupper, R.J. (1994) Separation and characterization of pigments from bleached and deodorized canola oil. *Journal of American Oil and Chemists' Society*, **71**, 401–407.

Cho, F.A.Y. and de Man, J.M. (1989) Sulphur and Chlorophyll content of Ontario canola oil. *Canadian Institute of Food Science Technology Journal*, **22**, 222–226.

Clark, J. (1996) Tocopherols and sterols from soybean. *Lipid Technology*, **8**, 111–114.

Cmolik, J., Schwarz, W., Svoboda, Z., Pokorny, J., Reblova, Z., Dolezal, M. and Valentova, H. (2000) Effects of plant-scale alkali refining and physical refining on the quality of rapeseed oil. *European Journal of Lipid Science Technology*, **102**, 15–22.

Codex Alimentarius Commission (2001) Vol. 8. *Fats, oils and related products*. Rome: Food and Agriculture Organization of the United Nations.

Coleman, M.H. (1960) Further studies on the pancreatic hydrolysis of some natural fats. *Journal of American Oil and Chemists' Society*, **38**, 685–688.

Daun, J.K. (1984) Composition and use of canola seed, oil and meal. *Cereal Foods World*, **29**, 291–296.

Daun, J.K. (1987) *Chlorophyll in Canadian Canola and Rapeseed and its Role in Grading* (ed. J. Krzymanski). *Proceedings of the 7th International Rapeseed Congress*, 11–14 May 1987; Poznan, Poland. Poznan Poland: Plant Breeding and Acclimatization Institute, pp. 1451–1456.

Daun, J.K. and Adolphe, D.F. (1997) A revision to the canola definition. *Bulletin of GCIRC*, **14**, 134–141.

Daun, J.K. and Hougen, F.W. (1977) Identification of sulfur compounds in rapeseed oil. *Journal of American Oil and Chemists' Society*, **54**, 351–354.

Daun, J.K. and Jeffrey, L.E. (ed. E.E. McGregor) (1991) *Sediment in Canola Oil*. #87–44, Winnipeg: Canola Council of Canada.

de Clercq, D.R., Daun J.K., Loutas, P., Marianchuk, M.N. and Moody, M. (1991) Sulfur levels in canola oils from Canadian crushing plants: analysis by Raney nickel reduction and inductively coupled plasma atomic emission spectroscopy (ed. D.I. McGregor). Rapeseed in a Changing World, *Proceedings of the Eighth International Rapeseed Congress*, 9–11 July 1991; Saskatoon. Saskatoon: Organizing Committee of the Eighth International Rapeseed Congress, pp. 1396–1401.

de Man, J.M., Porgorzelska, E. and de Man, L. (1983) Effect of the presence of sulfur during the hydrogenation of canola oil. *Journal of American Oil and Chemists' Society*, **60**, 558–562.

Devinat, G., Biasini, S. and Naudet, M. (1980) Composés soufrés des huiles de colza. *Revue Française des corps gras*, **27**, 229–236.

Diosady, L.L., Sieggs, P. and Kaji, T. (1983) Degumming of canola oils. In: *7th Progress Report, Research on Canola Meal, Oil and Seed* (ed. E.E. McGregor), pp. 186–199. Winnipeg: Canola Council of Canada.

Dolde, D., Vlahakis, C. and Hazebroek, J. (1999) Tocopherols in breeding lines and effects of planting location, fatty acid composition, and temperature during development. *Journal of American Oil and Chemists' Society*, **76**, 349–355.

Drexler, H.-J. (1975) Bestimmung schwefelhaltiger Verbindungen in Rapsolen. Bundesanstal fur Fettforschng in Munster/Westf. Jahresbericht ed. Munster: Bundesanstal fur Fettforschng in Munster/Westf.

Drozdowski, B., Niewiadomski, H. and Szukalska, E. (1973) Dezaktywacja kontaktu niklowego zwiazkomi siarki podczas uwodorniania oleju rzepakowego. *Przemsl Chem.*, **2**, 556–557.

Ekman, R. and Pesner, G. (1973) Studies on components in wood Suom. *Kemistiseuran Tied*, **82**, 48–59.

Elson, C.M., Hynes, D.L. and MacNeil, P.A. (1979) Trace metal content of rapeseed meals, oils and seeds. *Journal of American Oil and Chemists' Society*, **56**, 998–999.

Endo, Y., Daun, J.K. and Thorsteinson, C.T. (1992) Characterization of chlorophyll pigments present in canola seed, meal and oils. *Journal of American Oil and Chemists' Society*, **69**, 564–568.

Eskin, N.A.M. (1996) Canola oil: physical and chemical properties. In: *Bailey's Industrial Oil and Fat Products*, Vol. 2, 5th edition (ed. Y.H. Hui) John Wiley & Sons, Inc., Toronto, pp. 43–55.

Eskin, N.A.M., Durance-Todd, S., Vaisey-Genser, M. and Przybylski, R. (1989) Stability of low linolenic acid canola oil to frying temperatures. *Journal of American Oil and Chemists' Society*, **66**, 1081–1084.

Evershed, R.P., Male, V.L. and Goad, L.J. (1987) Strategy for the analysis of steryl esters from plant and animal tissues. *Journal of Chromatography*, **400**, 187–205.

Federal Register USA (1985) *Direct Food Substances Affirmed as Safe: Low Erucic Acid Rapeseed Oil*. Federal Register, 28 January 1985, Washington, DC, **50**, 3745–3755.

Franzke, C., Göbel, R., Noacke, G. and Siffert, I. (1975) Über den Gehalr an Isothicyanaten und Vinylthiooxazolidon in Rapssaat und Rapsöl. *Nahrung*, **19**, 583–593.

Friedt, W. and Lühhs, W.W. (1999) Breeding of rapeseed (*Brassica napus*) for modified seed quality – synergy of conventional and modern approaches [CD ROM]. Organizing Committee of the Tenth International Rapeseed Congress, ed. New Horizons for an Old Crop, *Proceedings of the 10th International Rapeseed Congress*, 27–29 September 1999, Canberra, Australia.

Gao, Z. and Ackman, R.G. (1995) Gradient density isolation and fractionation of haze-causing solids of canola oil. *Journal of Science and Food Agriculture*, **68**, 421–430.

Garrido, M.D., Frías, I., Díaz, C. and Hardisson, A. (1994) Concentrations of metals in edible vegetable oils. *Food Chemistry*, **50**, 237–243.

Goffman, F.D. and Becker, H.C. (2002) Genetic variation of tocopherol content in a germplasm collection of *Brassica napus* L. *Euphytica*, **125**, 189–196.

Gordon, M.H. and Miller, L.A.D. (1997) Development of steryl ester analysis for the detection of admixtures of vegetable oils. *Journal of American Oil and Chemists' Society*, **74**, 505–510.

Grynberg, H., Ceglowska, K. and Szczepanska, H. (1966) Étude de la composition des glycérides de l'huile de colza. *Rev. Franc. Corps Gras*, **13**, 595–602.

Gunstone, F.D., Hussain, M.G. and Smith, D.M. (1974) Fatty acids. Part 42. The preparation and properties of methyl monomercaptostearates. Some related thiols, and some methyl epithiostearates. *Chemical and Physical Lipids*, **13**, 71.

Gunstone, F.D., Wijesundera, R.C. and Scrimgeour, C.M. (1978) The component acids of lipids from marine and freshwater species with special reference to furan-containing acids. *Journal of Science and Food Agriculture*, **29**, 539.

Hasma, H. and Subramaniam, A. (1978) The occurrence of furanoid fatty acid in Hevea brasilients latex. *Lipids*, **13**, 905–908.

Hazuka, Z. and Drozdowski, B. (1987) Major pigments in double-low rapeseed oils. In: *Proceedings of the 7th International Congress*, pp. 1457–1462.

Hermann, L., Mailer, R.J. and Robards, K. (1999) Sedimentation in canola oil: a review. *Australian Journal of Experimental Agriculture*, **39**, 103–113.

Hilditch, T.P., Laurent, P.A. and Meara, M.L (1947) The mixed unsaturated glycerides of liquid fats. Part IV. Low temperature crystallization of rape oil. *J. Soc. Chem. Ind.* **66**, 19–22.

Holmes, M.R.J. and Bennet, D. (1979) Effect of nitrogen fertilizer on the fatty acid composition of oil from low erucic acid rape varieties. *Journal of Science and Food Agriculture*, **30**, 264–266.

Hu, X., Daun, J.K. and Scarth, R.J. (1994) Lipids containing C18:1($n$-7) and C18:1($n$-9) fatty acids in canola seedcoat surface and internal lipids. *Journal of American Oil and Chemists' Society*, **71**, 221–222.

Institute of Medicine (2000) Dietary Reference Intakes for Vitamin C, Vitamin E, Selenium, and Carotenoids: a report of the Panel On Dietary Antioxidants and Related Compounds, subcommittees on Upper Reference Levels of Nutrients, and the Standing Committee on the Scientific Evaluation of Dietary Reference Intakes, Food and Nutrition Board, Institute of Medicine, National Academy Press, Washington, DC, pp. 186–283.

Jàky, M. and Kurnik, E. (1981)Verteilung der linolsaeure in den glyceriden unter besonderer beruecksichtigung der beta-stellung. *Fette, Seifen, Anstrichm*, **83**, 267–270.

Johansson, A. (1977) Nonvolatile sulfur content of zero-erucic acid rapeseed, sunflower and poppyseed oils. *9th Proc. Scand. Symp. Lipids* (ed. R. Marcuse), Stockholm. Goteborg: Lipidforum.

Jonsson, R. (1975) Breeding for improved fatty-acid composition in oilseed crops. 2. Investigation on the effect of some environmental factors on the fatty-acid pattern in winter rape (*Brassica napus* L. var. *biennis* L.) *Sver. Utsädestför. Tidskr.* **85**, 9–18.

Kamal-Eldin, A. and Appelqvist, L.-. (1996) The chemistry and antioxidant properties of tocopherols and tocotrienols. *Lipids*, **31**, 671–701.

Kaufmann, H.P. and Wessels, H. (1964) Die Dünnschicht-chromatographie auf dem Fettgebiet XIV: Die Trennung der Triglyceride durch Kombination der Adsorptions-und der Umkehrphasen-Chromatographie. *Fette, Seifen, Antrichmittel*, **66**, 81–86.

Kochhar, S.B. (1983) The influence of processing on sterols of edible vegetable oils. *Progress in Lipid Research*, **22**, 161–188.

Krygier, K. and Rutkowski, R. (1991) The content of sulphur compounds in solvent extraction products of winter type low and high glucosinolate rapeseed. *Fat Science Technology*, **93**(4), 153–154.

Latta, S. (1990) New industrial uses of vegetable oils. *Informatics*, **1**, 434–443.

Law, M. (2000) Plant sterol and stanol margarines and health. *British Medical Journal*, **320**, 861–864.

Ling, W.H. and Jones, P.J.H. (1995) Dietary phytosterols: a review of metabolism, benefits and side effects. *Life Science*, **57**, 196–206.

Litchfield, C. (1972) *Analysis of Triglycerides*, Academic Press, NY.

Liu, H., Przybylski, R., Dawson, K., Eskin, N.A.M. and Biliaderis, C.G. (1996) Comparison of the composition and properties of canola and sunflower oil sediments with canola seed hull lipids. *Journal of American Oil and Chemists' Society*, **73**, 493–498.

Mag, T. (2001) *Canola Seed and Oil Processing.* Winnipeg, MB, Canola Council of Canada. http://www.canola-council.org/pubs/Oilprocessing.pdf

Marcuse, R. (ed.) (1977) Nonvolatile sulfur content of zero-erucic acid rapeseed, sunflower and poppyseed oils. Goteborg: Lipidforum.

Morchio, G., de Andreis, R. and Fedeli, E. (1987) Total sterols in olive oil and their variation during refining. *Riv. Ital. Sostanze Grasse,* **64**, 185–192.

Morris, L.J., Marshall, M.O. and Kelly, W. (1966) A unique furanoid fatty acid from exocarpus seed oil. *Tetrahedron Letters,* **36**, 4249–4253.

Neff, W.E., Selke, E., Mounts, T.L., Rinsch, W.M., Frankel, E.N. and Zeitoun, M.A.M. (1992) Effect of triacylglycerol composition and structures on oxidative stability of oils from selected soybean germplasm. *Journal of American Oil and Chemists' Society,* **16**, 11–118.

Neff, W.E., Mounts, T.L., Rinsch, W.M., Konish, H. and El-Agaimy, M.A. (1994a) Oxidative stability of purified canola oil triacylglycerols with altered fatty acid compositions as affected by triacylglycerol composition and structure. *Journal of American Oil and Chemists' Society,* **71**, 1101–1109.

Neff, W.E., El-Agaimy, M.A. and Mounts, T.L. (1994b) Oxidative stability of blends and interesterified blends of soybean oil and palm olein. *Journal of American Oil and Chemists' Society,* **71**, 1111–1116.

Neff, W.E., Mounts, T.L. and Rinsch, W.M. (1997) Oxidative stability as affected by triacylglycerol composition and structure of purified canola oil triacylglycerols from genetically modified normal and high stearic and lauric acid canola varieties. *Lebensmittelwissen-schaft und Technologie,* **30**, 793–799.

Ohlson, R., Podlaha, O. and Töregard, B. (1975) Stereospecific analysis of some Cruciferae species. *Lipids,* **10**, 732–735.

Persmark, U. (1972) Analysis of rapeseed oil. In: *Rapeseed: Cultivation, Composition, Processing and Utilization* (eds L.-. Appelqvist and R. Ohlson). Elsevier, Amsterdam, pp. 174–197.

Persson, C. (1985) High palmitic acid content in summer turnip rape (*Brassica campsetris* var. *annua* L.). *Cruc. News,* **10**, 137.

Plat, J. and Mensink, R.P. (2001) Effects of plant sterols and stanols on lipid metabolism and cardiovascular risk. *Nutritional, Metabolism, and Cardiovascular Diseases,* **11**, 31–40.

Pohl, P. and Wagner, H. (1972) Control of fatty acid and lipid biosynthesis in *Euglena gracilis* by ammonia, light and DCMU. *Fette, Seifen, Anstrichm.,* **74**, 53–61.

Przybylski, R. and Eskin, N.A.M. (1991) Phospholipid composition of canola oils during the early stages of processing as measured by TLC with flame ionization detector. *Journal of American Oil and Chemists' Society,* **68**, 241–245.

Przybylski, R. and Mag, T. (2002) Canola/rapeseed oil. In: *Vegetable Oils in Food Technology: Composition, Properties and Uses* (ed. F.D. Gunstone). Blackwell Publishing, CRC Press, pp. 100–126.

Przybylski, R., Biliaderis, C.G. and Eskin, N.A.M. (1993) Formation and partial characterization of canola oil sediment. *Journal of American Oil and Chemists' Society,* **70**, 1009–1015.

Rakow, G. (1973) Selektion auf Linol-und Linolensauregehalt in Rapssamen nach mutagener Behandlung. *Z. Pflanzenzucht.,* **69**, 62–82.

Ramamurthi, S. and Low, N.H. (1995) Effect of possible chlorophyll breakdown products on canola oil stability. *Journal of Agricultural and Food Chemistry,* **43**, 1479–1483.

Ratnayake, W.M.N. (1980) Ph.D. Thesis "Studies on fatty acids from Nova Scotian seaweeds and on the specificity of hydrazine reduction of unsaturated fatty acids", Dalhousie University, Halifax, Nova Scotia, Canada.

Richards, A., Wijesundera, C., Palmer, M. and Salisbury, P. (2002) Presented at the American Oil Chemists' Society Australasian Section Workshop, 4–5 November, Sydney, Australia.

Sauer, F.D. and Kramer, J.K.G. (1983) The problems associated with the feeding of high erucic acid rapeseed oils and some fish oils to experimental animals. In: *High and Low Erucic Acid Rapeseed Oils Production, Usage, Chemistry, and Toxicological Evaluation* (eds J.K.G. Kramer, F.D. Sauer, and W.J. Pigden). Academic Press, Toronto, pp. 254–292.

Sebedio, J.L. and Ackman, R.G. (1979) Some minor fatty acids of rapeseed oils. *Journal of American Oil and Chemists' Society,* **56**, 15–21.

Sebedio, J.L. and Ackman, R.G. (1981) Fatty acids of canola *Brassica campestris* var Candle seed and oils at various stages of refining. *Journal of American Oil and Chemists' Society*, **58**, 972–973.

Sonntag, N.O.V. (1991) Erucic, behenic feedstocks of the 21st century. *Informatics*, **2**, 449–463.

Sosulski, F.W., Zadernowski, R. and Babuchowski, K. (1981) Composition of polar lipids in rapeseed. *Journal of American Oil and Chemists' Society*, **58**, 561–564.

Strocchi, A. (1981) Determination of the fat blend in margarine based on fatty acid and unsaponifiable composition. *Riv. Ital. Sostanze Grasse*, **58**, 271–279.

Subbaram, M.R. and Young, C.G. (1967) Determination of the glyceride structure of fats. Glyceride structure of fats with unusual fatty acid compositions. *Journal of American Oil and Chemists' Society*, **44**, 425–428.

Suzuki, K. and Nishioka, A. (1993) Behavior of chlorophyll derivatives in canola oil processing. *Journal of American Oil and Chemists' Society*, **70**, 837–841.

Tautorus, C.L. and Low, N.H. (1993) Chemical aspects of chlorophyll breakdown products and their relevance to canola oil stability. *Journal of American Oil and Chemists' Society*, **70**, 843–847.

Taylor, D.C., Magus, J.R., Bhella, R., Zou, J., MacKenzie, S.L., Giblin, E.M., Pass, E.W. and Crosby, W.L. (1992) Biosynthesis of triacylglycerols in *Brassica napus* L. cv. Reston; Target: Trierucin. In: *Seed Oils for the Future* (eds S.L. MacKenzie and D.C. Taylor). AOCS Press, Champaign, IL, pp. 77–102.

Thelen, J.J. and Ohlrogge, J.B. (2002) Metabolic engineering of fatty acid biosynthesis in plants. *Metabolic Engineering*, **4**, 12–21.

Tso, P. (1985) Gastrointestinal digestion and absorption of lipid. *Advances in Lipid Research*, **21**, 143–186.

Unger, E.H. (1991) In: *Canola and Rapeseed: Production, Chemistry, Nutrition and Processing Technology*, van Nostrand Reinhold, New York, pp. 235–248.

van de Loo, F.J., Fox, B.G. and Somerville, C. (1993) Unusual Fatty Acids. In: *Lipid Metabolism in Plants* (ed. T.S. Moore). CRC Press, Boca Raton, pp. 92–126.

Vlahakis, C. and Hazebroek, J. (2000) Phytosterol accumulation in canola, sunflower, and soybean oils: Effects of genetics, planting location, and temperature. *Journal of American Oil and Chemists' Society*, **77**, 49–53.

Wada, S. and Koizumi, C. (1983) Influence of the position of unsaturated fatty acid esterified glycerol on the oxidation rate of triglycerides. *Journal of American Oil and Chemists' Society*, **60**, 1105–1109.

Warner, K., Frankel, E.N. and Mounts, T.L. (1989) Flavor and oxidative stability of soybean, sunflower and low erucic acid rapeseed oils. *Journal of American Oil and Chemists' Society*, **66**, 558–564.

Wijesundera, R.C. and Ackman, R.G. (1988) Evidence for the probable presence of sulfur-containing fatty acids as minor constituents in canola oil. *Journal of American Oil and Chemists' Society*, **65**, 959–963.

Wijesundera, R.C., Ackman, R.G., Abraham, V. and de Man, J.M. (1988) Determination of sulfur contents of vegetable and marine oils by ion chromatography and indirect ultraviolet photometry of their combustion products. *Journal of American Oil and Chemists' Society*, **65**, 1526–1530.

Willner, T., Jess, U. and Weber, K. (1997) Effect of process parameters on the balance of tocopherols in the production of vegetable oils. *Fett/Lipid*, **99**, 138–147.

Wolff, J.P. (1980) Recent progress in sterol analysis. *Riv. Ital. Sostanze Grasse*, **57**, 173–178.

Yukowski, M. (1989) Lipid content and fatty acid composition of muscle from some freshwater and marine fish from Central and Arctic Canada. In: *Health Effects of Fish and Fish Oils* (ed. R.K. Chandra). ARTS Biomedical Publishers & Distributors, St John's, Newfoundland, pp. 547–557.

Zadernowski, R. and Sosulski, F. (1979) Composition of fatty acids and structure of triglycerides in medium and low erucic rapeseed. *Journal of American Oil and Chemists' Society*, **56**, 1004–1007.

Zajic, J., Bares, M., Cmolik, I. and Volhejn, E. (1986) *The Conditioning Influence on the Content of Phospholipids in Pressed Rapeseed Oil*. Fette, Seifen, Anstrich, **88**, 67–69.

# 4 Chemical and physical properties of canola and rapeseed oil

Dérick Rousseau

## 4.1 Introduction

The physical and chemical properties of lipids strongly depend on their fatty acid diversity and structure as well as on their distribution in the glycerol backbone (deMan, 1964). As outlined in Chapter 3, fatty acids in canola and rapeseed oil vary in chain length and in the number, position and configuration of their double bonds. Furthermore, the fatty acid distribution within naturally occurring lipids is not random (Norris and Mattil, 1947; Hilditch, 1956). The taxonomic patterns of vegetable oils consist of triacylglycerols obeying the 1,3-random-2-random distribution, with saturated fatty acids being located almost exclusively at the 1,3-positions (*sn*-1,3) of triacylglycerols (Desnuelle and Savary, 1959; Rozenaal, 1992). Conversely, fats from the animal kingdom (e.g. tallow, lard) are quite saturated at the *sn*-2-position (Nawar, 1996).

This chapter focuses on canola oil and, to a lesser extent, on rapeseed oil (in this chapter, the term *rapeseed oil* refers only to oil containing high levels of erucic acid, whereas the term *canola oil* refers to rapeseed oil containing low levels of erucic acid). Canola oil is a genetically modified variant of rapeseed with increasing global importance, particularly in the Orient, due to its nutritional quality. In describing the physical properties of these oils, one must consider whether they are in the liquid or solid state. At room temperature, both processed oils will be liquid. For incorporation into processed foods (described in Chapter 6), the use of hydrogenation, interesterification and blending of these oils with other lipids may transform them into semi-solids or solids, with consequent changes in chemical and physical properties. While this chapter describes the primary chemical and physical properties of canola and rapeseed oil, it does not explore the properties of more specialised products such as high-lauric or low-linolenic canolas.

## 4.2 Chemical properties

Both canola and rapeseed oils are highly unsaturated vegetable oils. However, they have notable differences in composition, particularly the presence of erucic acid (22:1) in rapeseed oil (~45%), where there is little to none of this acid in canola oil (<1%). The reduction in erucic acid is reflected in substantial increases

**Table 4.1** Chemical properties of canola and rapeseed oils

| Property | Canola | Rapeseed |
|---|---|---|
| Saponification value (SV) | 182–193 | 168–181 |
| Iodine value (IV) | 110–126 | 94–120 |

*Source*: Firestone (1999).

in oleic (15–60%) and linoleic acids (14–20%), resulting in >95% 18 carbon fatty acids in canola, compared to >50% non-18 carbon fatty acids in rapeseed. These compositional differences will obviously lead to some notable differences in chemical and physical properties. Table 4.1 outlines two of their chemical properties – the saponification and iodine values.

### 4.2.1 Saponification value

The saponification value (SV) is defined as the number of milligrams of potassium hydroxide (KOH) required to saponify 1 g of oil or fat. A fat or oil's SV is inversely related to the mean molecular weight of its component fatty acids. Thus, a higher molecular weight will result in a lower SV. It is determined by measuring the amount of excess KOH remaining after saponification. During testing, the lipid is boiled under reflux in ethanolic KOH, and the excess KOH is back-titrated with HCl using phenolphthalein (pH turning point 8.4) as indicator. The presence of erucic acid in rapeseed oil compared to the presence of primarily 18 carbon fatty acids in canola oil is responsible for the different SVs between these two oils.

### 4.2.2 Iodine value

The iodine value (IV) of a lipid is a measure of its level of unsaturation. It is defined as the number of grams of iodine that is added to 100 g of fat (Allen, 1955). It is based on the ability of unsaturated carbon–carbon bonds to react with halogens (iodine). Thus, it expresses the concentration of the unsaturated fatty acids, together with the extent to which they are unsaturated, in a single number, and therefore is a simple and very useful quality parameter. The most common method of measuring IV is the Wijs' method, which involves dissolving the sample in carbon tetrachloride to which is added the Wijs' solution (iodine monochloride in glacial acetic acid and carbon tetrachloride). When the reaction is complete, KI solution is added and this reacts with the excess iodine monochloride liberating iodine for thiosulfate titration. For the sake of comparison, Table 4.2 shows the effect of unsaturation level on IV. Fully saturated lipids have an IV of zero. Canola's higher IV compared to rapeseed oil is primarily due to the replacement of erucic acid with unsaturated 18 carbon fatty acids (oleic, linoleic and linolenic acids). An obvious limitation of the IV is that it does not

Table 4.2 Iodine values of selected unsaturated acids

| Fatty acid | Iodine value |
|---|---|
| 16:1(n-9)cis | 99 |
| 18:1(n-9)cis | 89 |
| 18:2(n-6) | 181 |
| 18:3(n-3) | 273 |

define the specific fatty acids present. Even so, IV is extensively used in industry to follow the process of hydrogenation.

## 4.2.3 Oxidative stability

Oils and fats begin to decompose from the moment they are isolated from their natural environment. Oxidation of vegetable oils is the single most common cause of deterioration with the end result being a decline in the sensory and nutritive quality of the oil. For food applications, it is obviously undesirable, yet unavoidable. In fact, only a small proportion of oil needs to be oxidised before it results in objectionable smell and taste and becomes unacceptable from a sensory perspective (Coupland and McClements, 1997). Lipid oxidation proceeds via different pathways: autoxidation, photooxidation, thermal oxidation and/or due to the presence of enzymes, notably lipoxygenase. Each pathway depends on the presence of free radicals, although initiation differs in each mechanism. Enzymatic breakdown is much more common in processed foods, and not in refined canola nor rapeseed, and as such is not addressed here.

### 4.2.3.1 Mechanism

Autoxidation of unsaturated fatty acids occurs via a free radical mechanism and involves three steps: initiation, propagation and termination.

*Initiation.* In the presence of an initiator such as hydroxyl radical ($^{\bullet}$OH), a hydrogen atom from a methylene group ($-CH_2-$) is abstracted leaving behind an unpaired electron on the carbon ($-CH-$) or free lipid radical (L$^{\bullet}$). The radical can form a conjugated diene, which can combine with oxygen to form a peroxy radical (LOO$^{\bullet}$ or LO$_2^{\bullet}$).

$$LH \rightarrow L^{\bullet} + H^{\bullet}$$

*Propagation.* The peroxy radical can abstract hydrogen from another unsaturated fatty acid (LH) to form hydroperoxides (LOOH) as well as a new free lipid radical (L$^{\bullet}$), leading to an autocatalytic chain reaction via which lipid oxidation continues. It is important to note that hydrogen abstraction can occur at different points on the acyl chain and several isomeric hydroperoxides are formed. The alkyl radical also reacts with molecular oxygen to form a new peroxy radical (LOO$^{\bullet}$):

$$LO_2^{\bullet} + LH \rightarrow LOOH + L^{\bullet}$$
$$L^{\bullet} + O_2 \rightarrow LOO^{\bullet}$$

Hydroperoxides are the primary products of lipid oxidation. While not directly responsible for off-flavours and aromas, hydroperoxides decompose into secondary oxidation products, including aldehydes, ketones (carbonyls) and smaller amounts of epoxides and alcohols, with aldehydes being the most important source of the undesirable organoleptic properties (Xu et al., 1999). Propagation is accelerated by elevation in temperature, pressure and oxygen concentration, prior oxidation, presence of metal ions, lipoxygenases, removal of antioxidants (naturally occurring or otherwise), time and light (both visible and UV).

*Termination.* The termination of the free radical oxidative reactions occurs when two radical species react with each other to form a non-radical product. Reaction with an antioxidant may also terminate this reaction.

$$LOO^{\bullet} + LOO^{\bullet} \rightarrow LOOL + O_2$$
$$L^{\bullet} + L^{\bullet} \rightarrow LL$$
$$LOO^{\bullet} + L^{\bullet} \rightarrow LOOL$$

Numerous detailed reports on unsaturated fatty acid oxidation are available in the literature, namely Frankel (1998).

Photooxidation also involves the formation of hydroperoxides in a direct reaction of singlet oxygen and an unsaturated fatty acid without the formation of a free radical. Singlet oxygen is extremely reactive, being ~1500 times more reactive than its molecular counterpart. Light, and in particular ultra violet (UV) radiation, may be involved in the initiation of this pathway.

Oxidation is normally a slow reaction that only occurs to a limited degree. The rate of oxidation depends primarily on the fatty acid composition and less on the stereospecific distribution within the triacylglycerols. Oil oxidation will be roughly proportional to the degree of unsaturation. The most important components that are capable of reacting with oxygen are polyunsaturated fatty acids (PUFAs). Thus, linolenic acid (18:3) is more susceptible than linoleic acid (18:2), which in turn is more susceptible than oleic acid (18:1). Saturated fatty acids are not very susceptible to oxidation and *cis* isomers have lower oxidative stability than their *trans* counterparts.

The oxidative stability of canola and rapeseed may limit their applicability in foods. Given their moderately high polyunsaturated fatty acid content, both canola and rapeseed oils are susceptible to oxidation. Comparison of the relative oxidation rate of commonly used vegetable oils indicates that canola oil is mid-placed (Fig. 4.1).

Substantial research regarding the oxidation of canola oil has been performed. For example, Eskin *et al.* (1989) examined the stability of low-linolenic acid

Lowest stability ←——————————————————→ Highest stability

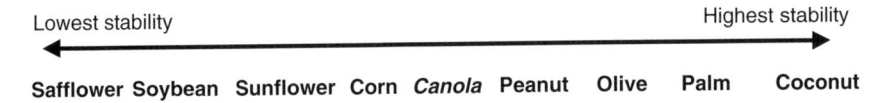

**Safflower  Soybean  Sunflower  Corn  *Canola*  Peanut  Olive  Palm  Coconut**

**Figure 4.1** Relative oxidative stability of some major vegetable oils. Scale is based on inherent oxidative stability, calculated by multiplying the decimal fraction of each unsaturated fatty acid present in a lipid by its relative oxidation rate and then summing these (Erickson and List, 1985).

canola oil for frying purposes. A significantly lower development of oxidation was evident for low 18:3 canola oil based on numerous tests, including peroxide value (PV), thiobarbituric (TBA) and free fatty acids. However, even low levels of 18:3 contributed to the development of *heated-room* odour. Warner *et al.* (1989) compared the flavour and oxidative stability of canola, soybean and sunflower oil establishing that the behaviour of each oil was unique, depending on the oxidation method (light vs dark storage, presence of citric acid and temperature (100 vs 60°C)). Lampi and Piironen (1999) examined the oxidative stability of blends of rapeseed and butter oils and found that rapeseed oil was more susceptible to oxidation than butter oil, as expected. From an oxidation standpoint, a key finding was that blends of these lipids behaved according to the dominant lipid in the mixture. Thus, addition of 10% rapeseed oil to butterfat did not lead to an appreciable increase in oxidation.

During the refining process (Chapter 2), removal of oxidation products takes place. However, once stored, canola and rapeseed will again be susceptible to oxidation. Oxidation is slowed down by packaging the oils in sealed, opaque containers. Furthermore, antioxidants are added to the oils.

Antioxidants hinder oxidation by reacting with the fatty acid peroxy radical to form unreactive radicals or non-radical products. There are two groups of antioxidants (1) chain-breaking antioxidants that react with the lipid radical to produce stable products (i.e. promote termination) and (2) preventative antioxidants that retard initiation. An antioxidant added to food must be effective at low concentrations. Typically, synthetic antioxidants are more effective than naturally occurring antioxidants. Often, mixtures of antioxidants are used to increase efficacy as oxidation is inhibited at various steps by various mechanisms.

Synthetic antioxidants that contain phenolic rings are the most common synthetic species added to these oils, namely butylated hydroxyanisole (BHA) and butylated hydroxytoluene (BHT). Vieira and Regitano-d'Arce (2001) examined the oxidative stability of refined, bleached and deodourised (RBD) canola during conventional and microwave heating, noting that BHA and BHT were effective at stabilising canola against oxidation in a conventional oven, but not in a microwave oven. While the use of synthetic antioxidants is widespread, there is growing concern over their potential toxicity in humans. Other synthetic antioxidants whose effects have been studied on canola oil include polymeric species (Vaisey-Genser and Ylimaki, 1985) and tert-butyl hydroquinone (TBHQ) (Hawrysh *et al.*, 1988). Slowly, synthetic antioxidants are being replaced by other more *natural* species, including vitamins C and E as well as plant phenolics.

Besides antioxidants, extensive research into the genetic modification of canola to enhance its oxidative stability has been carried out. In a study on the impact of fatty acid profile and stereospecific distribution of fatty acids on triacylglycerol molecules, Neff *et al.* (1997) found that canola oil with improved oxidative stability compared to normal canola resulted from two types of modification: (1) reduction in the content of linolenic and linoleic acids with consequent increase in oleic acid and saturated fatty acid content (e.g. stearic, lauric); and (2) an increase in oleic acid and a decrease in linoleic acid at the *sn*-2-position (Raghuveer and Hammond, 1967).

### 4.2.3.2 Susceptibility to oxidation

Lipid oxidation has been widely studied in bulk fats and oils, and now there is a relatively good understanding of the factors that affect oxidation in such systems (Frankel, 1991). However, canola oil is often used in complex, heterogeneous foods such as emulsions, where many different molecular species chemically and physically interact with one another (e.g. margarine, salad dressings and mayonnaise). The oxidation of vegetable oils in emulsion differs from that of bulk lipids, predominantly because of the presence of the emulsion droplet membrane, and the partitioning of ingredients between the oil, aqueous and interfacial regions (Coupland and McClements, 1996).

As mentioned, the most common method of retarding oxidation in fatty foods is the addition of antioxidants. Typically, the behaviour of antioxidants is more complex in emulsified foods than in bulk oils because more variables influence lipid oxidation, including the type and concentration of antioxidants and emulsifiers, and the properties of the aqueous phase (pH, ionic strength, etc.). For example, concentrations of emulsifier exceeding the critical micelle concentration have been shown to protect against oxidation (Ponginebbi *et al.*, 1999). The pH of the aqueous phase also influences antioxidant activity. Huang *et al.* (1996a) found that oxidation in corn oil emulsions stabilised with polysorbate 20 was accelerated by increasing pH. In multiphasic food systems, antioxidants partition into different phases according to their affinities towards and the concentration of these phases. This is due to the polarity and solubility of the antioxidants, which will determine their concentration in each phase of a multiphasic food system (Huang *et al.*, 1996b).

There exist a number of methods to assess the oxidative stability of canola and rapeseed oils. In the past, the active oxygen method (AOM) was the most frequently used method (AOCS official method Cd 12–57). During AOM, oil oxidation is accelerated by continuously bubbling air through a heated sample.

Peroxide values (PVs) are measured periodically to determine the time required for the oil to oxidise to a predetermined peroxide value. However, this method suffers from many deficiencies and difficulties (Jebe *et al.*, 1993) and has been declared obsolete in many circles. Alternative methods have been developed based on the changes in conductivity produced by volatile organic acids. The organic acids are stable tertiary-reaction products from the heated (oxidised) oil. The most common method based on this procedure is the oil stability index (OSI) (also called Rancimat) (Laubli and Bruttel, 1986; Matthaus, 1996). All official tests provide adequate information on oil shelf life, rancidity and oxidative stability at room temperature. However, they do not necessarily correlate with sensory analysis (Gertz *et al.*, 2000). Other tests have been designed to examine thermal stability of vegetable oils, particularly for heating and frying applications. Quantification of the extent of oxidation is based on several methods, the most common being described here.

*4.2.3.3   Peroxide value*
The usual method of assessment of hydroperoxides (primary oxidation products) is by the determination of PV. An elevated PV indicates that oxidation has taken place. PV depends on the amount of oxygen, light and number of double bonds in the substrate oil. The most common analytical methods are those based on an iodometric titration, which measures the iodine produced from potassium iodide by the peroxides present in the oil. PV is usually expressed as milli-equivalents of oxygen per kilogram of fat. PV has also been shown to correlate well with UV absorption at 232 nm (Wanasundara and Shahidi, 1994; Vieira and Regitano-d'Arce, 2001). A key limitation of PV is that since it measures only the primary products of oxidation (hydroperoxides), it may not be indicative of the full extent of oxidation (Sherwin, 1968).

*4.2.3.4   Thiobarbituric acid (TBA) test*
The TBA test is the classical method used to estimate secondary lipid-oxidation products (Gray, 1978). Oxidation of PUFAs results in malonaldehyde (OHCCH$_2$CHO) formation, which produces a pink colour with TBA. This can be determined spectrophotometrically, with the resultant pigment having an absorption maximum at 532 nm. TBA values should never be relied upon as the sole indicator of oxidative rancidity, as pigment development by other factors (e.g. browning reaction products, sugar decomposition products and amino acids) may skew results. Thus, it is more accurate to refer to TBA test as TBA-reactive substances (TBARS) test.

*4.2.3.5   p-Anisidine*
Anisidine values are used to determine the total amount of carbonyl compounds in oxidised vegetable oils. It is based on the reaction of *p*-anisidine with

unsaturated aldehydes, resulting in the formation of a yellowish pigment, measured at 350 nm. Fresh, fully refined oils should have values <10.

### 4.2.3.6 Conjugated dienes

The evaluation of conjugated dienes is usually performed to assess oxidation in heated oils. Oxidation products such as hydroperoxides of PUFAs contain conjugated dienes with a maximum UV absorption at 233 nm. Whilst sensitive, this method can only be applied to undegraded hydroperoxides.

### 4.2.3.7 Chromatography

Gas chromatography (GC) has been used to determine volatile oxidation products. Static headspace, dynamic headspace or direct injection methods are the more commonly used approaches. GC has been used extensively to study secondary oxidation products (e.g. alcohols, carboxylic acids, ketones, etc.) in canola oil (Snyder et al., 1985).

Liquid chromatography combined with mass spectroscopy detection can also be used for the determination of oxidation products. Byrdwell and Neff (2001) compared the autoxidation products of normal, high-stearic acid and high-lauric acid canola in the presence of air and were able to identify and characterise more oxidation products than had been reported to date, including two distinct types of previously unrecognised epoxides.

### 4.2.3.8 Electron-spin resonance

A recent technique that is gaining in popularity for the examination of lipid oxidation products is electron-spin resonance (ESR) (Thomsen et al., 2000). ESR relies on the paramagnetic properties of the unpaired electrons in radicals (Andersen and Skibsted, 2002). Concentrations as low as $10^{-9}$ M can be detected. A key feature of this technique is that it allows study of oxidation events prior to hydroperoxide formation.

### 4.2.3.9 Sensory analysis

Although chemical and physical measurements are useful in determining the oxidative stability of vegetable oils, sensory evaluation is considered the best way of assessing oil quality (namely odour and colour) as it provides information most closely associated with the quality of an oil. Defects in flavour or odour may be detected by trained panellists before they are recognised by instrumental means (AOCS, 1998). For example, flavours that have been described as rancid, metallic, soapy, fishy, beany, etc. can occur at very low levels of oxidation, detected only by a sensory approach. There are obvious limitations on sensory analysis, including poor reproducibility, lack of practicality and the high costs associated with panellist training and facilities. As a result, the recommended approach is to use reproducible chemical or instrumental means to complement/support sensory analyses (Frankel, 1998). For example, Coppin and

Pike (2001) correlated OSI with sensory analysis results and found good correlation between the two techniques.

## 4.3 Physical properties

Notable physical properties of canola and rapeseed oils are shown in Table 4.3.

### 4.3.1 Relative density

The density of a material is defined as the measure of its mass per unit volume (e.g. in g/ml). The density of vegetable oils is lower than that of water and the differences between vegetable oils are quite small, particularly amongst the common vegetable oils. Generally, the density of oils decreases with molecular weight, yet increases with unsaturation levels. Timms (1985) reported that between 20 and 60°C, density in vegetable oils changes by 0.3 kg/m$^3$ for each unit increase in $SV$, 0.14 kg/m$^3$ for each unit increase in $IV$, −0.68 kg/m$^3$ for each unit increase in temperature (°C), −0.2 kg/m$^3$ for every 0.1% increase in free fatty acids, and 0.8 kg/m$^3$ for every 1% increase in water. Over the range of temperatures that oils are usually exposed to during processing (~65–260°C), density exhibits a linear relationship with temperature, decreasing 0.64 kg/m$^3$ for every 1°C increase in temperature (Formo, 1979).

Several authors have developed empirical models to relate density to fatty acid composition and temperature. Density varies linearly with temperature according to the following relationship (Formo, 1979):

$$\rho = b + mT \tag{4.1}$$

where $\rho$ = density, $T$ = temperature and, $m$ and $b$ = constants (dependent on type of oil).

As mentioned, the density of a vegetable oil also depends on its SV, IV, free fatty acid content and moisture level. A common method used for the

**Table 4.3** Summary of the physical properties of canola and rapeseed oils

| Property | Canola | Rapeseed |
|---|---|---|
| Relative density (g/cm$^3$; 20°C) | 0.914–0.920 | 0.910–0.912 |
| Refractive index ($n_D$; 40°C) | 1.465–1.467 | 1.465–1.469 |
| Kinematic viscosity (20°C, mm$^2$/sec) | 78.2 | 84.6 |
| Surface tension (mN/m) | ~33 | ~33 |
| Cold test (15 h at 4°C) | Passed | Passed |
| Smoke point (°C) | 220–230 | 226–234 |
| Flash point, open cup (°C) | 275–290 | 278–282 |
| Specific heat (J/g at 20 °C) | 1.910–1.9116 | 1.900–1.911 |

*Source*: Firestone (1999); Przybylski and Mag (2002).

determination of density in a wide variety of vegetable oil is known as the Lund relationship, discussed by Halvorsen *et al.* (1993):

$$sg = 0.8475 + 0.0003SV + 0.00014IV \qquad (4.2)$$

where sg=specific gravity at 15°C, SV=saponification value and, IV=iodine value. More recently, Pantzaris *et al.* (1985) discussed the following relationship, which incorporated temperature besides IV and SV:

$$D = 0.8475 + 0.000308SV + 0.000157IV - 0.00068T \qquad (4.3)$$

where D=apparent density (g/ml), SV=saponification value, IV=iodine value and, T=temperature (°C).

Other models that have been used include the Rackett equation (Nourredini, 1992a) and the generalised model developed by Rodenbush *et al.* (1999) to estimate density based on fatty acid composition. Their model predicted density with a deviation of only 0.21% from experimental results.

Characteristic values of the density of canola and rapeseed oils are presented in Table 4.3. Rapeseed and canola oils do not differ appreciably in density, although the reduction in erucic acid and *cis*-11-eicosenic acid (20:1) and a subsequent increase in 18 carbon unsaturated fatty acids may play a role in density differences between the two oils (Ackman and Eaton, 1977). Density is an important parameter in fats and oils trade, as oil transactions are on a weight basis but measured on a volume basis (Gunstone, 2000).

### 4.3.2 Viscosity

Viscosity is defined as the resistance of a liquid to flow. From a molecular point of view, viscosity provides an indirect measurement of the internal friction between the molecules that constitute an oil. Given their long chain lengths, vegetable oils have rather high viscosities compared to other organic compounds. However, the viscosities of the most common vegetable oils do not differ significantly from one another. There are two types of viscosities in common use – dynamic (absolute) viscosity and kinematic viscosity (usually measured with falling ball, rising bubble, etc.), which is equal to the dynamic viscosity divided by the fluid density at the experimental temperature. Viscosity is an important parameter for numerous operations, including assessing the quality of frying oils. Viscosity increases with molecular weight but decreases with increasing unsaturated level and temperature. Nourredini *et al.* (1992b) investigated this relationship and found:

$$\ln \mu = A + \frac{B}{(T+C)} \qquad (4.4)$$

where $\mu$=viscosity, T=temperature (K) and, A, B, C=constants dependent on the oil.

Generally speaking, vegetable oils exhibit the flow behaviour of true Newtonian liquids. This is normally examined by measuring shear stress vs shear rate over a wide range of shear rates, where a linear relationship between the two indicates Newtonian flow. If an oil begins to crystallise (i.e. when the temperature is decreased below its crystallisation point), flow will become non-Newtonian.

The viscosity of RBD canola oil is substantially lower than that of rapeseed oil, namely due to the presence of erucic acid (Table 4.3). Lang *et al.* (1992) investigated the relation between temperature and kinematic viscosity for refined canola oil and found that viscosity was exponentially correlated with oil temperature. The kinematic viscosity of canola was modelled using:

$$\upsilon = \exp(C_0 + C_1 T + C_2 T) \tag{4.5}$$

where $\upsilon$ = kinematic viscosity, $T$ = temperature of the oil and, $C_0$, $C_1$, $C_2$ = constants determined experimentally.

Toro-Vazquez and Infante-Guerrero (1993) studied the viscosity of numerous vegetable oils (including canola) and found that multiple regression analysis was effective at establishing quantitative structure–function relationships in oils. Rather than other empirical approaches that did not necessarily incorporate terms with a physico-chemical significance, their approach included *meaningful* terms, namely SV and IV.

### 4.3.3 *Surface and interfacial tension*

The forces acting on the surface of a liquid against air, or at the interface between two liquids, will tend to minimise the area of the surface/interface in an effort to approach thermodynamic equilibrium. An understanding of surface and interfacial tension presents information on surface structure as well as adsorption of oil molecules at surfaces and/or interfaces, which may impact formation and stability of dispersed systems (i.e. foams or emulsions).

Very few reports have been published on the surface and interfacial tensions of oils. To the author's knowledge, there have not been any reports on the evolution of surface or interfacial tension of canola and/or rapeseed oils. As an approximation, the evolution in surface tension of tricaprylin and tripalmitin as a function of temperature is shown in Fig. 4.2. Surface tension decreases with increasing temperature essentially in a linear fashion (Chumpitaz *et al.*, 1999).

Surface tension also increases with an increase in chain length. Interfacial tension will be significantly reduced by free fatty acids, traces of soaps, monoacylglycerols and/or lecithin. For example, adding 2% monoacylglycerols to canola oil will reduce its interfacial tension by 70–80%. The interfacial tension of an oil, such as canola, against purified water should be ~30 mN/m, assuming that it is impurity-free (Johansson and Bergenståhl, 1995).

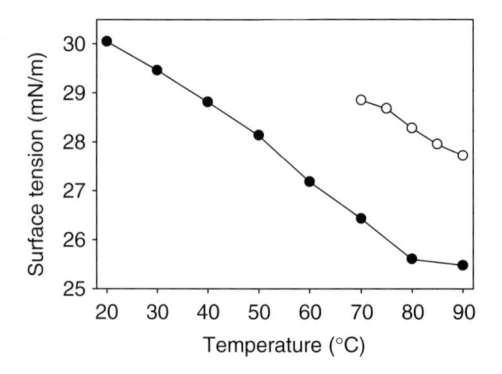

**Figure 4.2** Surface tension of tripalmitin and tricaprylin as a function of temperature (data from Chumpitaz et al., 1999); tricaprylin (●—●), tripalmitin (○—○).

## 4.3.4 Refractive index

The refractive index (RI) of a substance is the ratio between the speed of light *in vacuo* and that in the substance of interest. As the RI of a given fat or oil is quite constant, it can be used for identification purposes. The relationship between RI and oil composition is as follows: RI increases as acyl chain length increases, as the number of double bonds increases, and is higher for triacylglycerols than for fatty acids. It is an important characteristic given its ease of measurement, speed and small sample amount required. RIs decrease with an increase in temperature. Perkins (1995) mentioned the following relationship between RI and IV:

$$n^{25}_D = 1.45765 + 0.0001164 IV \tag{4.6}$$

Thus, when their IVs are known, the RI of canola and rapeseed oils can easily be determined. However, if using RI values for any hydrogenated samples of these oils, the presence of *trans* fatty acids can induce error as these affect RI but not IV. For example, with IVs less than 90–95, the presence of elaidic acid will make the fat harder than its RI would indicate (Gunstone and Norris, 1983). Furthermore, lipid oxidation and/or hydrolysis (resulting in free fatty acids) can affect the RI of a lipid.

## 4.3.5 Specific heat: heat of fusion or crystallisation

Specific heat capacity is defined as the measure of the energy required to raise the temperature of a unit of the material. For vegetable oils, it is largely independent of molecular weight, but is associated with unsaturation level (Timms, 1985). It is influenced by temperature (Formo, 1979) as follows:

$$C_p = 1.9330 + 0.0026T \qquad (4.7)$$

In general terms, there is little variation between all vegetable oils including canola and rapeseed.

### 4.3.6 Heat of combustion

The heat of combustion in saturated fatty acids increases with increase in chain length. Values for unsaturated fatty acids are slightly lower than that for their saturated counterparts (Formo, 1979). Triacylglycerols have about the same heat of combustion as the fatty acids they consist of. Thus, the heat of combustion for common fats and oils is ~9500 cal/g. A general equation linking the heat of combustion of vegetable oils to IV and SV was shown in Perkins (1995):

$$-\Delta H_c = 11\,380 - IV - 9.158 SV \qquad (4.8)$$

Thus, a higher degree of saturation and longer acyl chain length lead to a higher energy content in the oil. For canola oil, assuming $IV = 118$ and $SV = 187$, we obtain a heat capacity of 9550 cal/g. For rapeseed, assuming $IV = 107$ and $SV = 175$, we calculate a heat capacity of 9570 cal/g.

### 4.3.7 Smoke, flash and fire point

The smoke point is defined as the temperature at which a lipid produces a continuous veil of visible smoke when heated in the presence of air, and is a commonly used indicator of the suitability of a vegetable for frying purposes. As oil is heated, it breaks down into free fatty acids and glycerol. At the smoke point, glycerol is dehydrated to form acrolein ($CH_2CHCHO$), which is visible as bluish, acrid smoke. Given their higher vapour pressure compared to triacylglycerols, free fatty acids are also components of the smoke evolved. It has been observed that canola produces less smoke than soybean or sunflower oil, perhaps due to its lower IV (Chen and Chen, 2001). Shield et al. (1995) reported that the smoke from heated rapeseed oil was mutagenic vis-à-vis canola, which was not. The flash point is the temperature at which the volatile decomposition products formed in heated oils can be ignited, but cannot support combustion. An increase in the content of unsaturated fatty acids usually causes a decrease in the flash point. The fire point is the temperature at which the volatiles are produced in a quantity that supports a flame. All three of these parameters are strongly related to the free fatty acid content of oils. The smoke point of an oil will vary from ~230°C with a free fatty acid content of 0.01% to ~93°C for a free fatty acid content of 100%. Under similar conditions, flash points will vary from ~330°C to ~193°C and fire points from ~363°C to ~221°C (Formo, 1979).

*4.3.8 Solubility*

In the liquid state, fats and oils are soluble in most non-polar organic solvents (e.g. hydrocarbons, ethers, chlorinated solvents, etc.) and are essentially insoluble in water. A test designed to evaluate the miscibility of vegetable oils (and applied to canola), in a standard solvent mixture is the Crismer test, which uses a solvent mixture consisting of tert-amyl alcohol, ethanol and water (AOCS official method Cb 4-35). The Crismer value is used for international trade, particularly in Europe. The miscibility of a vegetable oil is primarily related to the chain length of its component fatty acids and their level of unsaturation. The Crismer values of canola and rapeseed range from 67 to 70 for canola and 80 to 82 for rapeseed (Przybylski and Mag, 2002). The solubility characteristics of oil in various solvents can be estimated from their dielectric properties (Sipos and Szuhaj, 1996).

*4.3.9 Cold test*

The ability of an oil to remain liquid at refrigerator temperature is determined by the cold test. Canola and rapeseed should be sediment-free during storage. However, even in properly winterised canola oil there is still the occasional development of a haze. This is an obvious problem for its quality and acceptability. Unfortunately, its occurrence is difficult to predict. It has been postulated that hot and dry weather during the growing season is at least partially to blame, as such climates lead to an increase in saturated fatty acid production in canola.

The chemical and physical properties of the haze closely resemble those of the winterisation sediment from bottled canola oil (Przybylski *et al.*, 1993; Liu *et al.*, 1996). Several investigators have examined the occurrence of haze formation in canola (Hu *et al.*, 1993; Lui *et al.*, 1993, 1995). These workers found that the major clouding substances were wax esters of fatty acids and fatty alcohols with saturated acyl chains, from 16 to 32 carbons, with minimal amounts of saturated triacylglycerols, free fatty acids, hydrocarbons and free fatty alcohols. Crystallisation conditions were found to significantly influence the final properties of the sediment (crystal size and shape) (Liu *et al.*, 1994).

*4.3.10 Spectroscopic properties*

The spectroscopic properties of fats and oils can provide an exceptional amount of data on composition, identification, purity and extent of degradation (e.g. via oxidation or heat). The main techniques in use in the fats and oils field include infrared (IR) (usually Fourier-transformed, hence FT-IR), Raman (often FT-Raman), and to a lesser extent, nuclear magnetic resonance (NMR) and near infrared (NIR) spectroscopy (AOCS, 1998). Spectra characterisation of a substantial number of fats and oils has led to the identification of specific bands or band ratios that may be useful for identification purposes. By spectroscopic means, it is possible to examine and quantify *cis–trans* isomerisation, IV, SV, PV, free fatty acid content, anisidine value, oxidation products and heat-induced polymerisation.

Fourier-transformed infrared is becoming more and more common in the fats and oils industry. It can be used to detect adulteration (Ozen and Mauer, 2002) and has been used for oxidation studies as it provides a quick and accurate means of evaluating degradation in vegetable oil (Moya Moreno *et al.*, 1999a). An extensive review on the uses of FTIR, and to a lesser extent NIR and Raman, in lipid analysis is given by Van der Voort *et al.* (2001). IR and Raman spectra provide complementary information. For example, in Raman spectra, the –C=C– stretching absorptions are readily visible, yet extremely weak in IR spectra and are observed at different Raman wave numbers (1656 vs 1670/cm) for *cis* and *trans* bonds respectively (Johnson *et al.*, 2002). The NIR spectrum of oils recorded between 10 000 and 45 000/cm provides substantial information on lipid properties. In fact in 2000, an FT-NIR method was adopted by AOCS for the determination of IV. NIR has been used for examining adulteration in vegetable oils, notably olive oil (Downey *et al.*, 2002).

With NMR, it is possible to evaluate the oxidation products (e.g. peroxides, epoxides) that result during lipid oxidation (Shahidi, 1992). Relative changes in the NMR absorption peaks of fatty acids and both primary and secondary oxidation products allow time-dependent changes in oxidation to be monitored (Saito and Udagawa, 1992). With proton NMR, Wanasundara *et al.* (1995) examined the ratio of aliphatic to olefinic protons in canola and found a steady increase over the storage duration, indicating increasing oxidation of unsaturated fatty acids. Subjecting a variety of vegetable oils including canola to temperatures up to 300°C, Moya Moreno *et al.* (1999b) followed and identified the evolution of oxidation products with proton NMR spectroscopy. Both primary and secondary oxidation products were identified, including hydroperoxides (primary oxidation products) and carbonyls (secondary oxidation products). Heating canola at 300°C for 40 minutes yielded ~6.6% carbonyl compounds in canola compared to 10.5% in olive oil and 3.0% in corn oil.

### 4.3.11  *Melting behaviour, polymorphism and crystal structure*

An oil is defined as that which is liquid at room temperatures, whereas a fat will be solid at room temperature. Compositional factors affecting the physical properties of vegetable oils include diversity in fatty acids (unsaturation and acyl chain length), stereoisomerism and positional distribution.

#### 4.3.11.1  *Unsaturation level*
Vegetable oils are composed of triacylglycerols that contain both saturated and unsaturated fatty acids. Triacylglycerols composed of saturated fatty acids (e.g. myristic, palmitic, stearic) have the highest melting points and are generally

solid at ambient temperature while those with unsaturated (monoene, polyene) fatty acids (e.g. oleic, linoleic, linolenic) are liquid at room temperature. Thus, oils with high levels of unsaturated fatty acids are liquid at room temperature.

*4.3.11.2 Acyl chain length*
As the chain length of a fatty acid increases, so does its melting point. Thus, a short fatty acid (e.g. caproic acid) will have a lower melting point than a long-chain fatty acid (e.g. stearic acid).

*4.3.11.3 Fatty acid isomers*
The *cis* and *trans* isomers of a fatty acid can substantially differ in physical properties. For example, oleic acid and its geometric *trans* isomer, elaidic acid, differ substantially in melting point. Oleic acid is liquid at room temperature (mp=4°C) whereas elaidic acid is solid (mp=46°C).

*4.3.11.4 Positional distribution*
The molecular configuration of fatty acids in triacylglycerols will substantially alter physical properties. A system consisting of monoacid triacylglycerols (e.g. tristearin) will have a sharp melting point. A lipid containing mixed triacylglycerols (e.g. 1-3-dioleoyl-2-stearin and 1-palmitoyl-2-oleoyl-3-behenin) will have a melting range rather than a melting point. Thus, edible fats and oils do not have a true melting point, but rather a melting range given their compositional variety and will pass through a gradual softening before becoming fully liquid. This solid–liquid conversion is normally examined using differential scanning calorimetry, dilatometry and/or pulsed nuclear magnetic resonance (pNMR). pNMR is now widely used compared to the older technique of dilatometry, which is less accurate (particularly at lower temperatures) and more time-consuming.

In the solid state, fats and oils exist as crystals, which consist of interacting triacylglycerol molecules with an asymmetrical tuning fork geometry (Jensen and Mabis, 1963). These crystals can exist in crystal forms of differing stabilities, a phenomenon known as polymorphism. The current crystal-polymorph nomenclature, proposed by Lutton (1950), consists of three main forms – alpha ($\alpha$), beta prime ($\beta'$) and beta ($\beta$). The $\alpha$-form chain packing is hexagonal, the $\beta'$-form is orthorhombic while the $\beta$-form is triclinic (Chapman, 1962). Crystal sub-forms include sub-$\alpha$ ($\gamma$), $\beta'_1$, $\beta'_2$, pseudo-$\beta'$, sub-$\beta$, $\beta_1$ and $\beta_2$ (D'Souza *et al.*, 1990). Most fats exhibit polymorphism monotropically (Manning and Dimick, 1985). When a transition occurs (e.g. from $\beta'$ to $\beta$), the molecular chain packing of the triacylglycerols becomes less motile resulting in a higher melting point. The most stable of the three structures is the $\beta$-form that results in rather large crystals (5–25 µm), whereas the $\beta'$-form results in small crystals (<3 µm). The $\alpha$-form is not commonly found in fats and oils, although it may concurrently exist in complex mixtures with both $\beta'$- and $\beta$-forms. X-ray diffraction, used to identify crystal polymorphs, is based on the determination of the long and short spacings of crystals. The α-form has a single short spacing near 4.15 Å, the β'-form consists of spacings at 3.8 and 4.2 Å or three at 4.27, 3.97 and 3.71 Å, whereas the β-form differs again and typically shows a single strong spacing at 4.6 Å (Larsson, 1966).

The rate at which fats and oils transform from one polymorph to another largely depends on composition and processing. In general, the more diverse the triacylglycerol composition of the highest melting portion of the fat, the less the β-form tendency will be. Triacylglycerols with similar fatty acids can pack closely, thus allowing crystals to transform to a larger, more stable, higher melting form. Molecules that differ in triacylglycerol and fatty acid composition cannot pack as closely, thus impeding polymorphic transformation. Hence, lipids consisting of a heterogeneous assortment of triacylglycerols will tend to remain in the less-stable polymorph. The short spacings of liquid and hydrogenated canola and rapeseed oils are shown in Table 4.4.

Both canola and rapeseed oils are liquid at room temperature. Properly processed canola will not crystallise at temperatures above ~−10°C (Riiner, 1970). Rapeseed can crystallise above 0°C, if its erucic acid content is quite high (>50%). In the liquid state, both canola and rapeseed will crystallise in the β-form; however, this crystallisation behaviour will strongly depend on the cooling and heating regime. Riiner (1970) found that in rapeseed oils with erucic acid contents >25%, the α-form could be detected only incidentally. In rapeseed with <8% erucic acid, the α-form could be more easily identified and there exists a β'-form polymorph, not visible in the higher erucic-acid content oils, where β-form crystals predominate. Characteristic crystallisation behaviour in rapeseed can be substantially influenced by interesterification, whereby the stereospecific distribution of the erucic acid can be changed (Kawamura, 1981).

The polymorphic behaviour of canola and rapeseed oils is strongly affected by hydrogenation. Partial hydrogenation introduces greater compositional variety (notably *trans* isomers), which slows the β'→β transformation of the lipid. In the case of canola, full hydrogenation results in an even more homogeneous fatty acid composition, resulting in a very strong tendency to form β-crystals. Hydrogenated canola oil's (HCO) most stable polymorph is the β-form (Hernqvist *et al.*, 1981). In Fig. 4.3, a polarised light microscopy image of the β-form found in HCO is shown. Naguib-Mostafa *et al.* (1985) investigated the crystal structure of HCO (IV=70) using scanning electron microscopy and found that quiescently crystallised HCO existed as β-crystals in the form of spherulites.

As a liquid added to solid fats, canola and rapeseed oils can substantially alter crystallisation properties. DeMan *et al.* (1989) found that quench-cooled fully hydrogenated rapeseed oil crystallised in α, β' and β-forms. The presence of canola oil in such a mixture would promote and accelerate the β'→β transition (deMan *et al.*, 1995). Likewise, Rousseau *et al.* (1996b) found that incremental

**Table 4.4** Short spacings for liquid and hydrogenated canola and rapeseed oils

| Product | IV | 5.20 | 5.00 | 4.8 | 4.6 | 4.5 | 4.4 | 4.3 | 4.2 | 4.1 | 3.9 | 3.8 | 3.6 | Polymorph |
|---|---|---|---|---|---|---|---|---|---|---|---|---|---|---|
| Canola (<−10°C) | 112 | 5.20 w | | 4.80 w | 4.60 s | | 4.38 m | | 4.15 w | | 3.90 m | | | $\beta' > \beta_1$ |
| Canola (non-selective hydrogenation) | 72.0 | | | | | | | 4.37 m | 4.20 s | 4.07 w | | 3.84 s | | $\beta'$ |
| Canola (selective hydrogenation) | 66.6 | | 5.05 vw | | | 4.55 w | 4.40 m | | 4.25 s | | | 3.84 s | | $\beta' > \beta$ |
| Canola (margarine) | | 5.20 w | | | 4.60 s | 4.51 w | 4.42 w | | | | 3.89 s | 3.8 m | 3.68 m | $\beta$ |
| Rapeseed (53% erucic) (<−22°C) | 102 | 5.20 w | | 4.80 w | | | 4.48 s | | | 4.10 m | | 3.80 m | 3.62 w | $\beta_2$ |
| Rapeseed (hydrogenated) | 70.2 | | | | | | | 4.36 s | 4.23 s | | 4.07 w | 3.86 s | | $\beta'$ |
| Rapeseed (non-hydrogenated) | 64.2 | | | | | 4.58 w | | 4.34 s | 4.21 s | | 4.07 w | 3.87 s | | $\beta' > \beta$ |

Legend: w, m and s designate the relative intensities of the spacings: w – weak; m – medium; s – strong.
*Source*: Rüner (1970); D'Souza *et al.* (1990).

CHEMICAL AND PHYSICAL PROPERTIES

**Figure 4.3** Image of hydrogenated canola oil as viewed with polarised light microscopy.

addition of canola oil to butter fat led to an increased proportion of β-crystals, likely due to the lower, continuous phase viscosity.

## 4.4 Modification strategies

The physical properties of canola and rapeseed oils can be altered by one of two key processing means – hydrogenation and interesterification.

### 4.4.1 Hydrogenation

Hydrogenation is an important tool for the fats and oils processor. The two main reasons to hydrogenate are: (i) to convert naturally occurring oils into fats with novel melting profiles and (ii) to improve oxidative stability. However, as a result of this process, there is a reduction in the nutritional quality of the oil (loss of essential fatty acids and an increase in *trans* fatty acid formation) (Bansal and deMan, 1982a,b).

This process relies on the addition of hydrogen molecules to the fatty acids present in triacylglycerols at high temperature and pressure in the presence of a catalyst, to saturate the double bonds present in unsaturated fatty acids, which increases their melting point and reduces their possibility of reaction with oxygen. Hydrogenation requires the use of a metal catalyst, high-purity hydrogen gas, well-refined vegetable oil, high temperatures (160–200°C) and high pressure (<25 bar). The usual catalyst is nickel, which is normally deposited on a carrier, such as kieselguhr. This process can substantially extend the application of canola, which is extensively used in margarine and shortening basestocks as well as in many processed foods (Nawar, 1996).

Three key factors will influence fatty acid properties during hydrogenation. The first is the rate of hydrogenation, usually expressed as the rate of decrease in IV, which as previously mentioned is essentially proportional to the level of unsaturation. During hydrogenation, it is usual to reduce the IV of canola oil from 110–120 to 75–100, depending on the final application. The second key factor is the selectivity of the reaction. Selectivity is defined as the preferential hydrogenation of polyunsaturated acid groups relative to monounsaturated groups (linolenic and linoleic vs oleic acid respectively). The third key factor is isomerisation. Unsaturated fats may be geometrically isomerised from the natural *cis* form to the *trans* form. This isomerisation will have an effect on the melting behaviour of the fat as *trans* containing fats are generally more solid than their *cis* counterparts. The double bonds in a fatty acid may also migrate along the chain.

The rate of the reaction depends on the level of unsaturation. In Fig. 4.4, the rate constants for the reduction of linolenic, linoleic and oleic acids, are $k_1$, $k_2$ and $k_3$ respectively (Koseoglu and Lusas, 1990a,b), whereas $k_a$, $k_b$ and $k_c$ are the rate constants for *trans* isomerisation, shown with relative rates (Kellens, 2000).

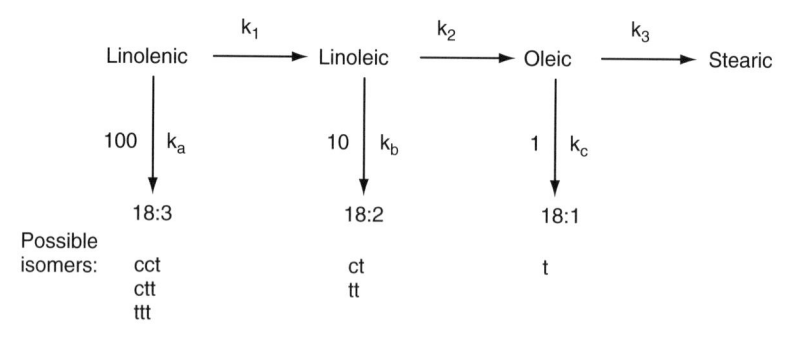

$$\text{Linolenic} \xrightarrow{k_1} \text{Linoleic} \xrightarrow{k_2} \text{Oleic} \xrightarrow{k_3} \text{Stearic}$$

| | | |
|---|---|---|
| 100 ↓ $k_a$ | 10 ↓ $k_b$ | 1 ↓ $k_c$ |
| 18:3 | 18:2 | 18:1 |

Possible isomers:
- 18:3: cct, ctt, ttt
- 18:2: ct, tt
- 18:1: t

**Figure 4.4** Hydrogenation reaction sequence (adapted from Koseoglu and Lusas, 1990a,b and Kellens, 2000). Values of 100, 10 and 1 are the relative reaction rates during hydrogenation.

Selectivities are used to explain the extent of each reaction by using ratios of rate constants for each step, defined as follows:

Linolenate selectivity (Ln) = $k_1/k_2$
Linoleate selectivity (Lo) = $k_2/k_3$
Oleate selectivity (O) = $k_1/k_3$

Linolenate selectivity is the ratio of the reaction rate of hydrogenation of linolenic to linoleic acid. Linoleate and oleate selectivities are the ratios of the reaction rates of the hydrogenation of linoleic acid to oleic acid, and of linolenic acid to oleic acid respectively.

The rate and selectivity of hydrogenation will be affected by type of catalyst, catalyst concentration, feedstock type and quality, purity of hydrogen, hydrogen pressure in the reactor, temperature of hydrogenation and degree of agitation. Table 4.5 outlines the effect of the key parameters on hydrogenation selectivity and isomerism.

A number of different metals will function as catalysts for the hydrogenation of triacylglycerols. The commonality between these metals is their interatomic spacings, which closely correlate with that of the double bond spacing in the fatty acid acyl chain. While research has been done on the use of palladium, platinum and copper as catalysts, nickel is used in essentially all industrial applications of edible oil hydrogenation as it is relatively inexpensive, and performs efficiently and consistently. It takes only small amounts of poisons to inactivate the catalyst, as the catalyst itself is present in the reactor at very low concentrations. Poisons may be introduced either with the hydrogen gas supply or with the oil. Poisons in the gas phase include sulfur compounds (e.g. hydrogen sulfide, carbon sulfide, sulfur dioxide and carbon oxysulfide) that bind irreversibly to the catalyst active sites. Other gases that affect the reactor performance are carbon dioxide, methane and nitrogen, which do not poison the catalyst *per se*, but will accumulate in the headspace of a dead-end batch reactor and decrease the partial pressure of the hydrogen.

Oil being fed into the reactor must be previously purified (Chapter 2). It is generally more economical to purify the oil than to add extra catalyst. Compounds that are particularly important in their effect on catalyst performance are sulfur, phosphorus, bromine and nitrogen (being found in proteinaceous material). Alkali

**Table 4.5** Effect of parameter change on hydrogenation

| Parameter | Selectivity | Isomerisation |
|---|---|---|
| High temperature (200°C) | Higher | Higher |
| High pressure (3–25 bar) | Lower | Lower |
| High agitation | Lower | Lower |
| High catalyst concentration (0.01–0.1%) | Higher | Higher |
| Low catalyst poisoning | Higher | Lower |

Adapted from Kellens (2000).

soaps and phosphorus act as inhibitors by blocking catalyst pores and reducing mass transfer of reactants to and from the catalyst active sites (Hastert, 1996).

Most commercial hydrogenation is performed in large-batch slurry reactors of 20–30 t. Purified heated oil is introduced into the reactor along with a slurry containing the catalyst. Hydrogen is introduced into the bottom of the tank and the mixture is agitated with rotating paddles or helices. The hydrogenation reaction is exothermic, so the tank is equipped with cooling coils to control the reaction temperature. Hydrogen is admitted into the reactor until the desired degree of saturation is achieved. The extent of the reaction is controlled either by monitoring RI and IV, or simply by measuring the amount of hydrogen that has been introduced into the reactor.

Continuous slurry reactors have found very limited use in industry as they are more dependent on insufficiently developed instrumentation and are subject to contamination when changing products. Continuous, fixed-bed systems are even less well suited to this application, as impurities in the feedstock tend to clog the catalyst bed. As a result, the selectivity of the reaction, and hence the physical properties of the product, will alter as the bed becomes gradually poisoned (Hastert, 1996).

In practice, hydrogenation reactors are under-agitated, and mass transfer of hydrogen and triacylglycerol to the catalyst site becomes the rate-limiting steps (Coenen, 1976; Allen, 1981). Hydrogen must diffuse through two stagnant liquid layers – one surrounds the hydrogen bubble while the second surrounds the catalyst particle – into the pores of the catalyst. This is obviously one of the key factors in selectivity.

Canola oil is often hydrogenated for incorporation into foods, particularly in the countries where it is grown. Table 4.6 shows some IV values and resultant changes in composition.

Depending on the final application, different levels of hydrogenation can be achieved (Eskin et al., 1996). Lightly (i.e. IV = 90) hydrogenated canola oil is used as a pourable frying fat. Highly hydrogenated canola will not be used in

**Table 4.6** Fatty acid composition of canola before hydrogenation and after mild and extensive hydrogenation

| Fatty acid | Non-hydrogenated (IV = 112) | Mild hydrogenation (IV = 97) | Extensive hydrogenation (IV = 7) |
|---|---|---|---|
| 16:0 | 4.3 | 4.3 | 4.6 |
| 18:0 | 2.0 | 2.9 | 11.7 |
| 18:1 (cis + trans) | 59.3 | 68.7 | 72.4 |
| 18:2 | 20.7 | 14.9 | 5.8 |
| 18:3 | 8.1 | 1.7 | 0.5 |
| Others | 5.6 | 7.5 | 5.0 |

Adapted from Kramer and Sauer (1983).

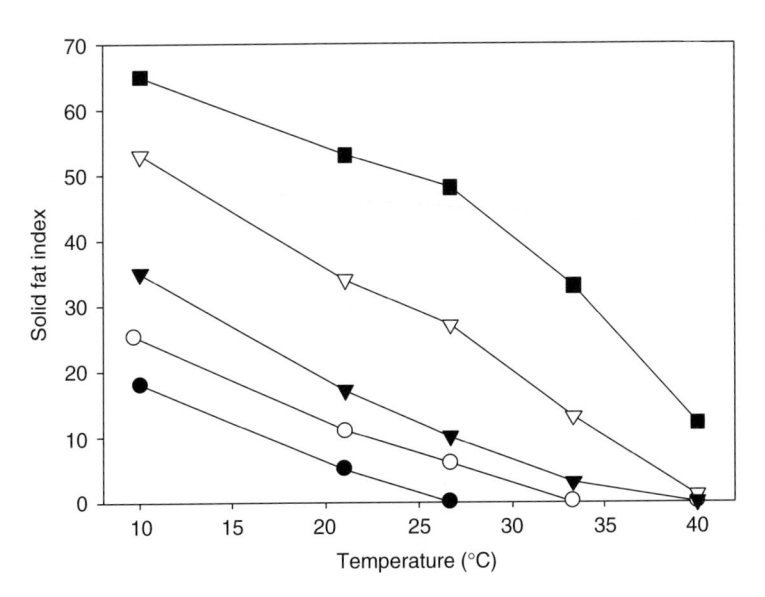

**Figure 4.5** Solid fat index of hydrogenated canola as a function of IV and temperature; IV = 62 (■—■); IV = 68 (△—△); IV = 72 (▲—▲); IV = 77 (○—○); IV = 82 (●—●) (adapted from Przybylski and Mag, 2002).

margarines, unless blended with a β′-stabilising agent such as palm oil, or emulsifiers such as sorbitan tristearate. Finally, Fig. 4.5 shows the change of solid fat index as a function of IV during canola oil hydrogenation.

### 4.4.2 Interesterification

Interesterification has been used industrially for many years to modify the physical properties of lard (Hoerr and Waugh, 1950; Luddy et al., 1955; Lutton et al., 1962). In the 1970s, there was renewed interest in this process, particularly as a hydrogenation replacement for the manufacture of zero-*trans* margarines (List et al., 1995), and as a result, vegetable oils such as canola began to be interesterified. Today, it plays a key role in the production of low-calorie fat replacers such as Olestra and Salatrim (Jandacek and Webb, 1978; Smith et al., 1994). This process causes the fatty acids within an oil or fat to be redistributed within and among triacylglycerol molecules, leading to substantial changes in lipid functionality. Most importantly, the fatty acid distribution is changed without changing the fatty acids' inherent properties. Unsaturation levels stay constant and there is no *cis–trans* isomerisation (Kaufmann and Grothues, 1959) as encountered during hydrogenation.

Two types of interesterification are presently in commercial use – chemical and enzymatic. Enzymatic modifications rely on the use of random or *sn*-1,3 regiospecific, fatty acid-specific lipases as catalysts, while for chemical modifications usually metal alkali catalysts are employed. Enzymatic interesterification does not find wide use in canola/rapeseed applications. As a result, only chemical interesterification will be discussed.

Typically, chemical interesterification is used for the following purposes:

- To modify the melting profile of a fat, typically linearising its solid fat content (SFC) vs temperature evolution.
- To change the melting point of a fat mixture (usually resulting in a decrease in melting point).
- To improve the plasticity of a fat by changing its crystallisation behaviour.
- To improve the compatibility of different solid fats (e.g. minimise eutectic formation).

Existing reviews on chemical interesterification include Kaufmann *et al.* (1958), Going (1967), Hustedt (1976), Sreenivasan (1978), Rozenaal (1992) and Marangoni and Rousseau (1995).

Mechanistically, during chemical interesterification, the ester bonds linking fatty acids to the glycerol backbone are split; the newly liberated fatty acids are randomly shuffled within a fatty acid pool and are then re-esterified onto a new position either on the same glycerol (intraesterification) or onto another glycerol (interesterification) (Sreenivasan, 1978). Intraesterification occurs at a faster rate than interesterification due to kinetic considerations (Freeman, 1968). Once the reaction has reached equilibrium, a complex, random mixture of triacylglycerol species is obtained.

Regardless of the type of interesterification used, the extent of its effects on the physical properties of a fat will depend on the composition of the starting material. If a single starting material (e.g. canola oil) is randomised, the effects will not be as great as if blended, dissimilar lipids are randomised (e.g. palm stearin combined with a vegetable oil) (Rousseau *et al.*, 1996a). Furthermore, if a material has a quasi-random distribution prior to randomisation, it will not lead to notable modifications. Thus, canola oils does not undergo substantial changes as a result of interesterification.

Chemical interesterification can occur without a catalyst at high temperatures (<200°C), although the reaction is slow and the lipid can break down. In the presence of a catalyst, the reaction is much more rapid (in as little as 15 minutes for certain applications) and takes place under much milder conditions. There are three groups of catalysts used (acids, bases and their corresponding salts and metals), which can be subdivided into high- and low-temperature groups. High-temperature catalysts include metal salts such as chlorides, carbonates, oxides, nitrates and acetates of zinc, lead, iron, tin and cobalt (Mattil and Norris, 1953). Others include alkali metal hydroxides of sodium and lithium. Most commonly used are sodium methoxide or sodium ethoxide; however, other bases, acids and metals are also available. These catalysts are simple to use, inexpensive, require only small quantities and are active at low temperatures (<50°C). Hence they can be used for directed interesterification, where newly formed high-melting triacylglycerols are removed from the reaction mixture through controlled crystallisation. Normally, interesterification reactions performed at temperatures above the melting point of the highest melting component in the mixture result in complete randomisation of fatty acids among all triacylglycerols according to the laws of probability (Going, 1967).

With a dry oil devoid of impurities, only small amounts of catalyst (<0.4% [w/w]) are required (Laning, 1985). Catalyst concentration should be minimised to prevent excessive losses resulting from saponification (Braun, 1960). Experience has shown that the addition of each additional 0.1% of catalyst, above 0.4% catalyst, results in the loss of 1% neutral fat.

For chemical interesterification of mixtures such as canola oil and palm stearin, the energy differences between the various possible triacylglycerols are insignificant and do not appear to lead to fatty acid selectivity. Hence, random interesterification is entropically driven leading to randomisation of fatty acids among all possible triacylglycerol positions (Coenen, 1974).

When the fatty acid distribution is fully randomised, the resulting triacylglycerol structure can be predicted from the overall fatty acid composition of the mixture. For example, 1-stearoyl-2-oleoyl-3-linoleoyl glycerol results in the fully randomised equilibrium mixture detailed in Table 4.7. The results of chemical interesterification of rapeseed and canola oils are shown in Table 4.8. Comparison of the fatty acid composition at the *sn*-2 position indicates substantial changes in positional distribution.

More often than not, canola will be interesterified in the presence of a hardstock. In a study on the crystallisation of lard, Rousseau *et al.* (1998c) chemically

**Table 4.7** Theoretical composition of fully randomised 1-stearoyl-2-oleoyl-3-linoleoyl glycerol. (Each triacylglycerol represents the sum of multiple isomers: 1 component fatty acid (e.g. SSS) = 1 isomer; 2 component fatty acids (e.g. SSO) = 3 isomers; 3 component fatty acids (e.g. SOL) = 6 isomers.)

| Triacylglycerol | Proportion (%) |
|---|---|
| SSS | 3.7 |
| OOO | 3.7 |
| LLL | 3.7 |
| SSO | 11.1 |
| SSL | 11.1 |
| SOO | 11.1 |
| SLL | 11.1 |
| OOL | 11.1 |
| OLL | 11.1 |
| SOL | 22.2 |

**Table 4.8** Effect of chemical interesterification on the fatty acid distribution at the *sn*-2 position in canola and rapeseed oils

| Fatty acid | Canola oil | | Rapeseed oil | |
| --- | --- | --- | --- | --- |
| | Non-interesterified | Interesterified | Non-interesterified | Interesterified |
| 16:0 | 0.6 | 7.3 | 0.7 | 4.2 |
| 18:0 | 0.3 | 2.2 | 0.4 | 2.0 |
| 18:1 | 45.7 | 56.7 | 40.3 | 26.2 |
| 18:2 | 37.8 | 24.9 | 38.9 | 16.4 |
| 18:3 | 14.4 | 8.0 | 16.5 | 6.8 |
| 20:1 | 0.6 | 0.7 | 2.4 | 10.3 |
| 22:1 | Trace | Trace | 1.0 | 33.3 |

Adapted from Kramer and Sauer (1983).

interesterified lard and canola oil. Native lard consisted of grainy crystals with large spherulitic aggregates. Chemical interesterification of native lard resulted in fine crystals and reduced aggregation whereas interesterification of lard–canola oil mixtures leads to lower-density aggregates. This had a direct effect on the hardness index of these fat blends. Other research by Rousseau and his co-workers has investigated the role of canola oil in modifying the properties of hard fats via blending and interesterification (Rousseau *et al.*, 1996a,b,c, 1998a,b,c,d).

Many authors have shown that chemical interesterification can alter the oxidative stability of fats and oils. Ledóchowska and Wilczyńska (1998) examined the oxidative stability of chemically interesterified tallow–canola oil blends. They found that triacylglycerols extracted from the randomised mixture had poorer oxidative stability compared to the initial mixture. They presumed that the inferior stability of the randomised mixture was due to the transfer of PUFAs from the *sn*-2 position towards the *sn*-1 and *sn*-3 positions, exposing them to easier oxygen access (Wada and Koizumi, 1983; Neff *et al.*, 1992). Lau *et al.* (1982) demonstrated that randomised corn oil oxidised 3–4 times faster than native corn oil. They concluded that triacylglycerol structure was probably implicated, but were unable to develop a working model. An important contribution to the literature was made by Zalewski and Gaddis (1967), who investigated the effect of interesterification on the stability, antioxidant-synergist efficiency and rancidity development in commercial lipids. They determined that there was no appreciable effect on initiation of oxidation and autoxidation rates, due to the position of unsaturated fatty acids at 1,3-positions or due to randomisation towards the 2-position in triacylglycerols. Tautorus and McCurdy (1990) demonstrated the effects of chemical and enzymatic randomisation on the oxidative stability of vegetable oils stored at different temperatures. Non-interesterified and interesterified oils (canola, linseed, soybean and sunflower) stored at 55°C made little difference to lipid oxidation, whereas non-interesterified samples were more stable at 28°C. Samples at 55°C underwent much greater oxidation than the samples at 28°C. These changes could perhaps be explained by the loss of pro-oxidants or antioxidants during the interesterification process and subsequent refining.

## References

Ackman, R.G. and Eaton, C.A. (1977) Specific gravities of rapeseed and canbra oils. *Journal of American Oil and Chemists' Society*, **54**, 435.

Allen, J.C. (1955) Determination of unsaturation. *Journal of American Oil and Chemists' Society*, **32**, 671–674.

Allen, R.R. (1981) Hydrogenation. *Journal of American Oil and Chemists' Society*, **58**, 166–174.

Andersen, M.L. and Skibsted, L.H. (2002) Detection of early events in lipid oxidation by electron spin resonance spectroscopy. *European Journal of Lipid Technology*, **104**, 65–68.

AOCS, Official Methods and Recommended Methods of the AOCS (1998) (ed. D. Firestone) AOCS Press, Champaign, IL.

Bansal, J. and deMan, J.M. (1982a) Effect of hydrogenation on the chemical composition of canola oil. *Journal of Food Science*, **47**, 1338–1344.

Bansal, J. and deMan, J.M. (1982b) Effect of hydrogenation on some physical properties of canola oil. *Journal of Food Science*, **47**, 2004–2007, 2014.

Braun, W.Q. (1960) Interesterification of edible fats. *Journal of American Oil and Chemists' Society*, **37**, 598–601.

Byrdwell, W.C. and Neff, W.E. (2001) Autoxidation products of normal and genetically modified canola oil varieties determined using liquid chromatography with mass spectrometric detection. *Journal of Chromatography A*, **905**, 85–102.

Chapman, D. (1962) The polymorphism of glycerides. *Chemical Review*, **62**, 433–456.

Chen, B.H. and Chen, Y.C. (2001) Formation of polycyclic aromatic hydrocarbons in the smoke from heated model lipids and food lipids. *Journal of Agricultural and Food Chemistry*, **49**, 5238–5243.

Chumpitaz, L.D.A., Coutinho, L.F. and Meirelles, A.J.A. (1999) Surface tension of fatty acids and triglycerides. *Journal of American Oil and Chemists' Society*, **76**, 379–382.

Coenen, J.W.E. (1974) Fractionnement et interestérification des corps gras dans la perspective du marché mondial des matières premières et des produits finis, II – interestérification. *Rev. franç. corps gras*, **21**, 403–413.

Coenen, J.W.E. (1976) Hydrogenation of edible oils. *Journal of American Oil and Chemists' Society*, **53**, 382–389.

Coppin, E.A. and Pike, O.A. (2001) Oil stability index correlated with sensory determination of oxidative stability in light-exposed soybean oil. *Journal of American Oil and Chemists' Society*, **78**, 13–18.

Coupland, J.N. and McClements, D.J. (1996) Lipid oxidation in food emulsions. *Trends in Food Science Technology*, **7**, 83–91.

Coupland, J.N. and McClements, D.J. (1997) Physical properties of liquid edible oils. *Journal of American Oil and Chemists' Society*, **74**, 1559–1564.

deMan, J.M. (1964) Physical properties of milk fat. *Journal of Dairy Science*, **47**, 1194–1200.

deMan, L., deMan, J.M. and Blackman, B. (1989) Polymorphic behavior of some fully hydrogenated oils and their mixtures with liquid oil. *Journal of American Oil and Chemists' Society*, **66**, 1777–1780.

deMan, L., deMan, J.M. and Blackman, B. (1995) Effect of tempering on the texture and polymorphic behavior of margarine fats. *Fat Science Technology*, **97**, 55–60.

Desnuelle, P. and Savary, P. (1959) Sur la structure glycéridique de quelques corps gras naturels. *Fette, Seifen, Anstrichmittel*, **61**, 871–876.

Downey, G., McIntyre, P. and Davies, A.N. (2002) Detecting and quantifying sunflower oil adulteration in extra virgin olive oils from the Eastern Mediterranean by visible and near-infrared spectroscopy. *Journal of Agricultural and Food Chemistry*, **50**, 5520–5525.

D'Souza, V., deMan, J.M. and deMan, L. (1990) Short spacings and polymorphic forms of natural fats and commercial solid fats: a review. *Journal of American Oil and Chemists' Society*, **67**, 835–843.

Erickson, D.R. and List, G.R. (1985) Fat degradation reactions. In: *Bailey's Industrial Oil and Fat Products*, Vol. 3, 3rd edition (ed. T.H. Applewhite), John Wiley & Sons, Inc., Toronto, Canada, pp. 275–277.

Eskin, N.A.M., Vaisey-Genser, M., Durance-Todd, S. and Przybylski, R. (1989) Stability of low linolenic acid canola oil to frying temperatures. *Journal of American Oil and Chemists' Society*, **66**, 1081–1084.

Eskin, N.A.M., McDonald, B.E., Przybylski, R., Malcolmson, L.J., Scarth, R., Mag, T., Ward, K. and Adolph, D. (1996) Canola oil. In: *Bailey's Industrial Oil and Fat Products*, Vol. 4, 4th edition (ed. Y.H. Hui), John Wiley & Sons, Inc., Toronto, Canada, pp. 1–95.

Firestone, D. (1999) *Physical and Chemical Characteristics of Oils, Fats, and Waxes*, AOCS Press, Champaign, IL.

Formo, M.W. (1979) Physical properties of fats and fatty acids. In: *Bailey's Industrial Oils and Fats* (ed. D. Swern), John Wiley & Sons, Toronto, Canada, pp. 177–232.

Frankel, E.N. (1991) Recent advances in lipid oxidation. *Journal of Science and Food Agriculture*, **54**, 495–511.

Frankel, E.N. (1998) *Lipid Oxidation*, The Oily Press Ltd, Dundee, Scotland.

Freeman, I.P. (1968) Interesterification. I. Change of glyceride composition during the course of interesterification. *Journal of American Oil and Chemists' Society*, **45**, 456–460.

Gertz, C., Klostermann, S. and Kochhar, S.P. (2000) Testing and comparing oxidative stability of vegetable oils and fats at frying temperature. *European Journal of Lipid Science Technology*, **102**, 543–551.

Going, L.H. (1967) Interesterification products and processes. *Journal of American Oil and Chemists' Society*, **44**, 414A–422A, 454A–456A.

Gray, J.I. (1978) Measurement of lipid oxidation: a review. *Journal of American Oil and Chemists' Society*, **55**, 539–546.

Gunstone, F.D. (2000) Composition and properties of edible oils. In: *Edible Oil Processing* (eds W. Hamm and R.J. Hamilton), CRC Press, Toronto, Canada, pp. 1–33.

Gunstone, F.D. and Norris, F.A. (1983) *Lipids in Foods: Chemistry, Biochemistry and Technology*, Pergamon Press, Toronto, Canada.

Halvorsen, J.D., Mammel, Jr, W.C. and Clements, L.D. (1993) Density estimation of fatty acids and vegetables based on their fatty acid composition. *Journal of American Oil and Chemists' Society*, **70**, 875–880.

Hastert, R.C. (1996) Hydrogenation. In: *Bailey's Industrial Oil and Fat Products*, Vol. 4, 4th edition (ed. Y.H. Hui), John Wiley & Sons, Inc., Toronto, Canada, pp. 213–300.

Hawrysh, Z.J., Shand, P.J., Tokarska, B. and Lin, C. (1988) Effects of tertiary butylhydroquinone on the stability of canola oil. I. Accelerated storage. *Canadian Institute of Food Science Technology*, **21**, 549–554.

Hernqvist, L., Herslöf, B., Larsson, K. and Podhala, O. (1981) Polymorphism of rapeseed with a low content of erucic acid and possibilities to stabilise the ß'-crystal form in fats. *Journal of Science and Food Agriculture*, **32**, 1197–1202.

Hilditch, T.P. (1956) *The Chemical Constitution of Foods*, 3rd edition, Chapman and Hall Ltd, London, pp. 1–24.

Hoerr, G.W. and Waugh, D.F. (1950) Some physical characteristics of rearranged Lard. *Journal of American Oil and Chemists' Society*, **32**, 37–41.

Hu, X., Daun, J.K. and Scarth, R. (1993) Characterization of wax sediments in refined canola oils. *Journal of American Oil and Chemists' Society*, **70**, 535–537.

Huang, S.-W., Frankel, E.N., Schwarz, K., Aeschbach, R. and German, J.B. (1996a) Antioxidant activity of carnosic acid and methyl carnosate in bulk oils and oil-in-water emulsions. *Journal of Agricultural and Food Chemistry*, **44**, 2951–2956.

Huang, S.-W., Hopia, A., Schwarz, K., Frankel, E.N. and German, J.B. (1996b) Antioxidant activity of ∀-tocopherol and trolox in different lipid substrates: bulk oils vs oil-in-water emulsions. *Journal of Agricultural and Food Chemistry*, **44**, 444–452.

Hustedt, H.H. (1976) Interesterification of edible oils. *Journal of American Oil and Chemists' Society*, **53**, 390–392.

Jandacek, R.J. and Webb, M.R. (1978) Physical properties of pure sucrose octa-esters. *Chemistry Physics of Lipids*, **22**, 163–176.

Jebe, T.A., Matlock, M.G. and Sleeter, R.T. (1993) Collaborative study on the oil stability index analysis. *Journal of American Oil and Chemists' Society*, **70**, 1055–1061.

Jensen, L.H. and Mabis, A.J. (1963) Crystal structure of ß-tricaprin. *Nature*, **197**, 681–682.

Johansson, D. and Bergenståhl, B. (1995) Wetting of fat crystals by triglyceride oil and water. 2. Adhesion to the oil/water interface. *Journal of American Oil and Chemists' Society*, **72**, 933–938.

Johnson, G.L., Machado, R.M., Freidl, K.G., Achenbach, M.L., Clark, P.J. and Reidy, S.K. (2002) Evaluation of Raman spectroscopy for determining *cis* and *trans* isomers in partially hydrogenated soybean oil. *Organic Proceedings of Research Development*, **6**, 637–644.

Kaufmann, H.P. and Grothues, B. (1959) Umesterungen auf dem Fettgebiet. III. Über den Einfluss verschiedener Umesterungs-Katalysoren auf ungesättigte Fettsäuren. *Fette, Seifen, Anstrichmittel*, **61**, 425–429.

Kaufmann, H.P., Grandel, F. and Grothues, B. (1958) Umesterungen auf dem Fettgebiet I: Theoretische Grundlagen und Schriftum; die Hydrier-Umesterung. *Fette, Seifen, Anstrichmittel*, **60**, 919–930.

Kawamura, K. (1981) The DSC thermal analysis of crystallization behavior in high erucic acid rapeseed oil. *Journal of American Oil and Chemists' Society*, **58**, 826–829.

Kellens, M. (2000) Oil modification strategies. In: *Edible Oil Processing* (eds W. Hamm and R.J. Hamilton), CRC Press, Boca Raton, FL, pp. 129–173.

Koseoglu, S.S. and Lusas, E.W. (1990a) Hydrogenation of Canola Oil. In: *Canola and Rapeseed: Production, Chemistry, Nutrition and Processing Technology* (ed. F. Shahidi), Van Nostrand Reinhold, New York, NY, pp. 129–148.

Koseoglu, S.S. and Lusas, E.W. (1990b) Recent advances in canola oil hydrogenations. *Journal of American Oil and Chemists' Society*, **67**, 39–47.

Kramer, J.K.G. and Sauer, F.D. (1983) Results obtained with feeding low erucic acid rapeseed oils and other vegetable oils to rats and other species. In: *High and Low Erucic Acid Rapeseed Oils* (eds J.K.G. Kramer, F.D. Sauer and W.J. Pigden), Academic Press, Toronto, Canada, pp. 414–474.

Lampi, A.-M. and Piironen, V. (1999) Dissimilarity of the oxidations of rapeseed and butter oil triacylglycerols and their mixtures in the absence of tocopherols. *Journal of Science and Food Agriculture*, **79**, 300–306.

Lang, W., Sokhansanj, S. and Sosulski, F.W. (1992) Modelling the temperature dependence of kinematic viscosity for refined canola oil. *Journal of American Oil and Chemists' Society*, **69**, 1054–1055.

Laning, S.J. (1985) Chemical interesterification of palm, palm kernel and coconut oils. *Journal of American Oil and Chemists' Society*, **62**, 400–405.

Larsson, K. (1966) Classification of glyceride crystal forms. *Acta Chemica Scandinavica*, **20**, 2255–2260.

Lau, F.Y., Hammond, E.G. and Ross, P.F. (1982) Effect of randomization on the oxidation of corn oil. *Journal of American Oil and Chemists' Society*, **59**, 407.

Laubli, M.W. and Bruttel, P.A. (1986) Determination of the oxidative stability of fats and oils: comparison between the Active Oxygen Method (AOCS Cd 12–57) and the Rancimat method. *Journal of American Oil and Chemists' Society*, **63**, 792–795.

Ledóchowska, E. and Wilczyńska, E. (1998) Comparison of the oxidative stability of chemical and enzymatically exteresterified fats. *Fett/Lipid*, **100**, 343–348.

List, G.R., Mounts, T.L., Orthoefer, F. and Neff, W.E. (1995) Margarine and shortening oils by interesterification of liquid and trisaturated triglycerides. *Journal of American Oil and Chemists' Society*, **72**, 379–382.

Liu, H., Biliaderis, C.G., Przybylski, R. and Eskin, N.A.M. (1993) Phase transitions of canola oil sediment. *Journal of American Oil and Chemists' Society*, **70**, 441–448.

Liu, H., Biliaderis, C.G., Przybylski, R. and Eskin, N.A.M. (1994) Effects of crystallization conditions on sedimentation in canola oil. *Journal of American Oil and Chemists' Society*, **71**, 409–415.

Liu, H., Biliaderis, C.G., Przybylski, R. and Eskin, N.A.M. (1995) Physical behaviour and composition of low- and high-melting fractions of sediment in canola. *Food Chemistry*, **53**, 35–41.

Liu, H., Przybylski, R., Dawson, K., Eskin, N.A.M. and Biliaderis, C.G. (1996) Comparison of the composition and properties of canola and sunflower oil sediments with canola seed hull lipids. *Journal of American Oil and Chemists' Society*, **73**, 493–498.

Luddy, F.E., Morris, S.G., Magidman, P. and Riemenschneider, R.W. (1955) Effect of catalytic treatment with sodium methylate on glycerine composition and properties of lard and tallow. *Journal of American Oil and Chemists' Society*, **32**, 522–525.

Lutton, E.S. (1950) Review of the polymorphism of saturated even glycerides. *Journal of American Oil and Chemists' Society*, **27**, 276–281.

Lutton, E.S., Mallery, M.F. and Burgers, J. (1962) Interesterification of lard. *Journal of American Oil and Chemists' Society*, **39**, 233–238.

Manning, D.M. and Dimick, P.S. (1985) Crystal morphology of cocoa butter. *Food Microstructure*, **4**, 249–265.

Marangoni, A.G. and Rousseau, D. (1995) Engineering triacylglycerols: the role of interesterification. *Trends in Food Science Technology*, **6**, 329–335.

Matthaus, B.W. (1996) Determination of the oxidative stability of vegetable oils by Rancimat and conductivity and chemiluminescence measurement. *Journal of American Oil and Chemists' Society*, **73**, 1039–1043.

Mattil, K.W. and Norris, F.A. (1953) U.S. patents 2 625 478-2 625 482.

Moya Moreno, M.C.M., Mendoza Olivares, D., Amézquita López, F.J., Gimeno Adelantado, J.V. and Bosch Reig, F. (1999a) Determination of unsaturated grade and *trans* isomers generated during thermal oxidation of edible oils and fats by FTIR. *Journal of Molecular Structure*, **482–483**, 551–556.

Moya Moreno, M.C.M., Mendoza Olivares, D., Amézquita López, F.J., Peris Martínez, V. and Bosch Reig, F. (1999b) Study of the formation of carbonyl compounds in edible oils and fats by H-NMR and FTIR. *Journal of Molecular Structure*, **482–483**, 557–561.

Naguib-Mostafa, A., Smith, A.K. and deMan, J.M. (1985) Crystal structure of hydrogenated canola oil. *Journal of American Oil and Chemists' Society*, **62**, 760–762.

Nawar, W.W. (1996) Lipids. In: *Food Chemistry*, 3rd edition (ed. O.R. Fennema), Marcel Dekker, Inc., New York, NY, pp. 225–319.

Neff, W.E., Selke, E., Mounts, T.L., Rinsch, W., Frankel, E.N. and Zeitoun, N.A.M. (1992) Effect of triacylglycerol composition and structures on oxidative stability of oils from selected soybean germplasm. *Journal of American Oil and Chemists' Society*, **69**, 111–118.

Neff, W.E., Mounts, T.L. and Rinsch, W.M. (1997) Oxidative stability as affected by triacylglycerol composition and structure of purified canola oil triacylglycerols from genetically modified normal and high stearic and lauric acid canola varieties. *Lebensmittelwissenschaft und Technologie*, **30**, 793–799.

Norris, F.A. and Mattil, K.F. (1947) A new approach to the glyceride structure of natural fats. *Journal of American Oil and Chemists' Society*, **24**, 274–275.

Nourredini, H., Teoh, B.C. and Clements, L.D. (1992a) Densities of vegetable oils and fatty acids. *Journal of American Oil and Chemists' Society*, **69**, 1184–1188.

Nourredini, H., Teoh, B.C. and Clements, L.D. (1992b) Viscosities of vegetable oils and fatty acids. *Journal of American Oil and Chemists' Society*, **69**, 1189–1191.

Ozen, B.F. and Mauer, I.J. (2002) Detection of hazelnut adulteration using FT-IR spectroscopy. *Journal of Agricultural and Food Chemistry*, **50**, 3898–3901.

Pantzaris, T.P. (1985) The density of oils in the liquid state. *PORIM Technology*, number 12.

Perkins, E.G. (1995) Physical properties of soybeans. In: *Practical Handbook of Soybean Processing and Utilization* (ed. D.R. Erickson), AOCS Press, Champaign, IL, pp. 29–38.

Ponginebbi, L., Nawar, W.W. and Chinachoti, P. (1999) Oxidation of linoleic acid in emulsion: effects of substrate, emulsifier and sugar concentration. *Journal of American Oil and Chemists' Society*, **76**, 131–138.

Przybylski, R. and Mag, T. (2002) Canola/rapeseed. In: *Vegetable Oils in Food Technology: Composition, Properties and Uses* (ed. F.D. Gunstone), CRC Press, Boca Raton, FL, pp. 98–127.

Przybylski, R., Biliaderis, C.G. and Eskin, N.A.M. (1993) Formation and partial characterization of canola oil sediment. *Journal of American Oil and Chemists' Society*, **70**, 1009–1015.

Raghuveer, K.G. and Hammond, E.G. (1967) The influence of glyceride structure on the rate of autoxidation. *Journal of American Oil and Chemists' Society*, **44**, 239–244.

Riiner, Ü. (1970) Investigation of the phase behavior of *Cruciferae* seed oils by temperature programmed X-ray diffraction. *Journal of American Oil and Chemists' Society*, **47**, 129–133.

Rodenbush, C.M., Hsieh, F.H. and Viswanath, D.S. (1999) Density and viscosity of vegetable oils. *Journal of American Oil and Chemists' Society*, **76**, 1415–1419.

Rousseau, D., Forestière, K., Hill, A.R. and Marangoni, A.G. (1996a) Restructuring butter fat through blending and chemical interesterification. 1. Melting behavior and triacylglycerol modifications. *Journal of American Oil and Chemists' Society*, **73**, 963–972.

Rousseau, D., Hill, A.R. and Marangoni, A.G. (1996b) Restructuring butter fat through blending and chemical interesterification. 2. Morphology and polymorphism. *Journal of American Oil and Chemists' Society*, **73**, 973–981.

Rousseau, D., Hill, A.R. and Marangoni, A.G. (1996c) Restructuring butter fat through blending and chemical interesterification. 3. Rheology. *Journal of American Oil and Chemists' Society*, **73**, 983–989.

Rousseau, D. and Marangoni, A.G. (1998a) Tailoring the textural attributes of butter fat-canola oil blends via *Rhizopus arrhizus* lipase-catalyzed interesterification. 1. Compositional modifications. *Journal of Agricultural and Food Chemistry*, **46**, 2368–2374.

Rousseau, D. and Marangoni, A.G. (1998b) Tailoring the textural attributes of butter fat-canola oil blends via *Rhizopus arrhizus* lipase-catalyzed interesterification. 2. Modifications of physical properties. *Journal of Agricultural and Food Chemistry*, **46**, 2375–2381.

Rousseau, D., Jeffrey, K.R. and Marangoni, A.G. (1998c) The influence of chemical interesterification on the physicochemical nature of two complex fat systems. 2. Morphology and polymorphism. *Journal of American Oil and Chemists' Society*, **75**, 1833–1840.

Rousseau, D., Hill, A.R. and Marangoni, A.G. (1998d) The effects of interesterification on physical and sensory attributes of butterfat and butterfat-canola oil spreads. *Food Research International*, **31**, 381–388.

Rozenaal, A. (1992) Interesterification of fats. *Inform*, **3**, 1232–1237.

Saito, H. and Udagawa, M. (1992) Application of nuclear magnetic resonance to evaluate the oxidation deterioration of brown fish meat. *Journal of the Science of Food and Agriculture*, **58**, 135–137.

Shahidi, F. (1992) Current and novel methods for stability testing of canola. *Inform*, **3**, 543.

Sherwin, E.F. (1968) Methods for stability and antioxidant measurement. *Journal of American Oil and Chemists' Society*, **45**, 632A–648A.

Shield, P.G., Xu, G.X., Blot, W.J., Fraumeni, Jr, J.F., Trivers, G.E., Pellizzari, E.D., Qu, Y.H., Gao, Y.T. and Harris, C.C. (1995) Mutation from heated Chinese and U.S. cooking oils. *Journal of the National Cancer Institute*, **87**, 836–841.

Sipos, E.F. and Szuhaj, B.F. (1996) Soybean oil. In: *Bailey's Industrial Oils and Fat Products, Edible Oil and Fat Products: Oils and Oilseeds*, Vol. 2, 5th edition (ed. Y.H. Hui), John Wiley & Sons, Inc., New York, NY, pp. 497–601.

Smith, R.E., Finley, J.W. and Leveille, G.A. (1994) Overview of SALATRIM, a family of low-calorie fats. *Journal of Agricultural and Food Chemistry*, **42**, 432–434.

Snyder, J.M., Frankel, E.N. and Selke, E. (1985) Capillary gas chromatography analyses of headspace volaties from vegetable oils. *Journal of American Oil and Chemists' Society*, **62**, 1675–1679.

Sreenivasan, B. (1978) Interesterification of fats. *Journal of American Oil and Chemists' Society*, **55**, 796–805.

Tautorus, C.L. and McCurdy, A.R. (1990) Effect of randomization on oxidative stability of vegetable oils at two different temperatures. *Journal of American Oil and Chemists' Society*, **67**, 525–529.

Thomsen, M.K., Kristensen, D. and Skibsted, L.H. (2000) Electron spin resonance spectroscopy for determination of the oxidative stability of food lipids. *Journal of American Oil and Chemists' Society*, **77**, 725–730.

Timms, R.E. (1985) Physical properties of oils and mixtures of oils. *Journal of American Oil and Chemists' Society*, **62**, 241–249.

Toro-Vazquez, J.F. and Infante-Guerrero, R. (1993) Regressional models that describe oil absolute viscosity. *Journal of American Oil and Chemists' Society*, **70**, 1115–1119.

Vaisey-Genser, M. and Ylimaki, G. (1985) Effects of a non-absorbable antioxidant on a canola oil stability to accelerated storage and to a frying temperature. *Canadian Institute of Food Science Technology Journal*, **18**, 67–71.

Van der Voort, F.R., Sedman, J. and Russin, T. (2001) Lipid analysis by vibrational spectroscopy. *European Journal of Lipid Science Technology*, **103**, 815–840.

Vieira, T.M.F.S. and Regitano-d'Arce, M.A.B. (2001) Canola oil thermal oxidation during oven test and microwave heating. *Lebensmittelwissenschaft und Technologie*, **34**, 215–221.

Wada, S. and Koizumi, C. (1983) Influence of the position of unsaturated fatty acid esterified glycerol of the oxidation of triglyceride. *Journal of American Oil and Chemists' Society*, 1105–1109.

Wanasundara, U.N. and Shahidi, F. (1994) Canola extract as an alternative natural antioxidant for canola. *Journal of American Oil and Chemists' Society*, **71**, 817–822.

Wanasundara, U.N., Shahidi, F. and Jablonski, C.R. (1995) Comparison of standard and NMR methodologies for assessment of oxidation stability of canola and soybean oils. *Food Chemistry*, **52**, 249–253.

Warner, K., Frankel, E.N. and Mounts, T.L. (1989) Flavor and oxidative stability of soybean, sunflower and low erucic acid rapeseed oils. *Journal of American Oil and Chemists' Society*, **66**, 558–564.

Xu, X.-Q., Tran, V.H., Palmer, M., White, K. and Salisbury, P. (1999) Chemical and physical analyses and sensory evaluation of six deep-frying oils. *Journal of American Oil and Chemists' Society*, **76**, 1091–1099.

Zalewski, S. and Gaddis, A.M. (1967) Effect of transesterification of lard on stability, antioxidant-synergist efficiency, and rancidity development. *Journal of American Oil and Chemists' Society*, **44**, 576.

# 5 High erucic oil: its production and uses
C. Temple-Heald

## 5.1 Introduction

High erucic oils have been used in the oleochemical industry for several decades. Erucic acid occurs in high levels in the oils of rape, mustard, crambe, nasturtium, lunaria, and meadowfoam. Although nasturtium (*Tropaeolum majus*) and lunaria (*Lunaria annua*) have high levels of erucic acid, they are unsuitable as erucic sources for industrial use. Nasturtium has over 80% erucic acid in the seed oil but contains only 7–10% oil. Lunaria has 45% erucic acid with an oil content of about 35%. However, it is still planted only in small acreages and is usually utilised for its nervonic acid content (24:1). Meadowfoam (*Limnanthes alba*) with only 10–12% erucic acid is not a useful source of erucic acid.

The main plant sources of erucic acid are from high erucic rapeseed oil (HERO) (*Brassica napus*), mustard, and crambe (*Crambe abyssinica*) – all crucifer oilseeds. There are several mustard varieties which will be discussed later. Fish oils also act as a cheap source of behenic acid (22:0) production but are subject to variability in supply and lipid profile.

This chapter reviews the production and uses of erucic acid derived from these oilseeds.

## 5.2 Crucifer oilseeds

Table 5.1 shows typical lipid profiles for HERO, crambe, and mustard oils. Meadowfoam and lunaria are included for comparison.

### 5.2.1 Brassica napus *(HERO)*

Council directive 76/621/EEC of 20 July 1976 stated that the maximum level of erucic acid in oils and fats intended for human consumption was limited to 5% maximum in the European Union (EU). This has since then been reduced to 2% maximum. It was also set at 2% maximum in Canada. However, most food-grade rapeseed now has <0.5% erucic acid. These levels were set because high levels of erucic acid in the diet were found to cause fatty deposits in the hearts of some laboratory animals. This occurred when high erucic oils with between 16 and 50% erucic acid were fed in very high amounts in the diet. However, significant quantities are still consumed in the diet in China and

**Table 5.1** Fatty acid composition (%) of HEAR oils

| Fatty acid | HERO[d] | Crambe[b] | Mustard (general)[a] | Lunaria[b] | Meadowfoam[c] |
|---|---|---|---|---|---|
| 14:0 | <0.1 | 0.1 | – | – | – |
| 16:0 | 3.1 | 2.3 | 2.6 | 1.4 | – |
| 16:1 | 0.1 | 0.2 | 0.1 | 0.1 | – |
| 18:0 | 0.9 | 0.7 | 0.8 | 0.2 | – |
| 18:1 | 12.8 | 13.6 | 23.2 | 23.7 | 1–2 |
| 18:2 | 13.6 | 10.5 | 8.9 | 5.5 | 0–0.5 |
| 18:3 | 8.5 | 6.2 | 10.4 | 0.9 | – |
| 20:0 | 0.6 | 0.8 | 0.7 | – | 0–0.5 |
| 20:1 | 8.2 | 3.9 | 8.1 | 0.9 | ($\Delta$5) 62–63 |
| 20:2 | – | 0.3 | – | – | – |
| 22:0 | 0.5 | 1.6 | – | 0.2 | – |
| 22:1 | 47.0 | 56.7 | 43.1 | 45.6 | ($\Delta$5) 2.5–4 |
|  |  |  |  |  | ($\Delta$13) 10–12 |
| 22:2 | – | 0.8 | – | – | 18 |
| 24:0 | 0.2 | 0.6 | – | – | – |
| 24:1 | 1.0 | 1.3 | – | 21.3 | – |
| Others | 3.5 | 0.4 | 2.1 | 0.3 | – |

[a] Gunstone *et al.* (1994a).
[b] Internal communication.
[c] Official methods (1996).
[d] NIAB (2001).

India, with no observed negative effects. Most low erucic rapeseed oil (LERO) crops now have <0.5% erucic acid.

Throughout the 1970s and 1980s, HERO was predominantly sourced from Eastern Europe, particularly Poland and East Germany, as it could not be grown for foodstuffs within the European Community. High erucic acid rapeseed (HEAR) crops were replaced by low erucic acid rapeseed (LEAR) crops and fear of cross-pollination effectively led to a ban on growing HEAR in the European Community. China is still a source of HERO but the oil quality does not meet the standards required by industrial users. India is a source of mustard oil but the Indian Government imposes tariffs on the oil because it is used internally as a food crop. This makes it prohibitively expensive to import. India does, however, export erucic acid.

In the early 1980s, political instability made the reliance on Eastern European sources unsatisfactory. One of the largest users of HERO in the UK (Croda Universal) decided that more secure sources were required. In fact, as a result of democratic changes, both Poland and East Germany quickly changed from growing HEAR to *double low* (low erucic acid and low glucosinolate) types. East German crops fell from 50 000 t oil in 1990 to less than 2000 t in 1991 and to effectively zero in 1992. Trials to grow HEAR were carried out in the UK and from an initial 16-ha trial, current production now extends right across the UK.

Re-establishing HEAR in the EU was no trivial task (Temple-Heald, 2000). The initial trials had to be carried out in remote parts of the UK to establish minimum separation distances. Cross-pollination could lead to a rapeseed variety that would be above the EU legislative limit of 5% maximum erucic acid content.

This would then be useless for both food and industrial applications. The legal requirements meant that a minimum isolation distance needed to be observed between high erucic and other rape oilseeds, grown in the same season. At Newcastle University, Bilsborrow *et al.* (1994) investigated the isolation requirements of oilseed rape. The work showed that limited cross-pollination occurs by bees but most is due to physical contact between plants. As a result of this work, the Ministry of Agriculture reduced the isolation distance to 50 m (Nicholls, 1996).

In association with issues surrounding isolation distances, work was also carried out to assess the problem of volunteers (Ramans, 1995). It was found that the maximum volunteers occurred out of HEAR production in the second year and that they were less than 1 plant/m$^2$. Advice could therefore be given to growers about the strategies available to minimise volunteer contamination.

Identity preservation is important when growing high erucic rapeseed in the EU (Hebard, 2002). The establishment of the HEAR crop in the UK was a good example of how it was implemented. This meant ensuring that the legislative issues were adhered to. These include seed quality, isolation distances, and sourcing. As rapid testing of the seed oil for fatty acid composition was not feasible then, reliance on the experience of both the crushers and the supply chain management company to provide excellent logistics was essential. Good collaboration existed throughout the supply chain and the success of establishing the UK-HERO crop was assured. As HERO is a commercial crop in the UK, the planted acreage is difficult to assess. HEAR planted in the USA for industrial uses occupies about 2500 acres. It is grown mainly in Dakota. It is also grown in Canada.

*5.2.1.1 HEAR agronomy*

Winter HEAR performs well over a wide range of soils across all regions in the UK. Good drainage and soil structure are essential for the early development of the tap root. Winter HEAR is well adapted to soil with a pH of 6.5–7.0 but can be grown over the range 6.0–8.0. Rotation is required no more than one year in five and winter beans can be grown to lengthen the rotation between susceptible crops. Ideally, winter HEAR is planted at the end of August, although both spring and winter seed types are available. Table 5.2 gives examples of the UK varieties available (NIAB, 2001).

Seed development for HEAR has fallen behind that of the double low rape, for obvious reasons, but a number of new varieties are being developed which combine high erucic acid and low glucosinolate levels. This will make utilisation of the meal easier. Yields from winter HEAR tend to be about 10–15% below

**Table 5.2** UK-HEAR seed varieties

| Seed variety | Type | Yield | Quality | Breeder |
|---|---|---|---|---|
| Askari | Winter | Below average compared to "00" varieties | High oil content, high erucic content, meal has high glucosinolate levels | Lembke, Germany |
| Cadwell | Winter | Below average compared to "00" varieties | High oil content, high erucic content, meal has low glucosinolate levels | Nickerson, UK |
| Martina | Winter | Below average compared to "00" varieties | Very high oil content, high erucic content, meal has high glucosinolate levels | Semundo, Germany |
| Steffi | Winter | Below average compared to "00" varieties | Very high oil content, high erucic content, meal has high glucosinolate levels | Semundo, Germany |
| Industry | Spring | Lower yielding than "00" types | Average oil content, high erucic content, meal has low glucosinolate levels | Danisco, Denmark |
| Sheila | Spring | Higher yielding than industry variety | Meal has low glucosinolate levels (no other data available) | Danisco, Denmark |

the best double low varieties. The spring HEAR varieties are early maturing types, with yields around 20% below the double low spring rape.

In Europe, HEAR is only grown under contract and can be grown on set-aside or on arable-area payment land, depending on the level of premium offered. The EU countries in which the majority of HEAR is grown are France, Germany, Italy, Netherlands, and the UK. Within the EU, the usage of high erucic rapeseed was 56 000 Mt of oil in 1998 (IENICA August 2000), with an expected increase to 79 000 Mt of oil by 2004. EU crop production statistics for 2000 for HEAR production is shown in Table 5.3 (IENICA, 2000).

It is anticipated that there will be a shortfall in the availability of HEAR oil in 2003 (De Guzman, 2003). China, Canada, and Europe had severe weather conditions during 2002 and world HEAR harvests were poor. As a result, the world stocks of HEAR oil are at a very low level for 2003. It is anticipated that there will not be enough HEAR oil to satisfy demand until the next harvest in late summer 2003.

**Table 5.3** HEAR production statistics for 2000

| Country | Area harvested $10^3$ ha | Yield (t/ha) |
|---|---|---|
| France | 4.51 | 2.0 |
| Germany | 28.0 | 2.9 |
| Italy | 0.54 | 1.8 |
| UK | 0.97 | 2.0 |

## 5.2.2 Crambe abyssinica

In a similar scenario to developing HEAR as an industrial crop in the UK, the US began developing crambe as an industrial crop in the late 1980s. The basis for developing the crop was that US companies requiring erucic acid and high erucic oils had been dependent upon the traditional rapeseed growing countries. As in the UK when the pool of HEAR was reducing rapidly, the US also needed a domestic supply of erucic acid.

Levels of $C_{18}$ polyunsaturates (diene and triene) are higher in HERO compared with crambe, but the level of erucic acid and oleic acid is higher in crambe. It is these fatty acids that make the crambe of interest as a feedstock and the lower levels of $C_{20}$ fatty acids add to its value.

Crambe originates from the Mediterranean region but has also been prevalent across Asia and western Europe. It was first introduced into the United States during the 1940s (Golz, 1993). Throughout the 1950s and 1960s, performance trials were carried out but it was not until much later that the crop became established. Trials were carried out in Indiana, Iowa, Kentucky, and Dakota. By 1993, 25 000 ha of crambe had been planted under contract in the US, mainly in North Dakota. In Europe, AIR-CT94-2480, which was part of the Research Programme of the EU Third Framework Programme, concentrated on all aspects of producing *Crambe abyssinica* in Europe. Over three years, the project evaluated the agronomy, breeding, production, and utilisation of crambe (BioMatNet, July 1998). The agronomic fundamentals were investigated in northeast Germany, southwest Germany, the PoValley in northeast Italy, south Italy, Scotland, southern England, northeast France, and Burgundy, France. The conclusion from the three years work was that the yields throughout the EU were very variable. For instance, in Scotland, two sites were being used for the crambe trials. At one site, the crop failed in both years due to drought and pest damage. At other sites, there were problems with weeds and the crop was of moderate height. The overall yields therefore were low. However, in southern England, seed yield averaged 3.5 t/ha with a seed oil content of 29.9–34.7%. On average, the erucic acid content in the oil was 57.3%. The conclusion was that crambe produced comparable yields to spring oil seed rape in southern England.

### 5.2.2.1 Crambe agronomy

*Crambe abyssinica* contains approximately 35% oil of which up to 60% may be erucic acid. The general agronomy (Coupland and Makin, 2003) is that it is a cool season crop requiring, on average, 100 days from the date of emergence to reach physiological maturity. It is well adapted to soils with a pH of 6.0–7.5 and is more drought tolerant than rapeseed at all stages of growth.

Crambe does not cross-pollinate with rapeseed and therefore the separation restrictions needed for LEAR and HEAR varieties do not apply. However, crambe should not follow on from other closely related crops such as rapeseed. Ideally it should follow after cereals or legumes.

Few disease problems have been associated with crambe. Alternaria can be seen as black speckling on the hull and can infect the seed, which may affect yield (Endres and Schatz, 1993). However, if treated with a fungicide towards the end of flowering, it will be controlled sufficiently not to cause a problem.

After flowering, crambe matures rapidly within one or two weeks, and should be harvested quickly after this to avoid seed shattering. The seeds should be harvested with the hulls intact. However, due to the presence of the hull, crambe seed is very lightweight and this makes it a crop with low bulk density. Prior to crushing it should be de-hulled. Table 5.4 gives some of the crambe seed varieties.

### 5.2.3 Mustard rapeseed

Mustard rapeseed is one of the most economically important oilseed crops in India (IACR-RI, 2002). *B. alba* (yellow mustard) and *B. juncea* (brown mustard) are the main varieties available today. Prior to 1950, *B. nigra* (black mustard) was grown in large quantities (Hemingway, 1995). Even though it had the largest area of cultivated oilseeds in the World (in 1998/99), India still imports over 5 Mt of oil to meet local demand. The mustard and rapeseed crops account for 20–30% of the national output, covering five to six million hectares. However, productivity is low compared to developed countries, at less than 1 t/ha.

*Brassica juncea* is also grown in China, southern Ukraine, Pakistan, and Bangladesh. However, the proportion of these oilseeds grown in all countries for industrial use is difficult to calculate. The majority is consumed domestically and not exported.

*Brassica carinata* (also a type of mustard) is grown in Ethiopia and Eritrea, and is used domestically. *B. carinata* was studied in the EU between 1994 and

**Table 5.4** Crambe seed varieties

| Seed variety | Yield | Quality |
|---|---|---|
| Galactica | Below average compared to "00" varieties | High oil content. Meal has high glucosinolate levels. Above average erucic contents compared to HEAR |
| Nebula | Below average compared to "00" varieties | High oil content. Meal has high glucosinolate levels. Above average erucic contents compared to HEAR |
| Carmen | Below average compared to "00" varieties | High oil content. Meal has high glucosinolate levels. Above average erucic contents compared to HEAR |
| Mario | Below average compared to "00" varieties | High oil content. Meal has high glucosinolate levels. Above average erucic contents compared to HEAR |
| BelAnn | Below average compared to "00" varieties | High oil content. Meal has high glucosinolate levels. Above average erucic contents compared to HEAR |

**Table 5.5** Typical fatty acid composition of *Brassica carinata* grown in central Italy (internal communication)

| Fatty acid | Oil composition (%) |
|---|---|
| 16:0 | 2.8 |
| 16:1 | Trace |
| 18:0 | 0.8 |
| 18:1 | 10.0 |
| 18:2 | 11.9 |
| 18:3 | 11.2 |
| 20:0 | 0.8 |
| 20:1 | 6.0 |
| 20:2 | 0.6 |
| 22:0 | 0.8 |
| 22:1 | 50.4 |
| 22:2 | 1.2 |
| 22:3 | 0.5 |
| 24:0 | 0.6 |
| 24:1 | 2.4 |

**Table 5.6** Levels of oil and erucic acid (%) in *Brassica carinata* grown in Mediterranean regions

| Country | Oil (%) | Erucic acid (%) |
|---|---|---|
| Spain | 34–37 | 47–50 |
| Greece | 35–48 | 45–55 |
| Italy | 38–43 | 47–55 |

1998 under the FAIR-CT96-1946 project (Final Report, 2000). One of the aims of the project was to see if the *B. carinata* was suitable for adoption in Mediterranean agriculture. Both high and low erucic acid genotypes were evaluated but it is only the high erucic varieties that are of interest here. Table 5.5 shows the lipid profile of some of the breeds of *B. carinata*. *B. carinata* was grown over three years in Italy, Greece, and Spain. Table 5.6 gives a summary of the qualities obtained from each country. The oil from *B. carinata* was processed to erucamide and was evaluated in comparison to high erucic rapeseed erucamide. The performance properties of the erucamide produced from *B. carinata* were similar to those produced by HEAR but the higher levels of polyunsaturated fatty acids gave poorer colour stability due to increased oxidation of the erucamide.

## 5.3 Processing of HEAR oils

There are several methods for splitting HEAR oil. Those most used by industry are the batch autoclave and continuous processes. With high quality oil, both processes can be used without further pre-treatment (Soap, Perfumery and Cosmetics, 1973).

### 5.3.1 Batch processes

The batch process is used little these days because of its limitations on throughput, the low degree of split obtained, and the post-treatment required. In order to increase the degree of split, the fat has to be treated twice. The oil and water are heated in the autoclave by direct injection of steam to a temperature of about 225°C. The contents are re-circulated and the degree of split is dependent on temperature, flow rate, and the ratio of oil to water. Initially the degree of split is 88–90% but increases to 95–98% with a second splitting process. The crude, aqueous glycerol mixture is called sweetwater. It contains 15–17% glycerol from the first split and decreases to about 12–14% when both sweetwaters are combined. The crude split fatty acids are then acid treated.

### 5.3.2 Continuous splitting processes

There are two types of continuous splitting processes, both developed by Lurgi:

(i) *Lurgi three-stage countercurrent splitting process* This process with more than one autoclave is operated in a continuous manner. The crude oil is fed to the first autoclave and the fresh water into the final autoclave. The raw materials are heated using heat exchangers that utilise the heat from the final fatty acid and crude sweetwater streams. The oil and water are fed through the autoclaves countercurrent to each other. The sweetwater is allowed to settle at each stage and is collected, with some being returned to the previous splitting stage. A split of 96–98.5% is usually obtained for this type of process. The glycerine content in the sweetwater is about 12%. Again, because of the throughput of about 40 t/day, this process is limited in its use.

(ii) *Lurgi single-stage high pressure countercurrent splitting process* This is the most widely used process used in industry (Fig. 5.1). This processes oils at a high throughput rate of about 100 t/day with a high degree of split (97–99%) and produces sweetwater containing about 12% glycerol. The reaction proceeds through the formation of intermediate di- and mono-glycerides, which help to solubilise the reactant water in the oil layer. The water content in this oil layer is further raised by an increase in temperature, which accelerates the reaction rate. A temperature of 250–255°C is used to split the rape oil, which is 98% complete in approximately 3–4 hours. Higher temperatures and the use of catalysts are avoided because of possible side reactions.

The oil is fed into the bottom of the tower, whilst the de-mineralised or condensate water is fed into the top of the tower. Splitting of the oil is pushed to completion by the continual addition of more water and steam to give a pressure

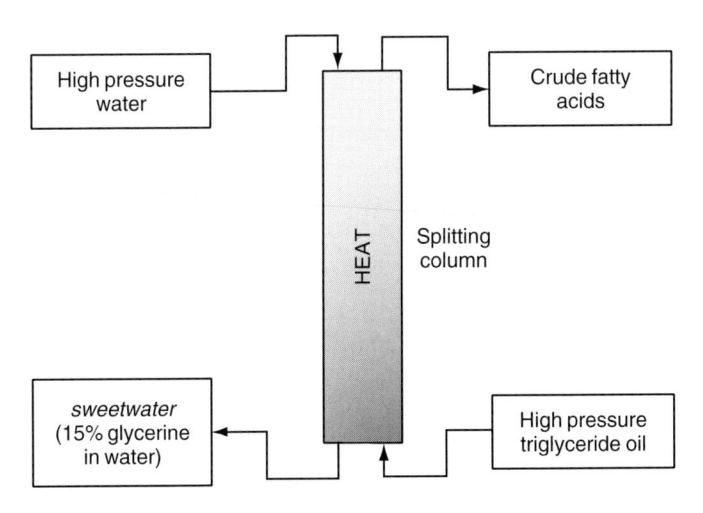

**Figure 5.1** Processing of high erucic oils – single-stage countercurrent splitting process.

of 750 psi in the splitting tower. This dilutes the concentration of the dissolved glycerol in the water phase, which in turn reduces the dissolved glycerol in the fat phase. The excess water prevents undue amounts of impurities formed by the reverse reaction, although there will be some present, predominantly in the form of monoacylglycerols. The sweetwater exits at the bottom of the tower, while the split fatty acids exit at the top. The hydrolysed fatty acids produced by any of the above processes will contain some unsplit material along with other unsaponifiable materials contained in the original oils.

### 5.3.3 Other splitting processes

(i) *Methanolysis*  Methanol is used to produce methyl esters of the high erucic rapeseed oil using alkaline catalysts, as in the production of biodiesel from LERO. The glycerol recovered will contain methanol, which must be removed before the crude glycerol can be processed. The crude methyl esters are then washed to remove methanol and the impurities before being fractionally distilled. However, this process is not used in large volumes for erucic acid production because of the difficulties in processing methanol and methanol-containing by-products.

(ii) *Enzymatic splitting*  Fats and oils are hydrolysed in the presence of natural enzymes (lipases) such as those from *Candida rugosa*, *Aspergillus niger*, and *Rhizopus arrhizus*. Total hydrolysis by enzymes is expensive and is not economically suitable for splitting HEAR oils. Henkel Corp. (1966) has patented a process for a more efficient way of enzymatic splitting. In the first stage, oil and lipase are mixed and effect partial splitting of the glyceride. The product from this stage is then passed through a high temperature, high pressure splitter to complete the hydrolysis. The combination of the two processes overcomes the disadvantages of each reaction. Enzymatic hydrolysis of the oil is a fast reaction initially but slows down at higher degrees of split. The high temperature, high pressure splitter has a slow initial hydrolysis induction period but the reaction increases substantially at higher degrees of split. The combination of the two processes increases the capacity of the splitting column.

## 5.4 Downstream processing of the split HEAR fatty acids

### 5.4.1 Fractional distillation

The main aim is to obtain high purity erucic acid for further derivatisation. A flow diagram of a typical process is shown in Fig. 5.2. The fatty acids must be fractionated to obtain the erucic acid. The main industrial process is fractional distillation in stills designed specifically for handling long-chain, high boiling fatty acids. The temperature required to distil split HEAR fatty acids is above 200°C and care must be taken to minimise side reactions. Rapeseed contains about 0.8% mixed sterols. These are high molecular weight cyclic alcohols, which can exist either as free alcohols or as their fatty acid esters. Even if sterol esters are hydrolysed during the splitting operation, they can re-combine during the distillation stage to re-form sterol esters. These will then have molecular weights over 600 and will be non-volatile under conditions of the distillation. They will then be removed in the residue stream.

In HEAR oil, unsaturation is at a maximum in the $C_{18}$ acids, of which over 20% contain either two or three such bonds per molecule. These acids, linoleic

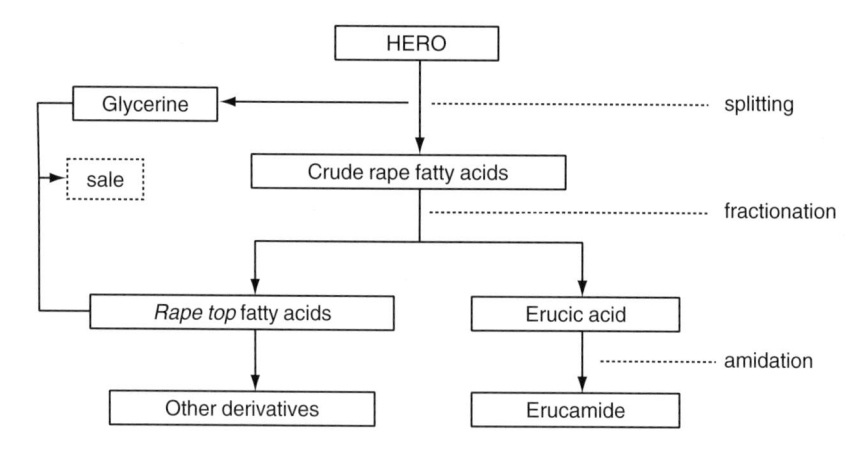

**Figure 5.2** Rapeseed oil processing.

and linolenic, tend to polymerise through the double bonds to give dimer and trimer acids. Such compounds are also of high molecular weight (~560 or 840) and again will exit the plant in the residue stream. The unsaturated fatty acids (particularly linolenic acid) can also react with oxygen to form peroxides. When these peroxides enter the heated zones of the distillation plant, they may be decomposed to liberate a range of low molecular weight organic compounds and higher molecular weight fragments of the original molecule. These need to be minimised during processing and must be removed from the erucic fraction.

Carboxylic fatty acids also tend to decarboxylate when distilled and the distillation plant should be designed to minimise this change. Once the erucic acid has been obtained, it can be converted to the required final product.

### 5.4.2 Dry or melt crystallisation

Dry or melt fractionation of fats and oils is a well-known process used to separate olein and stearin fractions from mixed glycerol esters or fatty acids (Gunstone et al., 1994b). This process comprises cooling the oil or fatty acid until a solid phase crystallises and separating the crystallised phase from the liquid phase usually by pressure filtration.

Fractional crystallisation of rape oil fatty acids to obtain erucic acid by this method is not feasible because of the difficulty with filtration. The problem arises from the fact that the liquid fraction is entrained in the inter- and intra-crystal spaces of the crystal mass of the erucic fraction. The crystal mass forms small crystals with a large surface area, and an efficient separation of the erucic fraction cannot be attained.

## 5.5 Quality problems associated with processing HEAR oils

High erucic rapeseed oil has several quality problems associated with it, the major ones being high contents of sulfur, phosphorus, and glucosinolates.

The high sulfur content in some plants is related to the fertiliser regimes used to increase growth and yields. The use of sulfurised fertiliser has increased as the levels of acid rain decrease. The required level of sulfur-based fertiliser is difficult to estimate and too much sulfur tends to be added. This has the effect of increasing the glucosinolate level. Chinese high erucic rapeseed has particularly high sulfur levels, which may be a result of local agronomic conditions.

The conditions used during crushing of the seed have a dramatic effect on the sulfur and phosphorus content. The oil is extracted from the seeds by a continuous two-stage process. The first stage involves expelling about 85% of the oil and the remaining cake is then extracted with hexane, which leaves 1–2% oil in the final meal. The extracted and expelled oils are combined (about 1:2 ratio) and the phospholipids are removed by degumming. This operation reduces the yield of the oil that would otherwise be obtained by 1–1.5% by weight. The crude HERO is then used without further refining. Depending on the efficiency of the extraction and degumming processes, the content of phosphorus and sulfur is usually in the range 70–400 ppm and 15–200 ppm respectively.

### 5.5.1 Downstream processing problems

Levels of phosphorus and sulfur in excess of 100 ppm can lead to downstream processing problems associated with high contents of phosphorus and sulfur in the oil and of high glucosinolate in the meal.

A high content of phosphorus in the oil is a problem because:

- phospholipids are good emulsifying agents and high levels of phosphorus produce oil emulsions at the splitting stage;
- some of the phospholipids are water-soluble and are retained in the *sweetwater* giving rise to problems during glycerol processing;
- high phosphorus residues from the fractional distillation process require disposal.

A high sulfur content in the oil leads to difficulties because:

- sulfur poisons the nickel catalysts used in the hydrogenation process. As a result, hydrogenation batch times may be doubled, thereby decreasing plant capacity;
- additional catalyst is required to overcome the high sulfur levels and may be as much as double the normal level;
- increased nickel content of the residues after distillation conflicts with environmental pressures to reduce heavy metals in waste streams.

## 5.6 Meal quality

A high glucosinolate content in the meal is an economic disadvantage. In recent years, plant breeders have developed cultivars which have a lower glucosinolate content for high erucic oils, e.g. Cadwell by Nickerson (Lincolnshire) and Industry by Danisco (Denmark). Typical composition of HEAR meal is shown in Tables 5.7, 5.8, and 5.9. High glucosinolate levels are not accepted into animal feed and the meal must be diluted with the meal from *double low* rape varieties to ensure they meet the low glucosinolate requirement. There is no set specification for low glucosinolates but it is generally accepted that 18–25 μmol/g is satisfactory as a low glucosinolate meal. Traditional cultivars of *B. napus* contain mainly aliphatic glucosinolates derived from tryptophan.

Crambe also contains high levels of glucosinolates (up to 8% in the crambe seed). The main glucosinolate in crambe is (3)-2-hydroxy-3-butenylglucosinolate

**Table 5.7** Typical composition of HEAR meal

| | |
|---|---|
| Protein | 30–45% |
| Oil | 4% max. |
| Fibre | 10–15% |
| Moisture | 10% max. |
| Sugars | 5–8% |
| Other carbohydrates | 5–17% |
| Mineral and vitamins | Remainder |

**Typical composition of fatty acids present**

| | |
|---|---|
| 16:0 | 6.5 |
| 16:1 | 0.9 |
| 18:0 | 1.1 |
| 18:1 | 16.5 |
| 18:2 | 19.0 |
| 18:3 | 7.0 |
| 20:0 | 0.8 |
| 20:1 | 11.6 |
| 20:2 | 0.6 |
| 22:0 | 0.7 |
| 22:1 | 33.0 |
| Other | 2.3 |

(epiprogoitrin) (Uppström, 1995). However, various strains of crambe are being bred for lower glucosinolate levels (Wang, 2000).

The mustard rapeseeds obviously contain high levels of glucosinolates as it is the hydrolysis products of the glucosinolates that give the mustard its flavour. The main glucosinolate of *B. alba* is 4-hydroxybenzylglucosinolate (sinalbin) and in *B. juncea* it is allylglucosinolate (sinigrin).

**Table 5.8** Amino acid profile of HEAR rapeseed meal (typical levels)

| Amino acid | % of total | mg/g of N |
|---|---|---|
| Lysine | 7.0 | 364 |
| Methionine | 2.1 | 111 |
| Isoleucine | 4.2 | 223 |
| Phenylalanine | 4.3 | 221 |
| Leucine | 7.7 | 395 |
| Tyrosine | 3.2 | 164 |
| Threonine | 4.6 | 240 |
| Valine | 5.8 | 301 |
| Histidine | 3.1 | 161 |
| Arginine | 6.8 | 351 |
| Glycine | 5.3 | 268 |
| Alanine | 4.8 | 241 |
| Aspartic acid | 7.5 | 386 |
| Glutamic acid | 20.1 | 1038 |
| Hydroxyproline | 1.3 | 66 |
| Proline | 7.7 | 399 |
| Serine | 4.5 | 234 |

**Table 5.9** Typical mineral content of low glucosinolate HEAR seed meal (ppm)

| Element | ppm |
|---|---|
| Magnesium | 900 |
| Phosphorus | 16 000 |
| Sulfur | 4200[a] |
| Potassium | 13 000 |
| Calcium | 8500 |
| Manganese | 65 |
| Iron | 180 |
| Molybdenum | 0.8 |
| Chromium | 5 |
| Nickel | 2 |
| Copper | 9 |
| Zinc | 65 |
| Iodine | 27 |

[a] Typically 10 500 ppm for high glucosinolate meal.

## 5.7 Users and producers of erucic acid

The major industrial use of erucic acid (*cis* 13-docosenoic acid) is as the primary amide derivative, erucamide ($RCONH_2$). This is used in the processing of polymers such as linear, low-density polyethylene (LLDPE) and polypropylene (PP) as a slip agent. It is a modifier that gives a reduced coefficient of friction to the surface of plastic films. This ensures that the film is able to slide over itself or another film. The erucamide molecule has polar and non-polar regions that allow it to migrate quickly to the surface of the film. Its main function is to impart surface lubrication, either during or immediately after the processing of the film. The reduction in surface coefficient of friction (CoF) over time achieved by amides in the film after extrusion is shown in Fig. 5.3.

The major producers of erucamide are shown in Table 5.10. Croda, Crompton, Akzo, and Uniqema all make their own erucic acid, Lutianhua is the main producer in China, and Fine Organics purchases erucic acid mainly from the

**Table 5.10** Major producers of erucamide for the industrial market

| Erucamide producer | Country | Seed source |
|---|---|---|
| Croda | UK | HERO/crambe |
| Crompton | USA | HERO/crambe |
| Akzo | USA, Korea, Italy | HERO |
| Ciba/Uniqema | Netherlands | HERO |
| Fine Organics | India | HERO |
| Lutianhua | China | HERO |
| Nippon Fine Chemicals | Japan | HERO |

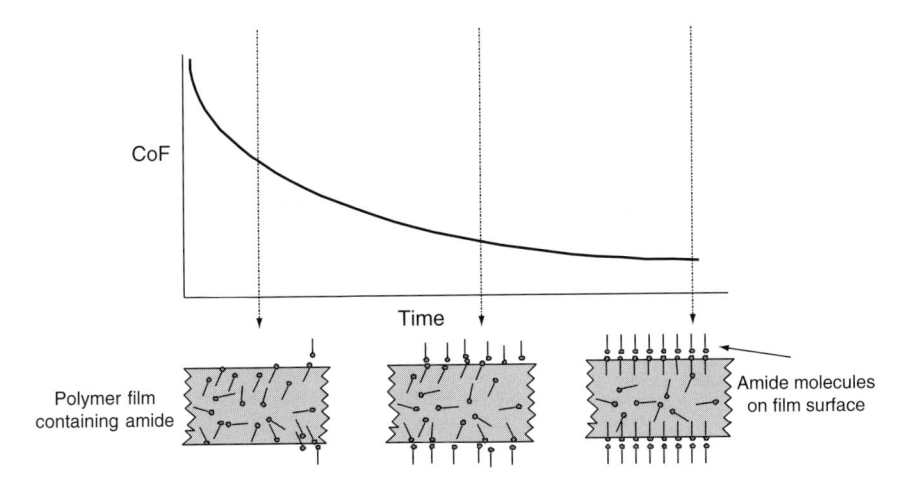

**Figure 5.3** Variation of coefficient of friction with time.

Indian local producers (VVF and Godrej). Cognis (GmbH) are also suppliers of erucic acid.

The market for erucic acid-based slip additive is generally put at approximately 30 000 t a year. Sonntag (1995) gave a detailed report of the world use of $C_{22}$ oleochemicals. Although he estimated that behenic (22:0) products would exceed the use of erucic acid and its derivatives by the twenty-first century, this has not yet occurred. Erucamide consumption is expected to grow around 3–5% per annum, in line with expected growth in the polymer (LLDPE and PP) industry. Figures 5.4 and 5.5 show the prediction of growth until the year 2010.

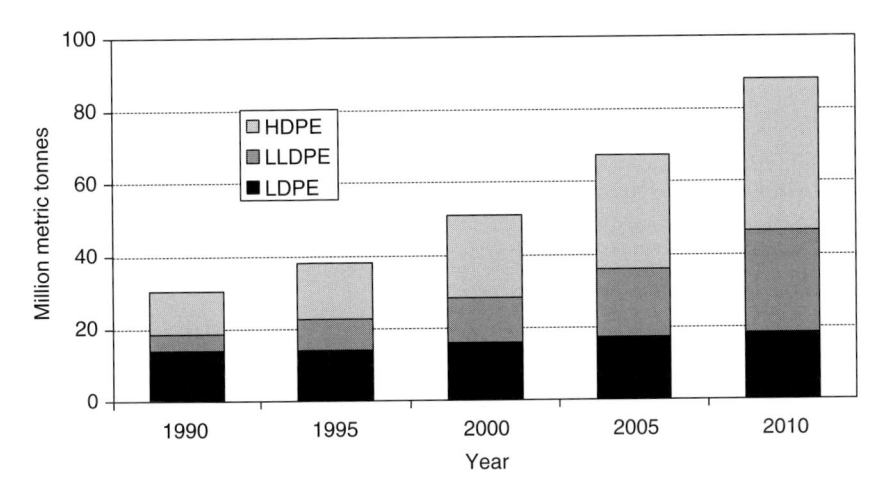

**Figure 5.4** Projected world growth for polyethylene by type to 2010.

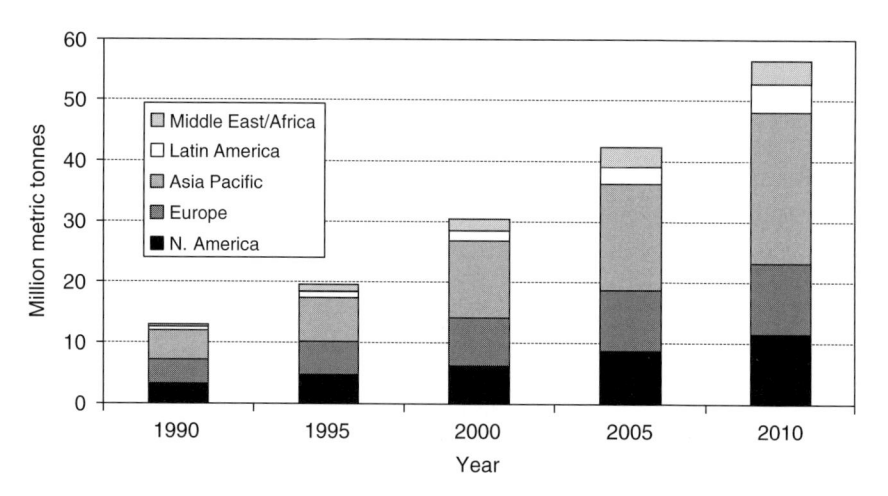

**Figure 5.5** Projected world growth for polymers by region to 2010.

Erucamide is not, as Sonntag (1995) predicted, becoming obsolete as a polyolefin film additive. Other amides produced from erucic acid are also utilised within the polymer industry, the prime one being stearyl erucamide (RCONHR'). It is produced from stearylamine and erucic acid and is used primarily in Nylon-6,6 and biaxially orientated polypropylene (BOPP). This is a niche speciality used because of its greater stability and slower migration effects compared to erucamide. This property gives the user time to modify the surface of the polymer before the surface becomes too slippy, i.e. lamination. It is estimated that stearyl erucamide holds less than 4% of the erucamide market.

## 5.8 Uses of erucic acid

The major use of erucic acid is still as erucamide for the polymer industry but erucic acid and its derivatives are also used in the detergent, photographic, food, cosmetic, pharmaceutical, ink, paper, textile, lubricant, household, and water treatment industries. Sonntag (1995) gives an estimate of the size of the markets for each application across the world.

An application of erucic acid that became the subject of a Hollywood film was Lorenzo's oil. The film dramatised the true story of Lorenzo Odone who was suffering from a rare lipid storage disease, adrenoleukodystrophy (ALD). The dietary lipid that was developed was high-purity glycerol trioleate and glycerol trierucate, which is now known as Lorenzo's oil (Coupland, 1999). A ten-year international study involving 12 laboratories has shown (Watkins, 2003) that boys who successfully followed a strict treatment/diet regime with

**Table 5.11** Utilisation of erucic acid and its derivatives

| Product | Uses | Patent, company, journal |
| --- | --- | --- |
| Erucic acid carbohydrate ester | Biodegradable non-toxic food-grade lubricating oils | Lgol Industries, France (1998) |
| Erucic acid sucrose esters | High quality, stable puff dough | Kokai Tokkyo Koho, Japan |
| Erucic acid, mixed ester with diol and triol (glycerol-mixed fatty acid esters) | Friction modifiers for diesel engines and turbines | Institut Français de Petrole (1996) |
| Erucic acid, reaction products with poly (isobutylmethacrylate) | Coagulating agents and their use for oil spill removal | W09722558 (1996) |
| Erucic acid, iodinated | Treatment of goitres and other disorders | W09703038 (1995) |
| Erucic acid, oxetanone derivatives | Sizing agents for paper | Hercules Inc. (1996) |
| Erucic acid | Lubricity of petroleum, well drilling fluids | Baker-Huges Incorp. USA (1994) |
| Erucic acid diglycerides | Diglycerides and sucrose esters for bitterness control of pharmaceuticals and foods | Kao Corp, Japan |
| Erucic acid esters with alkyl glycerol | Base oil for hydraulic fluids | Henkel (1993) |
| Erucic acid reaction products with rosin acids and hydroxyamines | Emulsifiers for oil–water emulsions | Lubrizol (1991) |
| Erucic acid diesters with dihydric alcohols | Edible fat substitutes | Nabisco Inc. (1991) |
| Erucic acid glycerol esters and erucic alkoxylate mixtures | De-inking agents, anti-foaming, in recycling of waste paper | Kao Crop, Japan |
| Erucic acid, glycolic acid derivatives | Anti-microbial activity | JAOCS (1989), **66**, 932–934 |
| Erucic acid monoglyceride | Emulsifying agent | Henkel (1988) |

10–15% of total calories from fat for two or more years were two-thirds less likely to develop neurological or structural brain abnormalities during the 2–4-year follow-up period.

A totally different use of HEAR oil is as a lubricant (Bently, private communication). The HEAR oil is first converted to *blown* oil. The blowing operation is usually carried out at high temperatures using sparged air, to produce an oxidised oil described as a *burnt rapeseed oil*. The oil is dark in colour with a reddish appearance due to the high processing temperature. This process improves the compatibility of the HEAR with mineral oils partly due to the oil having less polar properties than equivalent oils processed at lower temperatures. The oxidising process causes the following physical and chemical changes:

(i)   reduction in iodine value;
(ii)  increase in saponification value;
(iii) increase in specific gravity;
(iv)  increase in viscosity;
(v)   darkening of the oil.

Table 5.11 shows some of the less well-known uses of erucic acid and its derivatives.

## 5.9   Genetic modification of HEAR crops

There has been much research into producing very high erucic rapeseed oil with the justification that it is in the interest of the end user to buy more erucic acid per tonne of oil. The economics of utilising erucic acid as an industrial user is not that simple. The economic value of the by-products is an important factor to consider, i.e. glycerol, rapeseed meal, remaining rape fatty acids. Glycerol and meal are sold in the commodity markets at commodity prices. The remainder of the fatty acids is a commercially important source of $C_{18}$ fatty acids and command specialist fatty acid pricing. The economics therefore does not warrant the development costs of a very high erucic acid rapeseed variety. The theoretical value of an oil with increasing erucic content, if the value of the fatty acid by-products is taken into account, is shown in Table 5.12.

As there are no by-products to offset the cost of producing erucic acid at the 100% level, this makes for an expensive feedstock compared to the lower erucic varieties. There is a finite quantity of erucic required world-wide. If the 100% variety was available, the world-wide crop demand would be half of what it is today and growers would get less revenue. There is still a major reluctance for genetically modified crops to become established in Europe, and many customers of erucic acid and its derivatives are still insisting that the products they purchase are formally certified as being free of genetically modified organisms (GMO-free). This is of particular importance in the case of erucamide where, as an additive to plastic, it is in contact with food and can have food approval only if it has been assessed as being GMO-free.

**Table 5.12** The rise in theoretical cost of oils with increasing erucic content, taking into account the value of the fatty acid by-products (by-product at £300/t and erucic acid at £700/t)

| Erucic content of oil (%) | 50 | 60 | 70 | 100 |
|---|---|---|---|---|
| Yield of by-products (%) | 50 | 40 | 30 | None |
| Value of by-products (£/t) | 150 | 120 | 90 | None |
| Yield of erucic acid (%) | 50 | 60 | 70 | 100 |
| Value of erucic acid (£/t) | 350 | 420 | 490 | 700 |
| Total cost of oil (£/t) | 500 | 540 | 580 | 700 |

The use of the by-products will also be affected as the glycerol, rapeseed meal, and fatty acids must also be certified as GMO-free. Glycerol is used in food/cosmetics and rapeseed meal in animal feed. The debate on the growing of genetically modified crops in the EU continues at present.

## 5.10  Ideal crop for industrial users

The ideal erucic crop for industrial users would be the one that is high yielding and low in glucosinolates. Conventional ongoing breeding programmes have already produced low glucosinolate seeds (<20 µg/g) and yields are now much closer to the double low rape varieties than they were a few years ago (2.5 t/ha compared to 3.0–3.5 t/ha for "00" rapeseed). The ideal lipid profile shows 60% erucic (22:1) content and 40% oleic (18:1) content. Polyunsaturates are undesirable because of their oxidative instability and the 40% oleic acid content would be ideal as a high-purity vegetable oleic acid, the uses of which are numerous. The economics would be viable, as the maximum value could be achieved from both the erucic acid and the by-products.

## References

Bilsborrow, P.E., Evans, E.G. and Bland, B.F. (1994) Pollen transfer between high and low erucic acid oilseed rape crops. *Aspects of Applied Biology*, **35**, 163–166.
BioMatNet July 1998, http.//www.nf-2000.org/, AIR3-CT94-2480, *Crambe abyssinica*, Production and utilisation – a Comprehensive Programme, Final Year Report.
Coupland, D. and Makin, S. (2003) Springdale Crop Synergies Ltd, *Abyssinian* mustard (Crambe), Agronomy of a new oilseed crop for the UK, internal communication.
Coupland, K. (1999) Lorenzo's Oil – Six years later, *Inform*, 10, 8–9, 113.
De Guzman (2003) Chemical Marketing Reporter, 17 February, p. 11.
Endres, G. and Schatz, B. (1993) Crambe production, NDSU Extension Service, A-1010.
FAIR-CT96-1946 (2000) *Brassica carinata*: The outset of a new crop for biomass and industrial non-food oil, Final Report, March.
Golz, T. (1993) Crambe, NDSU Extension Service, Alternative Agriculture Series, No. 4, Jan.
Gunstone, F.D., Harwood, J.L. and Padley, F.B. (1994a) *The Lipid Handbook*, 2nd edition, Chapman and Hall, London, p. 78.
Gunstone, F.D., Harwood, J.L. and Padley, F.B. (1994b) *The Lipid Handbook*, 2nd edition, Chapman and Hall, London, pp. 225–226.
Hebard, A.B. (2002) Growing high erucic rapeseed oil as an example of identity preservation. *Lipid Technology Newsletter*, **8**, 53–55.
Hemingway, J.S. (1995) *The Mustard Species, Brassica Oilseeds, Production and Utilisation* (eds D. Kimber and D.I. McGregor), pp. 373–383.
Henkel Corp. (1996) Improved fat splitting process. Issued 5 December 1996. WO 96/38534.
IACR-R1-India Bulletin June 2002.
IENICA (August 2000) Summary Report for European Union, Oil Crops, Ref. 1495.
NIAB (2001) *Oilseeds Variety Handbook*, E&E Plumridge Ltd, Cambridge.
Nicholls, F.H. (1996) New crops in the UK: from concept to bottom line profits. In: *Progress in New Crops* (ed. J. Janick) ASHS Press, Alexandria, VA. pp. 21–26.
Official Methods on Recommended Practices of the American Oil Chemists Society (1996) *Physical and Chemical Characteristics of Oils, Fats and Waxes*, Section 1, p. 29.

Ramans, M. (1995) Winter HEAR volunteer study in East Anglia 1993/94 9th Int. Rapeseed Congress, **4**, 1369–1371.

Soap, Perfumery and Cosmetics, September 1973, pp. 525–532.

Sonntag, N.O.V. (1995) Industrial utilization of long-chain fatty acids and their derivatives. *Brassica Oilseeds, Production and Utilization* (eds D. Kimber and D.I. McGregor), CAB International, Wallingford, UK, pp. 339–352.

Temple-Heald, C.E. (2000) Sourcing erucic acid for the industrial market. *Lipid Technology Newsletter*, **6**, 59–63.

Uppström, B. (1995) *Seed Chemistry, Brassica Oilseeds, Production and Utilisation* (eds D. Kimber and D.I. McGregor), pp. 217–242.

Watkins, C. (2003) Lorenzo's oil vindicated. *Inform*, January, **14** (1), 38–39.

Wang, Y.P. (2000) *Industrial Crops and Products*, **12**, 47–52.

# 6 Food uses and nutritional properties
Bruce E. McDonald

## 6.1 Introduction

During the past decade, interest in dietary fat has switched from a concern about the amount in the diet to an emphasis on the type of fat in the diet. Several developments contributed to this change. Studies showed that it was not the total amount of fat in the diet that increased the risk factors for cardiovascular disease (CVD) but the amount of saturated fat in the diet. In addition, there was the recognition that substitution of unsaturated fat for saturated fat was more effective in reducing the risk factors for CVD than a similar substitution with carbohydrate. It was also established that dietary monounsaturated fat (viz., high in oleic acid) was equally as effective in reducing plasma total and low density lipoprotein (LDL) cholesterol levels as polyunsaturated fat (viz, linoleic acid). Another contributing factor was the demonstration that *trans* fatty acids, produced during the partial hydrogenation of vegetable oils, not only increased plasma LDL cholesterol levels but decreased plasma high density lipoprotein (HDL) cholesterol levels. Complementing these observations was the confirmation that *n*-3 (also referred to as omega-3) fatty acids were not only essential nutrients but played a role in reducing the risk of CVD.

Canola oil refers to low-erucic acid rapeseed oil and canola-type Brassica oils (e.g. canola quality *Brassica juncea* oil) and this designation will be used throughout the chapter. It has several of the properties that meet the nutritional attributes sought in a well-balanced diet. Canola oil is characterized by a low level of saturated fatty acids; it contains the lowest level of saturated fatty acids among common dietary fats and oils. Canola oil contains about half the level of saturated fatty acids present in corn oil, olive oil or soybean oil and about one-quarter the level in cottonseed oil. It is also characterized by a relatively high level of oleic acid. Among the common edible vegetable oils, canola oil is second only to olive oil in oleic acid content. In addition, it contains a significant amount of α-linolenic acid (α-LNA), an *n*-3 fatty acid. Canola and soybean oils are the only common edible vegetable oils that contain appreciable levels of α-LNA. Some seeds (e.g. flaxseed) and nuts (e.g. walnuts) and their oils contain significant levels of α-LNA. However, these sources make a relatively insignificant contribution to the α-LNA intake of the average diet. Furthermore, there is a relatively favorable balance between α-LNA and linoleic acid (LA) in canola oil; the ratio of α-LNA to LA is approximately 1:2, compared to a ratio of 1:7 for soybean oil and greater than 1:50 for most other common vegetable oils. It is doubtful that the developers of canola (Stefansson *et al.*, 1961), with due respect to the significance of their contribution, fully envisioned the nutritional quality of the oil that resulted from their efforts to reduce the level of erucic acid in rapeseed oil. Furthermore, it is unlikely that they would have set out to develop, on the basis of the then prevailing understanding of the nutrition of dietary fat, a cultivar with the fatty acid profile of current-day canola.

## 6.2  Food uses

Canola oil is a versatile vegetable oil that is widely used as a salad oil and in food formulations and food preparation in Europe, Canada, Australia and Japan. The refined, bleached and deodorized (RBD) oil, which is light in color and bland in flavor, is widely used as a salad oil and in the manufacture of salad oil products such as prepared salad dressings and mayonnaise. Its use in salad oil applications is the primary use of canola oil in the United States where it is promoted as a healthy oil. In some areas, for example parts of Asia, canola oil is blended with a variety of other oils and used as a *general purpose* salad and frying oil. Canola oil and partially hydrogenated canola oil also are extensively used in the manufacture of margarine, and baking and frying shortenings in Europe, Canada, Australia and Japan. Partially hydrogenated canola oil with somewhat higher levels of *trans* fatty acids, also, is used in the preparation of cream fillings and chocolate and other confectionery coatings.

### 6.2.1  *Salad oils, salad dressings and mayonnaise*

One of the current appeals of canola oil in salad oils is its nutritional quality; namely low level of saturated fatty acids, high level of oleic acid and good amount of α-LNA (see Chapter 3). The manufacturers of prepared salad dressings and mayonnaise look for oils that form strong emulsions. In order to ensure emulsion stability in mayonnaise and salad dressings, they seek oils that do not cloud and are free of crystalline materials, such as waxes and high-melting triacylglycerols, at refrigeration temperatures. They also seek oils that have good oxidative stability. Although canola oil contains an appreciable level of α-LNA, the introduction of plastic containers that cut out UV-light improves its shelf life by reducing the oil's susceptibility of photooxidation.

### 6.2.2  *Margarine*

Margarine was first produced to mimic the melting behavior of butter. This type of margarine has been largely replaced by the so-called *soft* margarine (a margarine that is spreadable at refrigeration temperatures). The latter is now produced largely by interesterification (see Chapter 4), a process that has been popular in Europe for some time but only relatively recently introduced to North America in response to the concern over *trans* fatty acids. Initially, soft margarine was produced in North America by blending partially hydrogenated oils with different melting points.

Interesterification allows the manufacturer to include liquid oil in the margarine without encountering problems with the development of a gritty texture (i.e. formation of β crystals as opposed to very small, polymorphic β′ crystals) and separation of the non-hydrogenated oil from the hydrogenated oils. In addition, the process preserves the nutritional advantages of the non-hydrogenated oil such as canola oil. Canola oil, like most of the oils used today, tends to produce β crystals due to the fact that these oils are high in 18 carbon fatty acids. Partial hydrogenation, interesterification and inclusion of inhibitors of crystal formation are used to overcome the tendency to β crystal formation.

Fat functions in the development of volume and texture of baked products. The role of fat in cakes and breads is to introduce a large volume of finely dispersed air into the batter. Fat is also required to lubricate and tenderize the structure of the baked product. Satisfactory incorporation of air into the fat requires that the fat be in very small crystals (β′ form) that are stable over a range of temperatures. Crystalline structure is also required to prevent the fat from coalescing.

Coalescence results in poor dispersion of the fat in the batter and, in turn, poor texture and structure of the baked product. Air incorporation and prevention of fat coalescence are the main reasons why liquid oils are poor substitutes for shortenings. A significant quantity of the shortening used in commercial and institutional baking is pourable at room temperature. The popularity of liquid or fluid shortenings stems from the increased control on product formulation, the use of less fat in baked products and increased economy in the handling of liquid shortening.

The primary characteristic sought in a frying shortening is good stability because of the high temperature used, the openness to air, and the introduction of moisture and other compounds by the food to be fried. Good stability means resistance of the fatty acids to oxidative changes and the triacylglycerols to hydrolysis. Oxidative stability is also important in packaged foods such as potato chips (crisps) and other savory snacks. Crystal structure is usually not a consideration in a frying fat. However, in some applications, e.g. savory snacks and donuts, a relatively high amount of solid fat is desired to avoid greasiness on aesthetic grounds.

*6.2.3 Other uses*

In addition to its use as a salad oil and in the production of salad oil products, margarine, and baking and frying shortenings, canola oil is also used in many other consumer products and food industry formulations. Examples include its use in anti-stick cooking sprays, the preparation of toppings and fillings for cookies (biscuits), thin-layer (spray) coatings on some crackers, Danish and puff pastry products, coatings for confectioneries and vacuum-packed canned foods.

## 6.3 Nutritional properties

The nutritional properties of canola oil, like most fats and oils, are based largely on its chemical characteristics, in particular its fatty acid composition and the latter's relationship to risk factors for CVD. Canola oil contains the lowest level of saturated fatty acids among common dietary fats and oils. It is also characterized by a high level of the monounsaturated fatty acid, oleic acid, and is second only to olive oil in oleic acid content among the major fats and oils. Avocados and high-oleic sunflower oil also contain high levels of oleic acid but they do not constitute major sources of dietary oleic acid. Canola oil is intermediate among vegetable oils in polyunsaturated fatty acid (PUFA) content. It contains higher levels of PUFA than olive oil or palm oil but lower levels than corn, cottonseed, soybean or sunflower oil. However, canola oil contains a relatively high level of the $n$-3 fatty acid, $\alpha$-LNA. In addition, there is a favorable balance between $\alpha$-LNA and the major PUFA in vegetable oils, LA an $n$-6 fatty acid. In fact, except for soybean oil, canola oil is the only common vegetable oil that contains any appreciable amount of $\alpha$-LNA but the ratio of $\alpha$-LNA to LA in canola oil is 1:2 compared to 1:7 in soybean oil.

Phytosterols and vitamin E are two other constituents of vegetable oils that may have an effect on the risk factors for CVD. Canola oil is relatively rich in both of these minor constituents (Przybylski and Mag, 2002). Although lower in total phytosterol content than corn oil (6.9 vs 9.7 g/kg), canola oil contains appreciably higher levels than soybean and sunflower oils (4.6 and 4.1 g/kg respectively). Similarly, canola oil contains higher levels of $\alpha$-tocopherol, the biologically active isomer of vitamin E, than soybean oil or corn oil but appreciably lower levels than sunflower oil (272, 116, 134 and 613 mg/kg respectively).

## 6.4 Dietary fat and cardiovascular disease

Cardiovascular disease is a multifactorial disorder, which is a major cause of mortality and morbidity among affluent populations worldwide. It is a disease that is characterized by a number of physiological conditions or events, including atherosclerosis, thrombosis and arrhythmia. Atherosclerosis, the formation of lipid-laden plaques rich in cholesterol on the intimal surface of blood vessels, is a slow process that develops over several years. It results in a narrowing of the lumen of blood vessels and the impedance of blood flow through the afflicted vessel. Thrombosis, the formation of a clot, can lead directly to a coronary attack or stroke, if it blocks a major vessel. In contrast to atherosclerosis, clot formation occurs relatively quickly, especially in vessels damaged by atherosclerotic plaque. Cardiac arrhythmia occurs suddenly and without warning and if not arrested, often are fatal. Although dietary fat has been implicated in all of these processes, the effect of fat on thrombosis and arrhythmia has only recently received the attention of researchers.

## 6.5 Effect of canola oil on plasma cholesterol and lipoproteins

Prior to the report by Mattson and Grundy (1985) that dietary monounsaturated fatty acids (MUFA) were equally as effective as PUFA in lowering plasma total and LDL cholesterol levels in hyperlipidemic patients, interest in dietary fat and CVD centered primarily on saturated fat and polyunsaturated fat. Shortly thereafter, Mensink and Katan (1989) confirmed these findings in normolipidemic subjects. These results provided a possible explanation for the earlier observation that canola oil was as effective as soybean oil in lowering plasma total cholesterol levels in normolipidemic men (McDonald *et al.*, 1974). They also provided impetus for further studies on the nutritional properties of canola oil. Interest in canola oil was also aroused by the proposal that $n$-3 fatty acids might play a role in CVD and other chronic diseases (Dyerberg, 1986). Appreciable research on the effect of canola oil on plasma cholesterol and lipoprotein levels has been reported over the past 10–12 years.

### 6.5.1 Studies with normolipidemic subjects

Carefully controlled studies in Canada, Finland, Germany and the United States found that canola oil was equally as effective as dietary fats high in PUFA in lowering plasma total and LDL cholesterol levels in healthy individuals when each replaced saturated fat in the diet (Table 6.1). In all of these studies, the canola oil and the PUFA oil diets resulted in statistically significant decreases in plasma total cholesterol (mean of −0.47 to −0.88 mmol/L) and LDL cholesterol (mean of −0.43 to −0.74 mmol/L) levels when substituted for the usual fat in the subjects' diets. As these values reveal, the lower levels of total cholesterol were primarily due to lower levels of LDL cholesterol. In addition, except for the study by Valsta *et al.* (1992), the lower plasma total and LDL cholesterol levels and apolipoprotein B levels (the lipoprotein characteristic of the LDL fraction) were the same on the canola and PUFA oil diets. Furthermore, except for the study by Kratz *et al.* (2002), there were no changes in HDL cholesterol levels in any of these studies. Mattson and Grundy (1985) had reported lower HDL cholesterol levels on a high PUFA diet but not on

**Table 6.1** Effect of canola oil vs PUFA dietary fat source on plasma lipid patterns of normolipidemic subjects

| Source of dietary PUFA | Plasma lipid parameter | Baseline[a] (mmol/L) | % change from baseline Canola diet | % change from baseline PUFA diet | Reference |
|---|---|---|---|---|---|
| Sunflower oil | Total cholesterol | 4.42 | −20 | −15 | McDonald et al. (1989) |
|  | LDL cholesterol | 2.76 | −23 | −21 |  |
| Sunflower oil | Total cholesterol | 5.35 | −15 | −12 | Valsta et al. (1992) |
|  | LDL cholesterol | 3.17 | −23[b] | −17[b] |  |
| Sunflower oil | Total cholesterol | 4.76 | −14 | −17 | Kratz et al. (2002) |
|  | LDL cholesterol | 2.70 | −18 | −20 |  |
| Safflower oil | Total cholesterol | 5.39 | −9 | −15 | Wardlaw et al. (1991) |
|  | LDL cholesterol | 3.71 | −12 | −15 |  |
| Soybean oil | Total cholesterol | 5.35 | −18 | −16 | Chan et al. (1991) |
|  | LDL cholesterol | 3.17 | −25 | −18 |  |

[a] Plasma levels on diets typical of usual fat intake.
[b] Significant difference ($p < 0.01$) between canola diet and PUFA diet.

a high MUFA diet. By contrast, Mensink and Katan (1989) found no effect of a high PUFA diet on plasma HDL levels. Kratz et al. (2002), who also included olive oil as one of the treatments in their study, observed slightly lower HDL cholesterol levels on both the canola oil and the sunflower oil diets but no change on the olive oil diet.

Substitution of canola oil or sunflower oil and products produced with these oils (margarines and salad dressings) for regular fat products in the diets of healthy subjects also resulted in significantly lower plasma total and LDL cholesterol levels. McDonald et al. (1995) replaced approximately 60% of the fat in the subjects' customary diet by either canola oil or sunflower oil and their respective products. Subjects on the canola oil or sunflower oil diet had significantly ($p < 0.05$) lower plasma total (−0.41 and −0.60 mmol/L respectively) and LDL (−0.40 and −0.54 mmol/L respectively) cholesterol levels than subjects (controls) who continued on their regular diet. Substitution of low-linolenic acid canola oil also resulted in lower plasma total and LDL cholesterol levels. Similarly, Nydahl et al. (1994) found that substitution of canola oil or sunflower oil and their respective margarines for regular oils and margarines in the diet resulted in significantly ($p < 0.001$) lower total and LDL cholesterol levels. However, the decreases on the canola oil and sunflower oil diets were less than those observed by McDonald et al. (1995) or by the studies summarized in Table 6.1. Regrettably, Nydahl et al. (1994) did not provide details on the degree of substitution or on the changes in fatty acid composition, when canola oil and sunflower oil were substituted for regular oils and margarines in the diets of their subjects. Becker et al. (1999) also found that replacing 65 g of fat in the regular diets of healthy, young individuals with a butter–canola oil (65:35) blend resulted in lower plasma total (−0.32 mmol/L, 8%) and LDL (−0.31 mmol/L, 12%)

cholesterol levels than with a butter–grapeseed oil (90:10) blend. The study was a crossover design with 3-week treatment periods separated by 3-week washout periods. Similarly, a 13-week diet intervention study (Matheson *et al.*, 1996) with a 23-member party on an Australian Antarctic research expedition found that substituting canola oil and a canola oil margarine for butter, a table margarine and variety of vegetable oils results in lower plasma total (−0.41 mmol/L, 7%) and LDL (−0.44 mmol/L, 10%) cholesterol levels.

The results of these studies with normolipidemic subjects are consistent with the primary goal of dietary intervention programs aimed at reducing the risk of CVD; namely, a lowering of plasma total and LDL cholesterol levels. All of these studies found that canola oil was equally as effective as high PUFA oils in lowering plasma total and LDL cholesterol levels when substituted for saturated fats in the diet.

### 6.5.2 Studies with hyperlipidemic subjects

Canola oil was also found to be effective in lowering plasma total and LDL cholesterol levels in subjects with elevated plasma lipids (Table 6.2). The results, however, have been less consistent than the studies with subjects with normal plasma lipid levels (Table 6.1). In well-controlled studies, decreases in plasma total and LDL cholesterol levels on canola oil or PUFA oil diets were similar to those reported for normolipidemic subjects (−12 to −17%). In addition, canola oil and PUFA oils were equally effective in lowering plasma cholesterol levels on high- (37% of energy, Södergren *et al.*, 2001) and low-fat diets (30% of energy, Gustafsson *et al.*, 1994; ≤27% of energy, Lichtenstein *et al.*, 1993). However, in contrast to studies with normolipidemic subjects, both Lichtenstein *et al.* (1993) and Gustafsson *et al.* (1994) found plasma HDL cholesterol levels decreased on the canola and PUFA diets. By contrast, the HDL cholesterol level on an olive oil-enriched diet did not differ from baseline (Lichtenstein *et al.*, 1993). Also noteworthy was the finding that the decrease in total cholesterol was less on the olive oil diet than on the canola or PUFA diets. However, the decreases in LDL cholesterol and apolipoprotein B were similar on all three diets. On the contrary, plasma HDL cholesterol levels did not differ on low-fat (30% of energy) diets in which the main source of fat was beef tallow, olive oil, canola oil, corn oil or rice bran oil (Schwab *et al.*, 1998). Total and LDL cholesterol levels, however, coincided with those reported in other studies; levels were lower on the canola oil, corn oil and rice bran oil diets than on the beef tallow diet. Total and LDL cholesterol levels on the olive oil diet were intermediate between those on the canola, corn and rice bran diets and those on the beef tallow diet, and did not differ statistically ($p<0.05$) from any of the other diets.

Intervention studies where canola oil and its products (e.g. margarine, cheese) were substituted for customary fat in the diet also resulted in lower

Table 6.2 Comparison of the effect of canola oil with other dietary fats on plasma lipid patterns of hyperlipidemic subjects

| Baseline diet | Fat source-test diet | Baseline values (mmol/L) Total chol. | Baseline values (mmol/L) LDL chol. | % change from baseline Total chol. | % change from baseline LDL chol. | Reference |
|---|---|---|---|---|---|---|
| NCEP Step 2 Diet[a] | Canola oil | 221 | 152 | −12 | −16 | Lichtenstein et al. (1993) |
| | Corn oil | | | −13 | −17 | |
| | Olive oil | | | −7[b] | −13 | |
| Typical Swedish diet | Canola oil | 6.79/7.28[c] | 4.85/5.23[c] | −15 | −16 | Gustafsson et al. (1994) |
| | Sunflower oil | | | −16 | −14 | |
| Subjects regular diet | Canola oil | 6.59/5.73[d] | 4.31 | −11 | −10 | Södergren et al. (2001) |
| | Saturated fat[e] | | | +6 | <+1 | |

[a] National Cholesterol Education Program Step 2 Diet – total fat <30% energy; saturated fat <7% energy; cholesterol <200 mg/day.
[b] Significantly different ($p < 0.01$) from canola oil and sunflower oil.
[c] Canola diet group/sunflower diet group.
[d] Canola diet group/saturated diet group.
[e] Saturated fatty acids provided 53% of total dietary fat (20% of total energy).

plasma cholesterol levels with moderately hyperlipidemic subjects. A relatively long-term study (6 months; Sarkkinen et al., 1994) found that a diet enriched with MUFA (canola oil and canola oil margarine) was as effective as an American Heart Association (AHA)-type diet in lowering plasma total and LDL cholesterol levels in hypercholesterolemic subjects (baseline levels of 6.5 mmol/L). The decrease in plasma cholesterol was 0.31 mmol/L on the AHA-type diet and 0.24 mmol/L on the MUFA-enriched diet, even though the latter contained higher levels of total fat (38 vs 30% of energy) and saturated fat (14 vs 10%) than the AHA-type diet. The decrease in LDL cholesterol level was 0.31 and 0.30 mmol/L on the AHA-type and MUFA-enriched diets respectively.

Substitution of canola oil at much lower levels was also found to favorably impact on serum lipid patterns. Karvonen et al. (2002) substituted 65 g of a low-fat, canola oil cheese, which provided 11 g of fat and less than 1 g of saturated fat, for regular cheese per day (15 g of fat, 10 g of saturated fat) in a 4-week crossover study. Plasma total cholesterol levels were significantly higher on the regular cheese than on the canola cheese at the end of the test period (6.28 vs 5.96 mmol/L; $p<0.001$). Plasma LDL cholesterol levels also were significantly higher on the regular cheese (4.23 vs 3.95 mmol/L; $p<0.002$). Although there was a slight difference in the amount of fat provided by the two cheeses, the major difference was in the amount of saturated fat (14.1% of total energy vs 9.6% on the regular and canola cheese respectively). This difference in saturated fat intake coincides with the finding that it is not the total amount of fat in the diet but the amount of saturated fat that is important (Barr et al., 1992). Replacing 50 g of regular fat in the diet by 50 g of canola oil mayonnaise also resulted in significant decreases from baseline values for both total and LDL cholesterol levels (−9 and −10% respectively) in subjects with a mean baseline serum cholesterol of 7.1 mmol/L (Miettinen and Vanhanen, 1994). Likewise, Bierenbaum et al. (1991) found that substituting 30 g of canola oil for the usual oils and spreads in the diet of hyperlipidemic subjects resulted in a significant decrease in plasma LDL cholesterol level (−0.18 mmol/L; $p<0.025$). However, total cholesterol did not differ from baseline.

Not all studies have found consistent decreases in plasma lipids in response to substitution of regular fat in the diet by canola oil and canola oil margarine. Seppänen-Laakso et al. (1992) also found that substitution of canola oil or a canola–sunflower oil margarine for butter resulted in significantly lower serum total (−0.49 and −0.39 mmol/L respectively) and LDL (−0.59 and −0.32 mmol/L respectively) cholesterol levels after three weeks. However, after six weeks, serum cholesterol levels had rebounded and only LDL cholesterol level for the canola oil group was significantly lower than baseline values. Changes in plasma fatty acid composition were similar for each of the diets at three and six weeks, which suggest adherence to the experimental protocol. However, estimated saturated fat intake, even for the subjects on the canola oil diet, was relatively high (14.0% of total energy). In a similar study with a cohort from the same study population, Seppänen-Laakso *et al.* (1993) substituted canola oil or olive oil for margarine in a select group of subjects. In this study, the decreases in plasma total or LDL cholesterol levels were not statistically significant at either three or six weeks for the group on the canola oil regimen. Although substitution with olive oil resulted in significantly lower LDL cholesterol level after three weeks, the level did not differ statistically from the baseline value after six weeks. However, the level of substitution in these studies may have been too low to produce a biological effect (15 and 18% of the dietary fat on the canola and olive oil diets respectively). Estimates of the decrease in saturated fatty acid intake were only 0.9 and 1.6% of total energy respectively, for the canola and olive oil groups.

### 6.5.3 Potential effect of phytosterols in canola oil on plasma cholesterol levels

Although the effect of unsaturated fats on plasma cholesterol levels is widely assumed to be primarily due to their fatty acid composition, there is evidence that the phytosterols (plant sterols) contained in vegetable oils also play a role (Ostlund *et al.*, 2002). Margarines and salad dressings containing added phytosterols and their hydrogenated derivatives (phytostanols) are available in several countries (Law, 2000; Jones and Raeini-Sarjaz, 2001; Yankah and Jones, 2001). Phytosterols lower plasma cholesterol levels by interfering with the absorption of dietary cholesterol and the reabsorption of endogenous biliary cholesterol. However, relatively high levels of purified phytosterol are required to produce significant lowering of plasma cholesterol levels. Although customary dietary intakes are appreciably lower than the supplemental levels needed to produce an effect, arguments can be made for the efficacy of naturally occurring dietary phytosterols (Law, 2000; Jones and Raeini-Sarjaz, 2001). Canola oil is a relatively rich source of phytosterols (see Chapter 3), which may account, in part, for the plasma cholesterol lowering effect found when canola oil is incorporated into the diet.

### 6.5.4 Effect of canola oil intake on lipid peroxidation

Although canola oil has been found to favorably alter plasma cholesterol patterns, it also has the potential to increase lipid peroxidation in the body. There is evidence that relatively high intakes of PUFA may produce unfavorable effects on the development of atherosclerosis (Eritsland, 2000). Oxidation of lipids in the LDL fraction (ox-LDL) has been proposed to play an important role in artherosclerosis (Parathasarathy and Rankin, 1992; Westhuysen, 1997). In addition, there is evidence that the LDL fraction of subjects fed a diet enriched with oleic acid was more stable to oxidation than the LDL of subjects fed a diet high in linoleic acid (Reaven et al., 1991; Abbey et al., 1993). Although canola oil contains a high level of oleic acid, it also contains a relatively high level of α-LNA, which might compromise the oleic acid. However, the fact that very little α-LNA is incorporated into tissue lipids would mitigate against this argument and research tends to support the premise that canola oil does not increase oxidative stress *in vivo*.

McDonald et al. (1995) compared the *in vitro* rate of conjugated diene formation (a measure of lipid oxidation) of the LDL fraction of subjects fed diets in which canola oil, sunflower oil and low-linolenic acid (L-LNA) canola oil replaced saturated fats in the regular diet. Oxidative stability of the LDL fraction of subjects fed the sunflower oil diet was significantly lower than that of subjects on the canola and L-LNA canola diets at the end of the 28-day study period. Moreover, the rate of conjugated diene formation in the LDL fraction from subjects on the canola oil and L-LNA canola oil diets did not differ from that in LDL from subjects who continued on their regular diet throughout the study. Turpeinen et al. (1995) and Kratz et al. (2002) reported similar findings for canola oil and sunflower oil-enriched diets. The rate of conjugated diene formation was significantly greater on a sunflower oil diet than on a canola oil diet. Likewise, the lag time before oxidation commenced was shorter for the sunflower oil group. However, Kratz et al. (2002) found that the lag time was longer and the rate of diene formation lower on an olive oil-enriched diet than on the canola oil diet. By contrast, Schwab et al. (1998) found no difference in *in vitro* oxidation of LDL from subjects fed diets varying appreciably in fatty acid composition. Mean lag times for LDL incubated with hemin and hydrogen peroxide did not differ when beef tallow, olive oil, canola oil, corn oil and rice bran oil provided two-thirds of the dietary fat. There is some question, however, concerning the relevance of *in vitro* indicators of lipid peroxidation to *in vivo* oxidative stress and, in turn, atherogenesis. In a separate study (Turpeinen et al., 1995) comparing sunflower oil and canola oil diets, no marked effect of diet on plasma levels of malondialdehyde (MDA) and conjugated dienes (*in vivo* lipid peroxidation) was observed. However, there are also questions about how well these measurements reflect *in vivo* lipid peroxidation *per se*.

A recent study by Södergren et al. (2001) assessed the effect of canola oil on *in vivo* lipid peroxidation by measuring plasma and urinary levels of 8-*iso*-prostaglandin $F_{2\alpha}$ (8-*iso*-PGF$_{2\alpha}$), a promising biomarker of *in vivo* lipid peroxidation. The effect of a canola oil-based diet was compared to that of a control diet containing a high level of saturated fatty acids (20% of total energy) in a 4-week crossover design. No differences in plasma or urinary levels of 8-*iso*-PGF$_{2\alpha}$ were found between the test diets. In addition, the levels after four weeks on the test diets did not differ from the baseline values. By contrast, Turpeinen et al. (1998) found that the urinary level of 8-*iso*-PGF$_{2\alpha}$ increased on a linoleic acid-enriched diet and was higher than that for subjects acting as controls, who consumed their habitual diet. The level of 8-*iso*-PGF$_{2\alpha}$ on an oleic acid-enriched diet did not differ from baseline value or from that of the control group. The results of these studies suggest that canola oil does not increase *in vivo* oxidative stress, whereas a diet containing a high level of PUFA increases oxidative stress.

In addition to the effect of dietary fat on the unsaturation and susceptibility of tissue lipids to oxidation, the level of antioxidants, in particular vitamin E, also could play a potential role in *in vivo* lipid oxidation. Although the intake of vitamin E (viz. α-tocopherol) was higher on the canola oil diet than on the saturated fat diet (15.3 vs 11.5 mg/day) in the study by Södergren *et al.* (2001), there were no differences in plasma α-tocopherol levels between baseline values and those after four weeks on the test diets. Similarly, McDonald *et al.* (1995) found no differences among treatment groups in plasma α-tocopherol levels nor did the values for the treatment groups differ from baseline values in the study by Turpeinen *et al.* (1995), even though canola oil and sunflower oil are rich sources of α-tocopherol. Although vegetable oils are the primary sources of vitamin E in the customary diet, the level ingested by the average person is appreciably less than that of individuals taking a vitamin E supplement. Furthermore, as pointed out by Schwab *et al.* (1998), variations in LDL α-tocopherol levels of subjects not taking α-tocopherol supplements do not correlate with *in vitro* LDL oxidation.

### 6.5.5 Canola oil and thrombogenesis

The effect of type and amount of dietary fat on the risk factors for atherosclerosis has been extensively studied over the past 40 years. By contrast, thrombogenesis, clot formation, and its relationship to coronary heart disease has received attention only over the past 12 to 15 years. The marked difference in CVD between Greenland Eskimos and Danes (Dyerberg, 1986) led to an interest in the antithrombotic effect of *n*-3 long chain PUFA (*n*-3 LC-PUFA) of fish oils, in particular, eicosapentaenoic acid (EPA; 20:5 *n*-3). This interest in *n*-3 LC-PUFA raised questions about the possible role of α-LNA in this process which, in turn, led to the finding that α-LNA was inversely correlated with platelet aggregation in a group of free-living subjects (Renaud *et al.*, 1986). However, there was considerable controversy surrounding the importance of α-LNA to platelet function (Kinsella, 1988); a recent study (Pawlosky *et al.*, 2001) confirmed the earlier contention that humans do not readily convert α-LNA to EPA. Nevertheless, the general hypothesis holds that the effect of EPA on thrombosis is related to its effect on eicosanoid metabolism, namely, the balance between thromboxane, a vasoconstrictor and platelet-aggregating agent, and prostacyclin, a vasodilator and inhibitor of platelet aggregation.

### 6.5.6 Effect of canola oil on fatty acid composition of plasma and platelet phospholipids

Metabolism of *n*-6 and *n*-3 fatty acids, namely, desaturation and elongation, share a common pathway (Fig. 6.1). In addition, the conversion of arachidonic acid (AA) (*n*-6) and EPA (*n*-3) to eicosanoids such as thromboxanes and prostacyclins also shares common metabolic pathways. Hence, competition among substrates for the enzymes in these pathways is a distinct possibility. In fact, Emken *et al.* (1994) found that a high dietary level of LA (30 vs 15 g) decreased the conversion of α-LNA to EPA. Other studies with human subjects bear out the influence of dietary LA on α-LNA metabolism. Mantzioris *et al.* (1994) found plasma EPA levels were severalfold higher on a high α-LNA, low LA diet than on a low α-LNA, high LA diet. In fact, there is evidence that the α-LNA/LA ratio had a greater influence on plasma EPA levels than the absolute levels of α-LNA and LA in the diet (Chan *et al.*, 1993).

A number of studies have shown that the ingestion of canola oil alters the fatty acid composition of plasma and platelet lipids, in particular the

**Diet**
(Vegetable fats and oils)

```
              /              \
      18:2 n-6            18:3 n-3
   (Linoleic acid)     (Linolenic acid)
          ↓    Δ⁶–Desaturase    ↓
      18:3 n-6             18:4 n-3
          ↓      Elongation    ↓
      20:3 n-6             20:4 n-3
          ↓    Δ⁵–Desaturase   ↓
```

| Diet | 20:4 *n*-6 | 20:5 *n*-3 | Diet |
|---|---|---|---|
| (Animal fats) → | (Arachidonic acid) | (Eicosapentaenoic acid) ← | (Fish oils) |

| Thromboxane A₂ | Prostaglandin I₂ | Thromboxane A₃ | Prostaglandin I₃ |
|---|---|---|---|
| (Strong platelet aggregating agent) | (Inhibitor of platelet aggregation) | (Weak platelet aggregating agent) | (Inhibitor of platelet aggregation) |

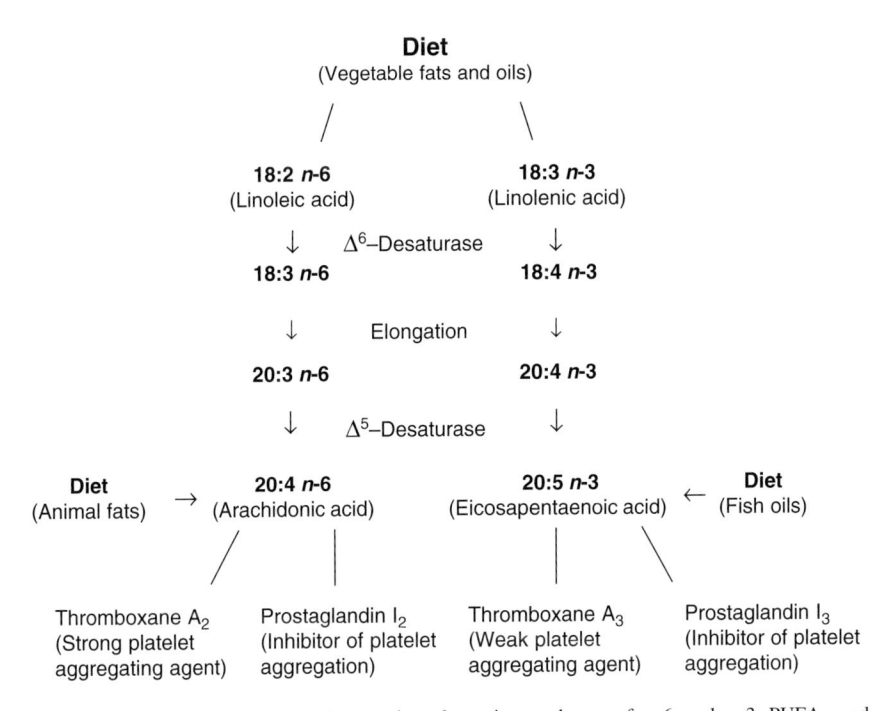

**Figure 6.1** Dietary sources and desaturation–elongation pathway of *n*-6 and *n*-3 PUFA, and thromboxane and prostacyclin species formed from arachidonic and eicosapentaenoic acids.

phospholipids, which contain the LC-PUFA required for eicosanoid synthesis. In general, phospholipid fatty acid patterns reflect the fatty acid composition of the diet (Chan et al., 1993; Lichtenstein et al., 1993; Nydahl et al., 1994; Schwab et al., 1998; Li et al., 1999). Canola oil and olive oil resulted in higher levels of oleic acid (OA; 18:1 n-9) whereas oils rich in LA, such as soybean oil, corn oil, sunflower oil and safflower oil, resulted in higher levels of LA. Canola oil and soybean oil also resulted in higher levels of α-LNA. However, very little α-LNA is incorporated into phospholipids. Even when the diet contained appreciable quantities of α-LNA, achieved by incorporating flaxseed oil, α-LNA made up less than 1% of the total fatty acids in plasma and platelet phospholipids (Sanders and Younger, 1981; Chan et al., 1993). By contrast, LA is readily incorporated in plasma or platelet phospholipids; it accounts for approximately 25% of the total fatty acids in the plasma phospholipid fraction. Even when subjects were fed relatively low LA diets (Valsta et al., 1996), it made up over 20% of the total fatty acids. On the other hand, dietary α-LNA has been found to alter the LC-PUFA composition of phospholipids, in particular AA and EPA. Several groups have reported lower levels of AA and higher levels of EPA on a canola oil diet (Table 6.3). In general, EPA levels increased from baseline values on the canola oil diet. Likewise, EPA levels were higher on a canola oil diet than on a PUFA-enriched diet. By contrast, similar comparisons for AA were less consistent. However, there was an overall trend to lower levels on a canola oil diet.

However, not all groups reported lower levels of AA or higher levels of EPA in response to canola oil. Corner et al. (1990), Mutanen et al. (1992), Nydahl et al. (1994), Seppänen-Laakso et al. (1992) and Li et al. (1999) did not find a change in baseline values for AA. Chan et al. (1993), Gustafsson et al. (1994), Nydahl et al. (1994) and Li et al. (1999) also found that the levels of AA on a canola oil diet did not differ from the levels on a PUFA-enriched diet. Similarly, Corner et al. (1990), Kwon et al. (1991) and Li et al. (1999) found that EPA levels did not differ from baseline values when canola oil was fed. Seppänen-Laakso et al. (1992) also found that EPA levels did not differ from baseline values after six weeks on a canola oil diet although the levels were higher than baseline values at three weeks. However, only Kwon et al. (1991) reported no differences in EPA levels between a canola oil diet and a PUFA-enriched diet. Part of the explanation for the seeming differences among studies may relate to the balance between fatty acids. Chan et al. (1993) found that the fatty acid composition of plasma and platelet phospholipid fractions varied not only with the level of α-LNA in the diet but with its ratio to LA (i.e. α-LNA/LA ratio). Other factors such as level of inclusion of the test fats in the diet, the length of time the diets were fed and the design of the study may also explain some of the observed variability among studies.

**Table 6.3** Effect of canola oil on levels of plasma and platelet phospholipids, arachidonic acid (20:4 n-6) and eicosapentaenoic acid (20:5 n-3), when compared to baseline values or levels on a PUFA-enriched diet

| Reference | Arachidonic acid (20:4 n-6) Baseline value | Arachidonic acid (20:4 n-6) PUFA diet | Eicosapentaenoic acid (20:5 n-3) Baseline value | Eicosapentaenoic acid (20:5 n-3) PUFA diet |
|---|---|---|---|---|
| Renaud et al. (1986) | ⇓[a] | Not det'd[b] | ⇑ | Not det'd |
| Corner et al. (1990) | ⇔ | ↓/↔[c] | ⇔ | ⇑⇑[d] |
| Weaver et al. (1990) | ⇒ | ↓/↔ | ⇐ | ⇑ |
| Kwon et al. (1991) | ⇒ | ⇓⇓ | ⇔ | ↔ |
| Mutanen et al. (1992) | ⇔ | ⇓⇓ | Not reported | Not reported |
| Chan et al. (1993) | Not reported | ↔ | Not reported | ⇑⇑ |
| Nydahl et al. (1994) | ⇔ | ↔ | ⇑ | ⇑⇑ |
| Seppänen-Laakso et al. (1992) 3 weeks | ⇔ | Mixed fat[e] | ⇔ | Mixed fat[e] |
| 6 weeks | ⇒ | Mixed fat[e] | ⇐ | Mixed fat[e] |
| Gustafsson et al. (1994) | ⇒ | ↔ | ⇐ | ⇑⇑ |
| Li et al. (1999) | ↔ | ↔ | ⇔ | ⇑⇑ |

[a] Change from baseline values on canola diet: ⇓ – decrease, ⇑ – increase, ⇔ – no change.
[b] Not determined.
[c] Canola vs PUFA diet: ↓/↔ – lower for phosphatidylethanolamine but no differences for other phospholipid fractions.
[d] Canola diet vs PUFA diet: ⇑⇑ – higher, ⇓⇓ – lower, ↔ – no difference.
[e] Compared to a diet enriched with a mixture of canola oil, sunflower oil, coconut oil and partially hydrogenated soybean oil; comparison between diets not reported.

### 6.5.7 Effect of canola oil on clotting time and factors involved in clot formation

In addition to changes in plasma and platelet fatty acid composition, canola oil has been found to alter several parameters related to clot formation (Table 6.4). McDonald *et al.* (1989) found that clotting time was increased over baseline values on a canola oil diet. Mean clotting time also was increased on a sunflower oil diet but the values did not differ statistically from baseline values. McDonald *et al.* (1989) also found that the production of prostacyclin (an antiaggregatory eicosanoid) during clot formation was increased on the canola oil diet, which coincided with the increased clotting time. Several groups also have assessed the effect of canola oil on thrombogenesis by measuring *in vitro* platelet aggregation. Renaud *et al.* (1986) and Kwon *et al.* (1991) found that canola oil reduced platelet aggregation whereas Freese *et al.* (1994) and Li *et al.* (1999) found that canola oil had no effect on *in vitro* platelet aggregation. Although Freese *et al.* (1994) found that platelet aggregation did not differ from baseline values, it was significantly lower than the values for a high-oleic sunflower oil diet, which resulted in enhanced aggregation. Kwon *et al.* (1991) also found that ATP production during platelet aggregation was decreased and the mean lag time before aggregation commenced was increased on a canola oil diet. They also found that the lag time in aggregation was negatively correlated with plasma thromboxane concentrations. Mutanen *et al.* (1992), on the other hand, reported enhanced platelet aggregation on both a canola oil and a sunflower oil diet. Of interest, they found that enhanced platelet aggregation was associated with increased *in vitro* platelet thromboxane production. The results of this study may have been confounded by a carryover effect associated with the crossover design of the study. Although the effect of canola oil on platelet activity and clot formation is not as well established as its favorable effects on plasma cholesterol and lipoprotein levels, there is evidence to suggest that canola oil may favorably alter thrombogenesis.

**Table 6.4** Effect of canola oil on clotting time and factors related to clot formation

| Parameter | Effect of canola oil on parameter | Reference |
| --- | --- | --- |
| Clotting time | Increased | McDonald *et al.* (1989) |
| *In vitro* platelet aggregation | Reduced | Renaud *et al.* (1986), Kwon *et al.* (1991) |
|  | No effect[a] | Freese *et al.* (1994), Li *et al.* (1999) |
|  | Enhanced | Mutanen *et al.* (1992) |
| Prostacyclin production | Increased | McDonald *et al.* (1989) |
|  | No effect[b] | Chan *et al.* (1993) |

[a] Freese *et al.* (1994) found that the levels did not differ from baseline values but aggregation was significantly lower on the canola oil diet than a high-oleic sunflower oil diet.
[b] Mean prostacyclin value on a canola oil diet was intermediate between values on a sunflower/olive oil diet and a sunflower/olive/flaxseed oil diet, which differed ($p < 0.05$).

## 6.5.8 Canola oil and cardiac arrhythmia

Sudden death due to fatal cardiac arrhythmia is a common accompaniment to myocardial infarction (MI). However, it is only relatively recently that the effect of diet on cardiac arrhythmia has received attention. There is evidence that $n$-3 LC-PUFA prevent induced arrhythmia in animals (McLennan *et al.*, 1988; McLennan *et al.*, 1992, 1993; Billman *et al.*, 1997) and that they may prevent fatal arrhythmia in humans (Burr *et al.*, 1989; Albert *et al.*, 1998). There is also evidence that α-LNA is equally as antiarrhythmic as the $n$-3 LC-PUFA in fish oils (viz. EPA and docosahexaenoic acid (DHA)), when infused intravenously as free fatty acids just before inducing ischemia in a dog model of cardiac sudden death (Billman *et al.*, 1999). Similarly, Siebert *et al.* (1993) found that canola oil was equally as effective as fish oil in reducing mortality and the pronounced arrhythmia produced in a rat model subjected to coronary occlusion and reperfusion. Canola oil was also found to be the only vegetable oil that effectively reduced arrhythmia in rats fed diets containing olive oil, sunflower oil, soybean oil and canola oil for 12 weeks prior to inducing cardiac arrhythmia (McLennan and Dallimore, 1995). Although soybean oil provided nearly the same amount of α-LNA as canola oil (7.3 vs 8.0% of total dietary fatty acids), it provided much higher levels of LA (50.2 vs 20.4% respectively). These differences were reflected in the $n$-3/$n$-6 ratios in the myocardial phospholipids. The sum of the $n$-3 fatty acid and the $n$-3/$n$-6 ratio in myocardial phospholipids was significantly higher ($p < 0.05$) on the canola oil diet than on the other three diets. These findings coincide with the observation by Chan *et al.* (1993) that the ratio of α-LNA/LA in the diet appears to play an important role in the incorporation of $n$-3 LC-PUFA into phospholipids. These results from animal studies suggest that inclusion of canola oil in the diet is effective in reducing cardiac events leading to life-threatening arrhythmia.

## 6.6 The Lyon Diet Heart Study: the canola oil connection

Cardiovascular disease is a complex, multifactorial disease and the manner in which dietary fat alters its development has not been completely resolved. Nonetheless, the benefit of reducing the level of saturated fat in the customary diet to plasma cholesterol and lipoprotein levels is well established. In addition, there is epidemiological evidence and evidence from controlled clinical studies that dietary MUFA (viz. oleic acid) favorably affect a number of risk factors for CVD (Kris-Etherton, 1999). There is also mounting evidence that higher intakes of α-LNA provide protection against CVD. Several epidemiological studies support the benefits of higher intakes of α-LNA. Dolecek (1992), e.g., in a re-evaluation of the multiple risk factor intervention trial (MRFIT) database, found that there was an inverse relationship between α-LNA intake and mortality from coronary heart disease (CHD), especially when α-LNA intakes were expressed as a percentage of total energy intake. Likewise, examination of the dietary intake data from the Nurses' Health Study (Hu *et al.*, 1999) indicated that higher intakes of α-LNA protect against fatal ischemic heart disease (IHD). The intakes of α-LNA for the 76 283 women who entered the study were divided into quintiles that ranged from a low of 0.71 g to a high of 1.36 g per day. During the 10-year follow-up from 1984 to 1994, 232 cases of fatal IHD and 597 cases of nonfatal MI were documented. The relative risks (RR) of fatal IHD for the lowest to highest quintile were: 1.0, 0.99, 0.90, 0.67 and 0.55 (for trend $p=0.01$). By contrast, only a weak, non-significant trend towards a reduced risk was found for non-fatal MI. The poor results of dietary intervention studies designed to prevent recurrence of MI following a first heart attack motivated de Lorgeril *et al.* (1994) to investigate the effect of the so-called *Mediterranean diet* on the prevention of recurrent MI. The Mediterranean diet was adopted because of the low mortality rate from CHD for the residents of the island of Crete in the Seven Countries Study. In addition to being a high-fat diet, the Cretan *Mediterranean* diet was also relatively high in α-LNA. The study, called the Lyon Diet Heart Study, was a randomized secondary prevention trial involving subjects who were enrolled within six months after a first MI. The study compared the effect of a Mediterranean-type diet, in which olive oil and canola oil were the prescribed dietary oils and butter was replaced by canola oil margarine, with the customary post-infarct, prudent diet. The subjects were followed for a mean of 27 months during which there were 16 cardiac deaths on the prudent diet compared to 3 on the Mediterranean diet and 17 non-fatal MI on the prudent diet vs 5 on the Mediterranean diet. Although there was a markedly lower recurrence of cardiac events and overall mortality on the Mediterranean diet, there were no differences between the groups in usual risk factor for CVD, such as plasma cholesterol and lipoprotein levels, blood pressure and body weight. Because of the statistical significance of the result, the study was stopped. However, de Lorgeril *et al.* (1999) conducted an extended follow-up of the subjects over an additional 19 months (hence, mean study period of 46 months). After nearly 4 years, there were three times as many major primary coronary events (cardiac deaths plus non-fatal MI) on the prudent diet as on the Mediterranean diet (44 vs 14; 19 vs 6 cardiac deaths; and 25 vs 8 non-fatal MI respectively). As in the original report, the marked decrease in the recurrence of cardiac events was independent of any differences between the groups in blood lipid or lipoprotein levels. Although one of the aims of the diet modification was to increase the α-LNA intake, there were several differences between the diets. In addition to a higher α-LNA intake, there was also a higher oleic acid and lower LA, saturated fat and cholesterol intakes on the Mediterranean diet. Nonetheless, only α-LNA, of the blood lipid analyses performed two months after randomization of the subjects to the diets, was associated with the improved prognosis on the Mediterranean diet. Although the results of this study are very encouraging and though canola oil was a major source of fat in the Mediterranean-type diet, there is need to confirm these findings in further clinical studies.

## 6.7 Summary

- The development of canola, which was aimed at preserving the favorable agronomic features of its parent, rapeseed, while eliminating the potential health concerns linked to the high-erucic acid level in rapeseed, resulted in an oil that meets many of the characteristics sought in dietary fats. The low level of saturated fatty acids, high level of oleic acid, relatively high level of α-LNA and a favorable α-LNA to LA balance are consistent with current dietary recommendations.
- Substitution of canola oil for saturated fat in the diet has been found to be equal to similar substitution with PUFA in lowering plasma total and LDL cholesterol levels. Incorporation of canola oil into the diet has also been found to favorably alter the fatty acid patterns of plasma and platelet phospholipids, in particular the production of higher levels of $n$-3 LC-PUFA, EPA.
- There is also evidence that canola oil results in a delay in clot formation, which has positive implication for CVD.
- In addition, studies have shown that canola oil prevents induced cardiac arrhythmia in animal models. This finding is of particular interest because cardiac arrhythmia is a frequent cause of death during MI.
- However, only the positive effect of canola oil on plasma lipid patterns has been well established by clinical trials. The other nutritional benefits attributed to canola oil need additional confirmation.

## References

Abbey, M., Belling, G.B., Noakes, M., Hirata, F. and Nestel, P. (1993) Oxidation of low-density lipoproteins: intraindividual variability and the effect of dietary linoleate supplementation. *American Journal of Clinical Nutrition*, **57**, 391–398.

Albert, C.M., Hennekens, C.H., O'Donnell, C.J., Ajani, U.A., Carey, V.J., Willett, W.C. and Manson, J.E. (1998) Fish consumption and risk of sudden cardiac death. *Journal of the American Medical Association*, **279**, 23–28.

Barr, S.L., Ramakrishnan, R., Johnson, C., Hollerman, S., Dell, R.B. and Ginsberg, H.N. (1992) Reducing total dietary fat without reducing saturated fatty acids does not significantly lower plasma cholesterol concentrations in normal males. *American Journal of Clinical Nutrition*, **55**, 675–681.

Becker, C.C., Lund, P., Hølmer, G., Jensen, H. and Sandström, B. (1999) Effect of butter oil blends with increased concentrations of stearic, oleic and linolenic acid on blood lipids in young adults. *European Journal of Clinical Nutrition*, **53**, 535–541.

Bierenbaum, M.L., Reichstein, R.P., Watkins, T.R., Maginnis, W.P. and Geller, M. (1991) Effects of canola oil on serum lipids in humans. *Journal of the American College of Nutrition*, **10**, 228–233.

Billman, G.E., Kang, J.X. and Leaf, A. (1997) Prevention of ischemia-induced cardiac sudden death by n-3 polyunsaturated fatty acids in dogs. *Lipids*, **32**, 1161–1168.

Billman, G.E., Kang, J.X. and Leaf, A. (1999) Prevention of sudden cardiac death by dietary pure ⍵-3 polyunsaturated fatty acids in dogs. *Circulation*, **99**, 2452–2457.

Burr, M., Fehily, A.M., Gilbert, J.F., Rogers, S., Holliday, R.M., Sweetnam, P.M., Elwood, P.C. and Deadman, N.M. (1989) Effects of changes in fat, fish, and fibre intakes on death and myocardial reinfarction: diet and reinfarction trial (DART). *Lancet*, **2**, 757–761.

Chan, J.K., Bruce, V.M. and McDonald, B.E. (1991) Dietary α-linolenic acid is as effective as oleic acid and linoleic acid in lowering blood cholesterol in normolipidemic men. *American Journal of Clinical Nutrition*, **53**, 1230–1234.

Chan, J.K., McDonald, B.E., Gerrard, J.M., Bruce, V.M., Weaver, B.J. and Holub, B.J. (1993) Effect of dietary α-linolenic acid and its ratio to linoleic acid on platelet and plasma fatty acids and thrombogenesis. *Lipids*, **28**, 811–817.

Corner, E.J., Bruce, V.M. and McDonald, B.E. (1990) Accumulation of eicosapentaenoic acid in plasma phospholipids of subjects fed canola oil. *Lipids*, **25**, 598–601.

de Lorgeril, M., Renaud, S., Mamelle, N., Salen, P., Martin, J.-L., Monjaud, I., Guidollet, J., Touboul, P. and Delaye, J. (1994) Mediterranean alpha-linolenic acid-rich diet in secondary prevention of coronary heart disease. *Lancet*, **343**, 1454–1459.

de Lorgeril, M., Salen, P., Martin, J.-L., Monjaud, I., Delaye, J. and Mamelle, N. (1999) Mediterranean diet, traditional risk factors, and the rate of cardiovascular complications after myocardial infarction: final report of the Lyon Diet Heart Study. *Circulation*, **99**, 779–785.

Dolecek, T.A. (1992) Epidemiological evidence of relationships between dietary polyunsaturated fatty acids and mortality in the Multiple Risk Factor Intervention Trial. *Proceeding of the Society for Experimental Biology and Medicine*, **200**, 177–182.

Dyerberg, J. (1986) Linolenate-derived polyunsaturated fatty acids and prevention of atherosclerosis. *Nutrition Reviews*, **44**, 125–134.

Emken, E.A., Adlof, R.O. and Gulley, R.M. (1994) Dietary linoleic acid influences desaturation and acylation of deuterium-labeled linoleic and linolenic acids in young adult males. *Biochimica et Biophysica Acta*, **1213**, 277–288.

Eritsland, J. (2000) Safety consideration of polyunsaturated fatty acids. *American Journal of Clinical Nutrition*, **71**, 197S–201S.

Freese, R., Mutanen, M., Valsta, L.M. and Salminen, I. (1994) Comparison of the effects of two diets rich in monounsaturated fatty acids differing in their linoleic/α-linolenic acid ratio on platelet aggregation. *Thrombosis and Haemostasis*, **71**, 73–77.

Gustafsson, I.-B., Vessby, B., Öhrvall, M. and Nydahl, M. (1994) A diet rich in monounsaturated rapeseed oil reduces the lipoprotein cholesterol concentration and increases the relative content of n-3 fatty acids in hyperlipidemic subjects. *American Journal of Clinical Nutrition*, **59**, 667–674.

Hu, F.B., Stampfer, M.J., Manson, J.E., Rimm, E.B., Wolk, A., Colditz, G.A., Hennekens, C.H. and Willett, W.C. (1999) Dietary intake of α-linolenic acid and risk of fatal ischemic heart disease among women. *American Journal of Clinical Nutrition*, **69**, 890–897.

Jones, P.J. and Raeini-Sarjaz, M. (2001) Plant sterols and their derivatives: the current spread of results. *Nutrition Reviews*, **59**, 21–24.

Karvonen, H.M., Tapola, N.S., Uusitupa, M.I. and Sarkkinen, E.S. (2002) The effect of vegetable oil-based cheese on serum total and lipoprotein lipids. *European Journal of Clinical Nutrition*, **56**, 1094–1101.

Kinsella, J.E. (1988) Food lipids and fatty acids: importance of food quality, nutrition and health. *Food Technology*, **42**(10), 126–142.

Kratz, M., Cullen, P., Kannenberg, F., Kassner, A., Fobker, M., Abuja, P.M., Assman, G. and Wahrburg, U. (2002) Effects of dietary fatty acids on the composition and oxidizability of low-density lipoprotein. *European Journal of Clinical Nutrition*, **56**, 72–81.

Kris-Etherton, P.M. (1999) Monounsaturated fatty acids and risk of cardiovascular disease. *Circulation*, **100**, 1253–1258.

Kwon, J.-S., Snook, J.T., Wardlaw, G.M. and Hwang, D.H. (1991) Effects of diets high in saturated fatty acids, canola oil, or safflower oil on platelet function, thromboxane $B_2$ formation, and fatty acid composition of platelet phospholipids. *American Journal of Clinical Nutrition*, **54**, 351–358.

Law, M. (2000) Plant sterol and stanol margarines and health. *British Medicinal Journal*, **320**, 861–864.

Li, D., Sinclair, A., Wilson, A., Nakkote, S., Kelly, F., Abedin, L., Mann, N. and Turner, A. (1999) Effect of dietary α-linolenic acid on thrombic risk factors in vegetarian men. *American Journal of Clinical Nutrition*, **69**, 872–882.

Lichtenstein, A.H., Ausman, L.M., Carrasco, W., Jenner, J.L., Gualatieri, L.J., Goldin, B.R., Ordovas, J.M. and Schaefer, E.J. (1993) Effect of canola, corn, and olive oils on fasting and postprandial lipoproteins in humans as part of a National Cholesterol Education Program Step 2 Diet. *Arteriosclerosis and Thrombosis*, **13**, 1533–1542.

Mantzioris, E., James, M.J., Gibson, R.A. and Cleland, L.G. (1994) Dietary substitution with an α-linolenic acid-rich vegetable oil increases eicosapentaenoic acid concentrations in tissues. *American Journal of Clinical Nutrition*, **59**, 1304–1309.

Matheson, B., Walker, K.Z., Taylor, D. McD, Peterkin, R., Lugg, D. and O'Dea, K. (1996) Effect on serum lipids of monounsaturated oil and margarine in the diet of an Antarctic expedition. *American Journal of Clinical Nutrition*, **63**, 933–938.

Mattson, F.H. and Grundy, S.M. (1985) Comparison of the effect of dietary saturated, monounsaturated, and polyunsaturated fatty acids on plasma lipids and lipoproteins in man. *Journal of Lipid Research*, **26**, 575–581.

McDonald, B.E., Bruce, V.M., LeBlanc, E.L. and King, D.J. (1974) Effect of rapeseed oil on serum lipid patterns and blood hematology of young men. *Proceedings 4. Internationaler Rapskongress*, Giessen, Germany, pp. 693–700.

McDonald, B.E., Gerrard, J.M., Bruce, V.M. and Corner, E.J. (1989) Comparison of the effect of canola oil and sunflower oil on plasma lipids and lipoproteins and on *in vivo* thromboxane $A_2$ and prostacyclin production in healthy young men. *American Journal of Clinical Nutrition*, **50**, 1382–1388.

McDonald, B.E., Bruce, V.M., Murthy, V.G. and Latta, M. (1995) Assessment of the nutritional properties of low-linolenic acid (LL) canola oil with human subjects. *Proceedings of the Ninth International Rapeseed Congress*, Cambridge, UK, pp. 849–851.

McLennan, P.L. and Dallimore, J.A. (1995) Dietary canola oil modifies myocardial fatty acids and inhibits cardiac arrhythmias in rats. *Journal of Nutrition*, **125**, 1003–1009.

McLennan, P.I., Abeywaradena, M.Y. and Charnock, J.S. (1988) Dietary fish oil prevents ventricular fibrillation following coronary artery occlusion and reperfusion. *American Heart Journal*, **116**, 709–717.

McLennan, P.I., Bridle, T.M., Abeywaradena, M.Y. and Charnock, J.S. (1992) Dietary lipid modulation of ventricular fibrillation threshold in the marmoset monkey. *American Heart Journal*, **123**, 1555–1561.

McLennan, P.I., Bridle, T.M., Abeywaradena, M.Y. and Charnock, J.S. (1993) Comparative efficacy of *n*-3 and *n*-6 polyunsaturated fatty acids in modulating ventricular fibrillation in marmoset monkeys. *American Journal of Clinical Nutrition*, **58**, 666–669.

Mensink, R.P. and Katan, M.B. (1989) Effect of a diet enriched in monounsaturated or polyunsaturated fatty acids on levels of low-density and high-density lipoprotein cholesterol in healthy women and men. *New England Journal of Medicine*, **321**, 436–441.

Miettinen, T.A. and Vanhanen, H. (1994) Serum concentration and metabolism of cholesterol during rapeseed oil and squalene feeding. *American Journal of Clinical Nutrition*, **59**, 356–363.

Mutanen, M., Freese, R., Valsta, L.M., Ahola, I. and Ahlostrôm, A. (1992) Rapeseed oil and sunflower oil diets enhance platelet *in vitro* aggregation and thromboxane production in healthy men when compared with milk fat or habitual diets. *Thrombosis and Haemostasis*, **67**, 352–356.

Nydahl, M., Gustafsson, I.-B., Öhrvall, M. and Vessby, B. (1994) Similar serum lipoprotein cholesterol concentrations in healthy subjects on diets enriched with rapeseed and with sunflower oil. *European Journal of Clinical Nutrition*, **48**, 128–137.

Ostlund, Jr, R.E., Racette, S.B. and Stenson, W.F. (2002) Effects of trace components of dietary fat on cholesterol metabolism: phytosterols, oxysterols, and squalene. *Nutrition Reviews*, **60**, 349–359.

Parathasarathy, S. and Rankin, S.M. (1992) Role of oxidized low density lipoprotein in atherogenesis. *Progress in Lipid Research*, **31**, 127–143.

Pawlosky, R.J., Hibbeln, J.R., Novotny, J.A. and Salem, Jr, N. (2001) Physiological compartmental analysis of α-linolenic acid metabolism in adult humans. *Journal of Lipid Research*, **42**, 1257–1265.

Przybylski, R. and Mag, T. (2002) Canola/rapeseed oil. In: *Vegetable Oils in Food Technology: Composition, Properties and Uses* (ed. F.D. Gunstone), Blackwell Publishing Ltd, Oxford, UK, pp. 98–127.

Reaven, P., Parathasarathy, S., Grasse, B.J., Miller, E., Almazan, F., Mattson, F.H., Khoo, J.C., Steinberg, D. and Witzum, J.L. (1991) Feasibility of using an oleate-rich diet to reduce the susceptibility of low-density lipoprotein to oxidative modification in humans. *American Journal of Clinical Nutrition*, **54**, 701–706.

Renaud, S., Morazain, R., Godsey, F., Dumont, E., Thevenon, G., Martin, J.L. and Mendy, F. (1986) Nutrients, platelet function and composition in nine groups of French and British farmers. *Atherosclerosis*, **60**, 37–48.

Sanders, T.A.B. and Younger, K.M. (1981) The effect of dietary supplements of ω3 polyunsaturated fatty acids on the fatty acid composition of platelets and plasma choline phosphoglycerides. *British Journal of Nutrition*, **45**, 613–616.

Sarkkinen, E.S., Uusitupa, M.I.J., Pietinen, P., Aro, A., Ahola, I., Penttilä, I., Kervinen, K. and Kesäniemi, Y.A. (1994) Long-term effects of three fat-modified diets in hypercholesterolemic subjects. *Arteriosclerosis*, **105**, 9–23.

Schwab, U.S., Vogel, S., Lammi-Keefe, C.J., Ordovas, J.M., Schaefer, E.J., Li, Z., Ausman, L.M., Gualtieri, L., Goldin, B.R., Furr, H.C. and Lichtenstein, A.H. (1998) Varying dietary fat type of reduced-fat diets has little effect on the susceptibility of LDL to oxidative modification in moderately hypercholesterolemic subjects. *Journal of Nutrition*, **128**, 1703–1709.

Seppänen-Laakso, T., Vanhanen, H., Laakso, I., Kohtamäki, H. and Viikari, J. (1992) Replacement of butter on bread by rapeseed oil and rapeseed oil-containing margarine: effects on plasma fatty acid composition and serum cholesterol. *British Journal of Nutrition*, **68**, 639–654.

Seppänen-Laakso, T., Vanhanen, H., Laakso, I., Kohtamäki, H. and Viikari, J. (1993) Replacement of margarine on bread by rapeseed and olive oils: effects on plasma fatty acid composition and serum cholesterol. *Annals of Nutrition and Metabolism*, **37**, 161–174.

Siebert, B.D., McLennan, P.L., Woodhouse, J.A. and Charnock, J.S. (1993) Cardiac arrhythmia in rats in response to dietary n-3 fatty acids from red meat, fish oil and canola oil. *Nutrition Research*, **13**, 1407–1418.

Södergren, E., Gustafsson, I.-B., Basu, S., Nourooz-Zadeh, J., Nälsen, C., Turpeinen, A., Berglund, L. and Vessby, B. (2001) A diet containing rapeseed oil-based fats does not increase lipid peroxidation in humans when compared to a diet rich in saturated fatty acids. *European Journal of Clinical Nutrition*, **55**, 922–931.

Stefansson, B.R., Hougen, F.W. and Downey, R.K. (1961) The isolation of rape plants with seed oil free from erucic acid. *Canadian Journal of Plant Science*, **41**, 218–219.

Turpeinen, A.M., Alfthan, G., Valsta, L., Hietanen, E., Salonen, J.T., Schunk, H., Nyyssönen, K. and Mutanen, M. (1995) Plasma and lipoprotein lipid peroxidation in humans on sunflower and rapeseed oil diets. *Lipids*, **30**, 485–490.

Turpeinen, A.M., Basu, S. and Mutanen, M. (1998) A high linoleic acid diet increases oxidative stress *in vivo* and affects nitric oxide metabolism in humans. *Prostaglandins, Leukotrienes and Essential Fatty Acids*, **59**, 229–233.

Valsta, L.M., Jauhianinen, M., Aro, M.B., Katan, M.B. and Mutanen, M. (1992) Effects of a monounsaturated rapeseed oil and a polyunsaturated sunflower oil diet on lipoprotein levels in humans. *Arteriosclerosis and Thrombosis*, **12**, 50–57.

Valsta, L.M., Salminen, I., Aro, A. and Mutanen, M. (1996) α-Linolenic acid in rapeseed oil partly compensates for the effect of fish restriction on plasma long chain n-3 fatty acids. *European Journal of Clinical Nutrition*, **50**, 229–235.

Wardlaw, G.M., Snook, J.T., Lin, L.-C., Puangco, M. and Kwon, J.S. (1991) Serum lipid and apolipoprotein concentrations in healthy men on diets enriched in either canola oil or safflower oil. *American Journal of Clinical Nutrition*, **54**, 104–110.

Weaver. B.J., Corner, E.J., Bruce, V.M., McDonald, B.E. and Hould, B.J. (1990) Dietary canola oil: Effect on the accumulation of eicosapentaenoic acid in the alkenylacyl fraction of human platelet ethanolamine phosphoglyceride. *American Journal of Clinical Nutrition*, 51, 594–598.

Westhuysen, J. (1997) The oxidation hypothesis of atherosclerosis: an update. *Annals of Clinical and Laboratory Science*, **27**, 1–10.

Yankah, V.V. and Jones, P.J.H. (2001) Phytosterols and health implications – commercial products and their regulation. *Inform*, **12**, 1011–1016.

# 7 Non-food uses
Kerr Walker

## 7.1 Introduction

This chapter considers the non-food uses of rapeseed oil. Conventional rapeseed varieties produce an oil well suited to a number of different non-food or industrial uses. In addition, some specialist varieties are available, which have been bred to produce oil with a fatty acid profile differing from the conventional food oil. These varieties have a number of differing specialist uses. In recent years, advanced breeding techniques, including genetic engineering, have permitted the development of new rape varieties with an increasingly wide range of oil types. In these cases, rape with its well-proven agronomy and cropping reliability is used as the *vehicle* for delivering new oils that might not be so easily produced from novel crops. In addition, oils normally not readily available from crop plants may also be produced (see Chapter 8).

Conventional rapeseed oil can be used in the manufacture of lubricants, surfactants, paints, inks, polymers and pharmaceuticals. Rapeseed oil can also be used as a fuel in diesel engines and it is this use that has seen the most dramatic expansion over the last ten years.

## 7.2 Biodiesel

Most vegetable oils can be used as fuels in the diesel engine (Diesel, 1895). Indeed, it was peanut (groundnut) oil that was used by Rudolf Diesel to run a small engine at the World Exhibition in Paris in 1900. At that time, vegetable oils were cheaper than mineral oils but the subsequent rapid development of the oil industry, the seemingly never-ending supply of consistent standardised mineral oil products, and the crucial factor of *cheapness* resulted in the domination of mineral products for the internal combustion engine.

Interest in vegetable oils as a fuel was rekindled in the 1970s, at the time of the oil crisis. The fragility of the world economy and its heavy dependence on cheap energy from politically unstable areas of the world became apparent at that time. In addition, public concerns over the rapid consumption of non-renewable, finite fossil reserves and the environmental impact of their use particularly in terms of global warming all contributed to the interest in developing renewable forms of energy.

Further stimulus for the development of biofuels came from the introduction of set-aside programmes in the US and Europe to combat the over-production of food and the excessive expenditure associated with disposal of surpluses (Ferdinand, 1992). In the US, this led to the development of bioethanol from corn in the first instance but over the last six years development of biodiesel from soybean oil has expanded. In the EU, biodiesel has become the biofuel of choice.

Many vegetable oils are suitable for use in diesel engines including corn oil, cotton oil, soybean oil and rapeseed oil. Some oils are less suitable, e.g. castor oil because of its extreme viscosity, and linseed oil because of its rapid oxidation and polymerisation. Rapeseed oil has been targeted for fuel use because it produces an oil with a close-to-optimum set of fuel characteristics.

Rapeseed oil can be used in a diesel engine in four different ways, viz.:

- as pure rapeseed oil;
- in mixtures of rapeseed oil and diesel;
- as rapeseed oil methyl ester (RME);
- in mixtures of RME and diesel.

Pure rapeseed oil can be used without modification only in indirect injection-compression ignition engines, i.e. diesel engines where the fuel is not injected directly into the firing cylinder. This is because its viscosity and gumming properties would lead to large droplets being sprayed from the injector resulting in incomplete combustion. This leads to inefficient fuel use, high exhaust emissions, excessive carbon deposits, and accumulation of gumming factors. Few diesel engines, other than those built for specialist situations, are now of the required indirect injection type. Indirect injection engines have a pre-combustion chamber that allows the oil to mix more thoroughly with air before combustion, helps the injectors to achieve a higher temperature (thus improving vapourisation), and burns off some of the glycerol-gum deposits.

Mixtures of rapeseed oil with diesel would partly overcome these problems (especially starting the engine in cold climates), but would negate some of the environmental advantages of rapeseed oil fuel. They also add complexity to the preparation and handling of mixed fuels. However, this is the preferred route in French production plants, in which the domestic petroleum industry is heavily involved. Finally, rapeseed oil can be converted to its methyl ester for use in most diesel engines. RME is the basis of the biodiesel industry in Europe and can be used either alone or in mixtures with mineral diesel.

*7.2.1 Biodiesel feedstocks*

Any vegetable oil is a potential feedstock for the production of a fatty acid methyl ester or biodiesel but the quality of the fuel will be affected by the oil composition. Ideally, the vegetable oil should have low saturation and low

**Table 7.1** Fatty acid composition of a range of selected oils and fats

| Fatty acid | Rapeseed | Sunflower | Soybean | Palm | Tallow |
|---|---|---|---|---|---|
| Palmitic (16:0) | 4 | 6 | 8 | 42 | 25 |
| Stearic (18:0) | 1 | 4 | 4 | 5 | 19 |
| Oleic (18:1) | 60 | 28 | 28 | 41 | 40 |
| Linoleic (18:2) | 20 | 61 | 53 | 10 | 4 |
| Linolenic (18:3) | 9 | – | 6 | – | 1 |
| Erucic (22:1) | 2 | – | – | – | – |
| Average number of double bonds | 1.33 | 1.49 | 1.51 | 0.59 | – |

*Source*: Batel (1980), Schűtt (1982).

polyunsaturation, i.e. be high in monounsaturated fatty acids. As can be seen in Table 7.1, rapeseed oil meets these criteria.

Vegetable oils with a high degree of saturation tend to have high freezing points as measured by cloud point or cold filter plugging point (CFPP) (Table 7.2). These oils have poor flow characteristics and may become solid (e.g. palm oil) at low temperatures though they may perform satisfactorily in hot climates (Schäfer, 1991). Vegetable oils with high levels of polyunsaturates (as reflected in the iodine number) are prone to oxidation and polymerisation. Fuels with these characteristics are likely to produce thick sludges in the sump of the engine, when fuel seeps down the sides of the cylinder into the crankcase.

Quality of the feedstock, therefore, has a considerable impact on the end-product quality of the biodiesel produced. An important factor in the success of biodiesel has been its acceptance by engine manufacturers. Key to achieving this has been the production of a biodiesel specification for which engine manufacturers have extended their warranties (Prankl and Wőrgetter, 1995). The biodiesel specification provided in Table 7.3 covers the factors influenced by both the feedstock used and the efficiency of the production process. In Europe, the specification is very much geared to RME. Both

**Table 7.2** Properties of a range of selected vegetable oils and fats

| Property | Rapeseed | Sunflower | Soybean | Palm | Tallow |
|---|---|---|---|---|---|
| Density at 15° (g/cm$^3$) | 0.915 | 0.925 | 0.930 | 0.920 | 0.937 |
| Flash point (°C) | 317 | 316 | 330 | 267 | – |
| Cloud point (°C) | 0 | −16 | −8 | 31 | – |
| Viscosity (mm$^2$/s) | 74 | 66 | 64 | 40 | solid |
| Iodine number | 113 | 131 | 130 | 49 | 45 |
| Calorific value (MJ/kg) | 41 | 40 | 40 | 35 | 39 |
| Cetane number | 44 | 36 | 39 | 42 | – |

*Source*: Batel (1980), Schűtt (1982).

**Table 7.3** Proposed EU biodiesel specification (EN 14214)

| Property | Unit | Limits minimum | Limits maximum | Test method |
|---|---|---|---|---|
| Ester content | % (m/m) | 96.5 | | prEN 14103[a] |
| Density at 15°C | kg/m³ | 860 | 900 | EN ISO 3676 |
| | | | | EN ISO 12185 |
| Viscosity at 40°C | mm²/s | 3,5 | 5,0 | EN ISO 3104 |
| Flash point | °C | above 101 | – | ISO/CD 3679[b] |
| Sulfur content[c] | mg/kg | – | 10 | |
| Carbon residue (on 10% distillation residue)[d] | % (m/m) | – | 0,3 | EN ISO 10370 |
| Cetane number | | 51,0 | | EN ISO 5165 |
| Sulfated ash content | % (m/m) | – | 0,02 | ISO 3987 |
| Water content | mg/kg | – | 500 | EN ISO 12937 |
| Total contamination[e] | mg/kg | – | 24 | EN 12662 |
| Copper strip corrosion (3 h at 50°C) | rating | | class 1 | EN ISO 2160 |
| Thermal stability[f] | | | | |
| Oxidation stability, 110°C | hours | 6 | – | prEN 14112[g] |
| Acid value | mg KOH/g | | 0,5 | prEN 14104 |
| Iodine value | | | 120 | prEN 14111 |
| Linolenic acid methyl ester | % (m/m) | | 12 | prEN 14103[a] |
| Polyunsaturated (≥4 double bonds) methyl esters[l] | % (m/m) | | 1 | |
| Methanol content | % (m/m) | | 0,2 | prEN 14110[h] |
| monoacylglycerol content | % (m/m) | | 0,8 | prEN 14105[i] |
| diacylglycerol content | % (m/m) | | 0,2 | prEN 14105[i] |
| triacylglycerol content | % (m/m) | | 0,2 | prEN 14105[i] |
| Free glycerol[d] | % (m/m) | | 0,02 | prEN 14105[i] |
| | | | | prEN 14106 |
| Total glycerol | % (m/m) | | 0,25 | prEN 14105[i] |
| Alkaline metal (Na + K)[j] | mg/kg | | 5 | prEN 14108 |
| | | | | prEN 14109 |
| Phosphorus content | mg/kg | | 10 | prEN 14107[k] |

[a] CEN/TC 307 publication of NF T 60-703: 1997.
[b] Apparatus equipped with a thermal detection device shall be used.
[c] Suitable test methods to be proposed by CEN/TC 19.
[d] ASTM D 1160 shall be used to obtain the 10% distillation residue.
[e] Pending development of a suitable method by CEN/TC 19, EN 12662 shall be used. The precision of EN 12662 is however poor for FAME products.
[f] Suitable test method and limit to be proposed by CEN/TC 19.
[g] CEN/TC 307 publication of ISO 6886 modified.
[h] CEN/TC 307 publication of NF T 60-701 (procedure A) and DIN 51608 (procedure B).
[i] CEN/TC 307 publication of NF T 60-704: 1997.
[j] Extension of this limit to cover additional elements, e.g. Ca and Mg to be considered.
[k] CEN/TC 307 publication of NF T 60-705: 1997.
[l] Suitable test method to be developed.

sunflower and soybean methyl esters would fail to meet this specification because of their high iodine number (high polyunsaturated feedstock).

### 7.2.2 Production of biodiesel

Oilseed rape can be processed by a range of different crushing and refining technologies as detailed in Chapter 2. Choice of technology depends upon the scale of operation and the end-use of the product. In essence, there are three major steps in the production of biodiesel. These are, first, crushing the seed to extract the oil, second, refining the oil, and third esterification. Crushing methods are identical whether the oil is to be used for food or fuel. Refining vegetable oil for food involves four processes – degumming, neutralisation, bleaching, and deodourising. Degumming involves the removal of natural phosphorus-based gums and pigments. Neutralisation involves the removal of free fatty acids (FFA). Both these processes are necessary to prepare oils for esterification and are often conducted together. Removal of phospholipids is important as these compounds can damage the engine. FFA should be removed to protect the esterification catalyst. The third step, bleaching, to improve oil colour and clarity is not necessary in biodiesel production as these criteria are not important for a fuel. Similarly, deodourisation to scrub out volatiles that would otherwise provide an unpleasant smell is not necessary for biodiesel production. As half the cost of refining rests on this final process (deodourisation), oil refining specifically for biodiesel is much less costly than for food use.

Esterification involves reaction between a monohydric alcohol and the oil in the presence of a catalyst. The triacylglycerols in the oil are transformed into fatty acid esters and glycerol. Normally methanol is the alcohol used in this reaction. The catalyst promoting the reaction may be acid or alkali. In most modern plants, the preferred catalyst is alkali for the main transesterification process but a pre-esterification step may be used with an acid catalyst for the conversion of FFA. This can be important where recycled oils are used as a feedstock as FFA levels may be high. For the main esterification process, KOH has superseded NaOH as the catalyst and is dissolved in the alcohol (Junek and Mittelbach, 1986). This reaction will take place at room temperature and the esterification reaction results in the separation of the heavier glycerol with a density of 1.26 g/ml from the lighter ester of density 0.88 g/ml. Separation can be conducted as a batch process in settling containers but in large plants it is usually a continuous process involving tube settlers or other separation technologies. Although this is a relatively simple chemical process, quality control is key if the biodiesel specification indicated in Table 7.3 is to be achieved. In addition, maximising the efficiency of conversion in the plant is key in terms of optimising the economics of the operation. The biodiesel may contain traces of soaps and some excess methanol, and these are removed by centrifuge for the former and by distillation for the latter. The biodiesel is then ready for use. However, being unsaturated, biodiesel can suffer oxidative decay when subjected to light and air, particularly at higher temperatures. Storage at low temperatures (e.g. 5°C) in underground tanks greatly reduces deterioration and the addition of antioxidants reduces this further (Kőrbitz, 1995).

Fuel quality of biodiesel is determined by the quality of the feedstock and the efficiency of operation of the biodiesel plant (Table 7.3). The density of biodiesel (0.88 g/ml) is very similar to that of mineral diesel (0.82–0.86 g/ml). Thus the biofuel has similar behaviour in the injectors and engine nozzles. Flash point of biodiesel is much higher (100–180°C) than for mineral diesel (55–60°C), so the biofuel is much safer to handle. However, biodiesel does have a higher CFPP (the temperatures at which the fuel will block filters) and in winter or cooler climates a CFPP depressant may have to be added. The cetane number measures the ignition property of the fuel. For biodiesel it is at least as high as for mineral diesel, resulting in smoother running of the engine. Methanol and glycerol contents are both indicators of the efficiency of the transesterification process. Although higher methanol levels do not adversely affect the performance of the engine, a level of less than 0.2% is desirable in order to achieve the low flash point. This safety characteristic may be important for certain markets. Both total glycerol and free glycerol content indicate production efficiency and levels higher than specification values will cause problems with engine performance. Rapeseed oil contains virtually no sulfur and consequently biodiesel contains almost no sulfur compared to mineral diesel. This is of considerable advantage in environmental terms as countries strive to reduce sulfur emission. In addition, this characteristic also prolongs the life of the catalytic converter now attached to most cars (Kőrbitz, 1995).

The mass balance for RME production is indicated in Table 7.4.

One tonne of rapeseed with an oil content of 43% will produce 0.41 t of crude oil and 0.58 t of meal. With slight losses during refining, 0.40 t of refined oil will produce approximately 0.38 t of RME. With a density of 0.88 g/ml, this represents 432 litres of biodiesel for each tonne of rapeseed. With seed yields

**Table 7.4** Mass balance for RME production

|  |  |  |  |
|---|---|---|---|
|  | 1 t rapeseed ↓ | → | 0.58 t meal |
|  | 0.41 t crude oil ↓ |  |  |
| 0.04 t methanol (recycled) → | 0.40 t refined oil ↓ | → | 0.04 t glycerol |
|  | 0.38 t RME @ 0.88 g/ml density = 432 litres |  |  |

of 3 t/ha, then approaching 1300 litres fuel/ha can be produced. In high crop-yielding situations, this can be 40% higher.

### 7.2.3 Fuel characteristics

When compared with mineral diesel, biodiesel is found to have a 13% lower energy value in terms of MJ/kg and a 7% lower level in MJ/dm$^3$ (as the density is higher) (Table 7.5).

On an analytical basis, biodiesel differs most markedly from mineral diesel in having virtually no sulfur present. Diesel, however, contains no oxygen compared with 10–11% in biodiesel. It is the presence of oxygen that is claimed to explain the improved combustion in terms of energy efficiency and reduced emissions as indicated in Table 7.5. There have been many studies conducted across the world comparing the efficiency of biodiesel with diesel and different results are produced according to the engine test cycle, the engine design, and the quality of the biodiesel (Havenith, 1993; Goetz, 1994). On balance, there appears to be little difference in potential power output but there is an increased fuel consumption of approximately 5% within the range of −5 to +14% (Schäfer, 1991; Sams and Schindlbauer, 1992; Walter, 1992). Other factors of note are dilution of sump oil ranging from +1 to +10%. The move to low-sulfur mineral diesel fuels results in poorer lubricity of the fuel causing adverse scoring and wear of the cylinder and piston. Here, biodiesel offers significant advantages and the addition of biodiesel to mineral diesel has significant benefits in this respect.

### 7.2.4 Emissions

In the same way that engine efficiency studies vary according to test cycle, engine, biodiesel, etc., so do emission comparisons vary (Table 7.6).

Generally speaking, results tend to indicate that most emissions are down, except for NO$_x$ which is generally up by around 3–5% (Mittelbach and Tritthard, 1988). As NO$_x$ is a greenhouse gas that is particularly damaging, more detailed studies have been carried out to show how engine timing adjustments specifically for biodiesel use can improve the NO$_x$ figure (Table 7.7).

**Table 7.5** Calorific value and fuel efficiency

| Fuel | Density (g/cm$^3$) | Energy value MJ/Kg | MJ/dm$^3$ | Efficiency degree (in % at 1200 rpm) |
|---|---|---|---|---|
| Diesel | 0.83 | 42.9 | 35.6 | 38.2 |
| Biodiesel | 0.88 | 37.2 | 32.9 | 40.7 |
| Variation on diesel | | −13.3% | −7.6% | +6.5% |

*Source*: Walter (1992).

**Table 7.6** Exhaust emissions* from direct injection engines of biodiesel compared to mineral-derived diesel

| Reference | SO$_2$ | CO | HC | PAH | NO$_x$ | Particulates | Smoke |
|---|---|---|---|---|---|---|---|
| Long | | >10% lower | | 60% lower | >10% lower | | 50% lower |
| Patcher | almost zero | lower or higher | 50% lower | | slightly higher | | |
| Austria | | | lower | much lower | slightly higher | | |
| FOP | | 65% lower | 12% lower | | 16% lower | | 57% lower |
| Koch | 90% lower | | lower | lower | | lower | lower |
| Wade | 90% lower | 10% lower | 40% lower | higher | 10–12% lower | | |

* HC = unburnt hydrocarbons; PAH = polycyclic aromatic hydrocarbons; NO$_x$ = nitrogen oxides; Smoke levels are measured on the Bosch Index.
*Source*: Culshaw and Butler (1992).

**Table 7.7** Emissions from Austrian bus trials, biodiesel relative to low-sulfur fossil diesel

| Emission | SO$_x$ | CO | NO$_x$ | NO$_x$* | PM | VOC | BS |
|---|---|---|---|---|---|---|---|
| % change with biodiesel | −99 | −20 | +1 | −23 | −39 | −32 | −50 |

−99 indicates a reduction by 99% (almost complete elimination), +1 indicates an increase of 1%.
* NO$_x$ result for an engine adjusted for biodiesel use.
*Source*: Sams (1996).

Carbon dioxide emissions from biodiesel combustion represent the release of $CO_2$ that was fixed during crop photosynthesis. One can therefore argue that biodiesel production and consumption is operating within a closed loop in environmental terms. However, $CO_2$ emissions will also occur during the preparation of inputs for growing and processing the crop, and a clearer view of the overall environmental benefit can be gained by examining the energy balance of crop production. The energy balance of a biofuel can be defined as the ratio of the energy used during its production to the energy value of the fuel produced and of any used by-products. For a biofuel to be sustainable, it is essential that the energy ratio is 1:>1. If the energy ratio is 1:<1, there will be a net loss of energy in the production of the fuel, thereby negating its status as a renewable energy source.

Table 7.8 provides the energy ratio for biodiesel and by-products from winter oilseed rape compared with spring oilseed rape (Culshaw and Butler, 1992).

Table 7.9 gives the energy ratio for biodiesel under a range of scenarios (Batchelor *et al.*, 1995).

In both the analyses, energy inputs include not just inputs to the field but also those for equipment manufacture, processing, commodity transport, etc. In both cases, the energy output for biodiesel without any by-product credits is 1.3–1.35 units of energy out for each unit supplied. The different values credited to the rapemeal reflect the varying approaches taken by the

**Table 7.8** Energy ratios for biodiesel from winter and spring oilseed rape

| Product (energy output/energy input) | Winter | Spring |
|---|---|---|
| Biodiesel only | 1.35 | 1.35 |
| Biodiesel + meal | 2.55 | 2.55 |
| Biodiesel + meal + glycerol | 2.62 | 2.61 |
| Biodiesel + meal + glycerol + straw | 3.77 | 3.77 |
| Biodiesel + straw | 2.50 | 2.50 |

*Source*: Culshaw and Butler (1992).

**Table 7.9** Energy ratios for RME production from winter rape

| | Good | | Poor | |
|---|---|---|---|---|
| Outputs included | Best-case scenario | Intermediate scenario | Intermediate scenario | Worst-case scenario |
| RME only | 1:2.23 | 1:1.58 | 1:1.12 | 1:0.674 |
| RME + rapemeal | 1:3.83 | 1:2.22 | 1:1.60 | 1:0.88 |
| RME + rapemeal + glycerol | 1:3.95 | 1:2.30 | 1:1.65 | 1:0.91 |
| RME + rapemeal + glycerol + straw | 1:9.18 | 1:5.46 | 1:3.92 | 1:2.22 |

*Source*: Batchelor *et al.* (1995).

two studies. In the Culshaw and Butler study, the full thermal credit of the straw is included whereas in the study by Batchelor *et al.* the metabolisable energy content is used for all but the best-case scenario. In both studies, the energy contribution from the glycerol is small. The inclusion of a credit to the straw is to a large extent academic as relatively little rape straw is used – most will be ploughed back into the soil. Consequently, in terms of RME + used by-products (meal + glycerol), the energy balance is in the region of 2–2.6 units of energy out per unit of energy in. By comparison, 1979 figures indicate that the energy used to extract, refine, and transport diesel products from crude oil is 5 units out for one unit in (Boustead, 1997). The very large differences between energy ratios calculated by Batchelor *et al.* (1995) under the best-case and worst-case scenarios highlight the fact that the energy ratio for RME should not be seen as a static value but is subject to variation depending on the prevailing conditions. The vast majority of energy used in the production of RME is consumed during the growth of the crop rather than during processing to RME. The input for nitrogen fertiliser varies from 10% in the best-case scenario to 27.5 and 20.7% in the intermediate scenarios and 43.5% in the worst scenarios. The importance of nitrogen fertiliser inputs to the energy balance suggests that modification of agricultural practice could have a notable effect on the energy ratio of RME production. In particular, the use of organic wastes, e.g. sewage sludge, could have a great significance for the energy balance (Batchelor *et al.*, 1995).

### 7.2.5 Economics of biodiesel production

The aim of this section is to provide an indication of the cost of producing biodiesel from commercial plants of differing scales. The impact of scale of production is also indicated. In preparing these costs, capital and operating quotes for all the three stages of production (extraction, refining, and esterification) have been obtained (Cook, personal communication). For the extraction stage, mechanical extraction via a full press process has been assumed. At crushing volumes of less than 500 t of rapeseed per day, it is generally accepted that the capital costs of solvent extraction facilities cannot be justified. It is more cost-effective to accept the loss of some oil in the meal than to incur the high capital cost. The costs of such facilities do not decline in line with the volume of seed processed. For example, in a quote for a 1000 t/day plant, hexane extraction accounted for $4.8 million of the capital costs. In a plant with half this capacity (500 t/day), the quote for the hexane extraction plant only fell to $3.3 m.

The following figures have been produced to indicate cost of production. Whilst interest on working capital, depreciation, and debt repayment are included in the calculations, no element of profit is included. The net RME cost is simply the cost of production.

**Option 1** 22 million litre biodiesel facility using rapeseed only. Full press seed mill (no hexane extraction) with 200 t/day capacity. 60 t/day RME plant

| | |
|---|---|
| **Process yield assumptions** | |
| 1 t rapeseed | by-product 660 kg meal (12% oil content) |
| ↓ | |
| 340 kg crude oil | |
| ↓ | |
| 332 kg refined oil | by-product 42 kg purified glycerol |
| ↓ | |
| 327 kg RME @ 0.88 g/ml density = 372 litres | |
| **Total quantities** | |
| 60 000 t rapeseed | by-product 39 600 t meal |
| ↓ | |
| 20 400 t crude oil | |
| ↓ | |
| 19 920 t refined oil | by-product 2520 t purified glycerol |
| ↓ | |
| 19 620 t RME @ 0.88 g/ml density = 22 295 454 litres | |

*Note*: This assumes 300 working days for the 200 t/day crushing plant, but approximately 330 working days for the 60 t/day RME facility.

Cost per litre estimate for 22 million litre facility

|  |  |  | Annual cost (£) | Cost per litre of RME |
|---|---|---|---|---|
| 1. | Capital cost | Euro |  |  |
|  | (A) 200 t/d full press mill | 2 650 000 |  |  |
|  | (B) Biodiesel and glycerol plant | 5 846 600 |  |  |
|  | (C) Tank farm and utilities | 1 312 000 |  |  |
|  | (D) Civil works | 1 640 000 |  |  |
|  |  | 11 448 600 |  |  |
|  | @ £0.63:£ | £7 212 618 |  |  |
|  | Annual charge to cover repayment and interest, over 5 years @ 9% interest |  | 1 853 643 | 8.3p |
| 2. | Operating costs |  |  |  |
| 2A. | Labour |  |  |  |
|  | Mill: 4 @ £25 000 | 100 000 |  |  |
|  | RME plant: |  |  |  |
|  | Chief operator 1 @ £40 000 | 40 000 |  |  |
|  | Shift workers 4 @ £25 000 | 100 000 |  |  |
|  | Laboratory 1 @ £25 000 | 25 000 |  |  |
|  |  | 265 000 | 265 000 | 1.2p |
| 2B. | Power |  |  |  |
|  | Mill: |  |  |  |
|  | 2000 kWH = 2000 units |  |  |  |
|  | = 7200 MJ @ 2p/MJ |  |  |  |
|  | = £144/day × 300 days | 43 200 |  |  |
|  | RME Plant: |  |  |  |
|  | 1000 kWH = 1000 units |  |  |  |
|  | = 3600 MJ @ 2p/MJ |  |  |  |
|  | = £72/day × 330 days | 23 760 |  |  |
|  |  | 66 960 | 66 960 | 0.3p |
| 2C. | Annual maintenance |  |  |  |
|  | Estimated 2.5% of investment cost | 180 315 | 180 315 | 0.8p |
| 2D. | Miscellaneous |  |  |  |
|  | Including business rates, environmental compliance, water charges, administration, audit, management board, insurance | 200 000 | 200 000 | 0.9p |
| 3. | Working capital interest |  |  |  |
|  | Mainly seed purchase and storage |  |  |  |
|  | Say £6m @ 9% for 8 months storage |  | 360 000 | 1.6p |
| 4. | Purchase of rapeseed |  |  |  |
|  | 60 000 t @ £140/t including any delivery cost |  | 8 400 000 | 37.7p |
| Total costs |  |  | 11 325 918 | 50.8p |
| 5. | By-product income |  |  |  |
|  | Meal 39 600 t @ £90/t ex-mill |  | 3 564 000 | 16.0p |
|  | Glycerol 2520 t @ £350/t* |  | 882 000 | 4.0p |
| Net cost of biodiesel |  |  | 6 879 918 | 30.8p |

* Estimated price for 78% pure refined glycerol. The 99.5% pure glycerol sells for over £600/t.

**Option 2** Small-scale 4500 t oilseed mill linked to 1500 t batch refining plant.
This option is included to illustrate the typical costs for a small-scale facility.

**Specification**
A 10–15 t/day crusher producing a 10% oil content meal. Batch refiner handling 2.5 t of oil per batch if refining to food quality. As each batch takes 10 h, throughput is 5 t/day if working in shifts. Slightly faster throughput for RME production as food quality refining not required.

**Total quantities**
4500 t rapeseed → 2970 t meal
↓
1530 t crude oil
↓
1494 t refined oil
↓
1400 t RME @ 0.88 g/ml = 1 590 909 litres

*Note*: This quantity of oil can be handled in a batch refiner if it operates two shifts per day over 300 days per year.

Cost per litre estimate

|  | Annual cost (£) | Cost per litre of RME (p) |
|---|---|---|
| 1. Capital cost | £ | |
| (A) Rapeseed mill | 130 000 | |
| (B) Biodiesel plant | 1 100 000 | |
| (C) Refinery | 25 000 | |
| (D) Site works, utilities, installation | 232 500 | |
|  | 1 487 500 | |
| Annual charge to cover repayment and interest, over 5 years @ 9% interest | 382 286 | 24.1 |
| 2. Operating costs | | |
| 2A. Labour | | |
| 2 skilled @ 25 000 | 50 000 | |
| 1 unskilled @ 15 000 | 15 000 | |
|  | 65 000 | 65 000 | 4.1 |
| 2B. Power | 15 000 | 1.0 |
| 2C. Annual maintenance | | |
| 2.5% of capital cost | 37 187 | 2.2 |
| 2D. Miscellaneous | 50 000 | 3.1 |
| 3. Working capital interest | | |
| Say £500 000 @ 9% for 8 months average | 30 000 | 1.9 |
| 4. Purchase of rapeseed | | |
| 4500 t @ £140/t including delivery cost | 630 000 | 39.6 |
| Total costs | 1 209 473 | 76.0 |
| 5. By-product income | | |
| Meal 2970 t @ £90/t ex-mill | 276 300 | 16.8 |
| Net cost of biodiesel | 933 173 | 59.2 |

Clearly, this small-scale option would not be competitive with larger-scale biodiesel facilities and more importantly will not produce a fuel that can compete with mineral diesel, even with excise duty concessions.

Currently, mineral diesel production cost before taxation and distribution costs is 10 p/litre. Biodiesel is therefore approximately three times this price when produced using modern facilities. It can compete with mineral diesel only if a tax reduction or exemption is given to the biodiesel (Walker and Kőrbitz, 1994). It is through this mechanism that biodiesel industries have been developed across the world.

### 7.2.6 Biodiesel market opportunities

There is a widespread misconception that biodiesel could substitute for mineral diesel in totality and so solve any future energy problems. This is not the case. Taking the UK as an example, the UK market for diesel is around 12.9 Mt per annum (Anon., 2003). At a rapeseed yield of 3 t/ha, it would require approximately 13.1 million hectares of rape to produce this quantity of biodiesel. There are only 5.2 million hectares of cropping land in the UK.

Clearly, potential market size is not a problem but given the higher cost of production of biodiesel, the biofuel is best placed in niche markets where either the consumer is prepared to pay more for the product, or where the environmental benefits of reduced emissions and biodegradability might justify a premium, relief from duty or a combination of both (Kasterine and Batchelor, 1998).

Examples of these markets are:

- inner-city vehicle use, especially taxis, buses, public utility vehicles, post office vehicles, local authorities, inner-city distribution vehicles;
- environmentally sensitive areas (ESA), e.g. water catchment areas;
- ski resorts;
- inland waterways;
- golf courses;
- underground tunnels, e.g. railway tunnels;
- mining operations;
- fire brigade practices, i.e. particularly inner-city brigades.

### 7.2.7 Biodiesel production in Europe

Most of the pioneering work on biodiesel was conducted in Austria. Historically, the first trials with RME were conducted in Austria in 1982 and they showed promising results. This was followed in 1985 by a pilot plant and

**Table 7.10** Biodiesel production (2001) and estimated production capacity (2003)

|  | Production (2001) | Capacity (2003) |
|---|---|---|
| Austria | 28 000 | 140 000 |
| Czech Republic | 50 000 | 70 000 |
| Denmark | 100 000 | 100 000 |
| France | 275 000 | 410 000 |
| Germany | 450 000 | 1 100 000 |
| Italy | 70 000 (2000) | 600 000 |
| Slovakia | 50 000 | 125 000 |
| Spain | – | 6 000 |
| Sweden | 7 000 | 8 000 |
| Switzerland | 2 000 | 2 000 |
| UK | – | 80 000 (estimated) |

*Source*: ABI (2003).

then in 1990 the first industrial biodiesel plant was constructed with a capacity of 10 000 t. From there capacity grew and spread across Europe. In France, in 1993 a demonstration plant in Compiegne with a capacity of 150 000 t RME/year was constructed. In 1995, commercial scale biodiesel production began in Germany. By 2001, production across Europe had risen to over 1 Mt (Table 7.10). Production capacity is estimated to reach over 3 Mt in 2003 (ABI, 2003).

## 7.3 Lubricants

The total EU lubricants market extends to over 4 Mt and is dominated by motor and gear oils which account for nearly 60% of consumption (Table 7.11). Prior to the development of the petrochemical industry, nearly all lubricants were based on animal fats or plant oils. However, their use was almost

**Table 7.11** EU lubricant market forecasts ('000 tonnes)

| Lubricants market sector | Total EU | Current RRM | RRM in 2010 (−) | RRM in 2010 (+) | RRM (% +) |
|---|---|---|---|---|---|
| Hydraulic fluids | 750 | 51 | 100 | 250 | 33 |
| Greases | 138 | 1 | 2 | 69 | 50 |
| Chainsaw lubricants | 40 | 29 | 30 | 38 | 96 |
| Mould release agents | 82 | 10.5 | 20 | 41 | 50 |
| Motor and gear oils | 2408 | 4.5 | 20 | 482 | 20 |
| Metal working fluids | 338 | 4.5 | 20 | 170 | 50 |
| Other applications | 486 | 0.5 | 10 | 240 | 50 |
| Total | 4242 | 101 | 202 | 1290 | 30 |

*Source*: Ehrenberg (2002).

totally extinguished by the development of mineral oils which offered the advantages of consistent quality, limitless quantity, low cost, and excellent functionality.

The opportunity for vegetable oils to substitute for mineral oils is principally based on the desire to reduce the dependence on non-renewable resources. In addition, vegetable oils offer the benefit of biodegradability and other environmental attributes, which may be important for certain markets. There has been a growing interest in the use of vegetable oils for these uses over the last 20 years. However, much of this interest is *producer push* (i.e. driven by the agricultural sector looking for new markets) rather than *consumer pull* where users actively seek out and prefer the vegetable oil option. Much of this lack of interest relates to existing users being content with the products currently available, coupled with a lack of knowledge of what attributes vegetable oils have to offer. Interest in the use of vegetable oils differs according to the part of the world and the dominant oil crop in that region. In the USA, most research centres on soybean. In northern Europe, rapeseed is the oil crop of interest whilst in southern Europe, it is sunflower. Consequently, most data of relevance is European in origin. At present, renewable raw materials (RRM) account for only 2.4% of the total EU lubricants market with less than 0.2% penetration of the largest market sector (motor and gear oils). However, it is noteworthy that over 70% of chainsaw lubricant use is with renewables.

An estimate of the market forecast and market share of RRM for 2010 is also presented in Table 7.1. The positive indicates potential share by 2010 should government intervene and actively encourage the use of renewables through legislation. The negative indicates market share where no active encouragement is made.

The key criteria for acceptability of renewables are cost and functionality. Environmental considerations are seen to be of very low priority by end-users – exceptions being in situations where contracts stipulate the use of renewables, e.g. chainsaw lubricants or hydraulic fluids in Forest Authority contracts.

The functionality of simple vegetable oils has always been regarded as good by their low evaporation rate, load carrying capacity, and low coefficient of friction, thereby often improving fuel consumption. They also have natural *multigrade* characteristics making them ideal for situations where there are significant variations in ambient temperatures. In addition, hydrolysis of fats and oils to their component fatty acids has the advantage that a wider range of natural sources (including fats, low-grade oils, or wastes such as tallow) can be used as a source of raw materials. Against these advantages are the concerns over thermal stability and cost. Thermal stability can be improved by converting oil into ester and this offers the advantages of vegetable oils without the disadvantages (Sala, 2000). Such processing adds to the cost but may improve performance beyond that of mineral oils. Both mineral and vegetable oils can provide the base oil for lubricants but a cocktail of additives, accounting for generally 10% but occasionally up to 25% by volume, may be required to raise functionality to an acceptable standard. This may counteract or at the very least confuse the environmental benefits as might otherwise be claimed that some additives are highly toxic. Oils made from RRM may not necessarily be better for the environment (Bondoli and Igartua, 2000).

In contrast, some synthetic oils based on mineral oils have been modified to improve their breakdown and thereby their environmental credentials. Consequently, assumptions of the comparative effect on the environment of oils made from renewable or non-renewable products should be made with caution, and more detailed comparative life-cycle assessment of products may be required to elucidate relative environmental merits.

## 7.3.1 Hydraulic fluids

Hydraulic fluids are seen as *high risk-loss* situations as the systems operate at high pressure, often in ESAs such as forests, watercourses, and skiing areas. Damage to exposed hoses and cylinders is not infrequent and leaks in seals and hose connections add to the risk of leaks to the environment.

The current EU market for renewable hydraulic fluids may be as high as 51 000 t (Table 7.11) or as low as 20 000 t (Battersby, 2002). It is assumed that the UK accounts for 10% of this use. The Shell company sees Sweden as the key market in Europe with the share of the market for environmentally acceptable hydraulic fluids (EAHF) having grown from 5% in 1996 to 18% by 2000. Use is concentrated on forest operations with EAHFs used for 70–80% of all harvest operations (Battersby, 2002).

Use of EAHFs is facilitated by international standards, e.g. Swedish Standard SS 15 54 34, American Standard D 6064 98, International Standard 150 1539- and eco-labels (e.g. Nordic White Swan, German Blue Angel), which can then be stipulated in contracts offered to contractors. Although vegetable oil-based hydraulic esters have advantages of high biodegradability, low eco-toxicity, and a high viscosity index, these are outweighed by poor oxidative stability, poor hydrolytic stability, giving operating problems at low temperatures, e.g. gel formation, loss of fluidity, and leaking hoses (Battersby, 2002).

Consequently, unmodified vegetable oils (of which rapeseed oil sets the standard) are not seen as ideal base fluids. Improvement can be achieved by using *winterised* base fluids (cooling the oil and filtering out the crystallised material). In effect, this increases the proportion of oleic acid in the base oil. This base oil can then be optimised using an antioxidant additive package together with a pour-point depressant and synthetic ester. Such a product is exemplified by Shell Naturelle HF-M46, which is currently under trial in a dedicated Caterpillar wheel loader. Continuous evaluation of the condition of the hydraulic fluid indicates no deterioration of the product, leading Shell to conclude that an optimised rapeseed oil-based hydraulic fluid can perform well under arduous field conditions (Battersby, 2002).

Clearly, biodegradable hydraulic fluids now meet the key criteria of functionality and environmental benefits but cost will restrain expansion of this sector. The raw material for the base oil (rapeseed oil) is readily available within the EU. The composition of model hydraulic fuels based on the vegetable oil in comparison with mineral oil is presented in Table 7.12 (Carruthers et al., 1999).

Generally, finished lubricant manufacturers are not prepared to release details of the additives used and only generic terms are used. The composition presented in Table 7.12 by Carruthers et al. (1999) is based on a number of sources including Crawford et al. (1997), Rasberger (1997), Caterpillar BIOHYDO Technical Information, RohnGMBH information (Internet), RohMax data sheets, and Copan and Haycock (1993).

The mineral oil base used is generally a severely solvent-refined and/or hydro-treated paraffinic mineral base oil. Base stocks used for lubricants are a mixture of $C_{20}$–$C_{45}$ hydrocarbons (Rasberger, 1997). Traces of N–, S–, O– containing heterocycles together with mercaptans (RSH), thioethers (R–S–R), and disulfides (R–S–S–R) are also an integral part of the complex composition of lubricating base oils. The base mineral oil contains 5% aromatic hydrocarbons and 95% paraffins (Mortensen et al., 1997) and also 0.03% sulfur, which is the median value for sulfur content given by Rasberger (1997). The assumed aromatic content is in close agreement with the value of 7.5% quoted by Betton (1993) for polyaromatic hydrocarbons (PAHs) in virgin lube oil. The nitrogen content of petroleum is <0.5% by weight, and residual oils from a Forties Field UK crude oil typically have contents of 20 ppm vanadium and 10 ppm nickel (CONCAWE, 1979). A survey of 54 lubricants sold in Europe showed chlorine contents ranging from 0.0008 to 0.039% (Hewstone, 1994). From this, it was assumed that the total chlorine content in mineral hydraulic fluid was 0.03%.

Table 7.12 Formulation of model hydraulic fluids – composition of 1 kg of hydraulic fluid

| Function | Mineral oil hydraulic fluid Components | g | Vegetable oil hydraulic fluid Components | g |
|---|---|---|---|---|
| Base oil | Mineral oil | 982 | Rapeseed oil | 988 |
| Antioxidant | Zinc di-2-ethylhexyl dithiophosphate (a primary alkyl ZDDP) | (12.9 Zn) | Alkylated diphenylamine | 3 |
| Antiwear | Zinc di-2-ethylhexyl dithiophosphate (a primary alkyl ZDDP) | | Trialkyl phosphate | 3 |
| Anti-corrosion | Overbase Ca sulfonate detergent | 0.5 | Overbased Ca sulfonate detergent | 0.5 |
| Antifoam | Polydimethyl siloxane (silicone) | 0.03 | Polyethylene glycol (PEG) (non-silicone) | 0.05 |

Rapeseed oil has a maximum sulfur content of 0.022% by weight (Kőrbitz, 1995). The only source of chlorine in rapeseed oil hydraulic fluid is the additives at an estimated figure of 0.005% (Hewstone, 1994).

The additives selected for the vegetable hydraulic fluid are typically chosen on the basis that they are less environmentally damaging than some of the additives in a typical mineral-based hydraulic fluid. Secondary aromatic amines such as diphenylamine are an important class of antioxidant additives in lubricants (Rasberger, 1997). The alkylated diphenylamine selected is a non water-extractable phenolic antioxidant cleared for use by the Food and Drug Administration (FDA) in the USA, and has very low toxicity.

The antiwear additive typically chosen for vegetable oil hydraulic fluid is a trialkyl phosphate. This is a phosphate ester with variable physical properties depending on the mix and type of organic substituents, molecular weights, and structural symmetry. The anti-foam additive, a polyethylene glycol (PEG) (non-silicone), is generally water-soluble and considered to be at a low risk to the environment.

### 7.3.2 Future market potential

The main driver for change towards environmental loss damaging hydraulic fluids has been Scandinavian environmental legislation and governmental agency stipulation (by many countries) to insist on EAHF use. This has provided the main impetus for the R&D to help produce acceptable products. Future market potential will depend on two criteria. First, extension of legislation (or public-contract specification) stipulating EAHF use, and secondly, reduction in the cost of production of EAHF to a level closer to that of the mineral-based product.

European Union figures (Table 7.11) indicate that without the first criterion, use of RRM-based hydraulic fluids will double by 2010 to 100 000 t, but with legislation this could increase to 250 000 t, i.e. one-third of total use. There is little in the way of restriction on the supply of rapeseed oil should such an expansion in demand take place.

Agriculture is well placed to supply such a need. It is noteworthy that industry would prefer a rapeseed oil higher in oleic acid and lower in saturated and polyunsaturated fatty acids. These objectives match with those of other industrial uses of rapeseed oil and, more importantly, match with the desires of nutritionists in terms of a healthy food oil. Plant breeders have high oleic breeding programmes underway.

### 7.3.3 Greases

Total market for greases in the EU is 138 000 t, with greases based on RRMs accounting for less than 1% of this. Technically, it is possible to make greases using vegetable oils such as rapeseed as a base fluid but it is necessary to include additives (e.g. TMP esters) and thickeners to achieve an acceptable specification. The main weaknesses of rapeseed as a base oil for this purpose are resistance to temperature, too low a dynamic viscosity, and low resistance to oxidation and hydrolysis (Muntada, 1999). A biodegradable grease has been developed by ROWE GmbH using a calcium-based thickener (as opposed to the more toxic lithium- and aluminium-based thickeners) (Zehe, 2000). However, this oil is a stearate-based oil.

Opportunities for RRMs for such a product therefore rest with tallow-based products (rather than rapeseed oil). Whilst biodegradability objectives can be achieved with such a product, functionality parity is achieved only with difficulty and products are expensive. At present, there is no legislation forcing the use of biodegradable greases (other than in food production lines) and there is little in the way of environmental impact pressurising industry to change.

Without legislation forcing industry to use biodegradable greases, the prospects for such products are not good, with as little as 2000 t being the possible European market value by 2010 (see Table 7.11). With legislation, use could be lifted to 50% (69 000 t). Target areas for such legislation could be railways, vehicle chassis and gears, mining, forestry, and agriculture. If biodegradable greases were produced along the lines of the ROWE product currently available, then tallow would readily supply the stearic acid required.

### 7.3.4 Mould release agents

From a total annual European lubricant market of 5 million tonnes, approximately 15% is accounted for by loss lubricants (750 000 t). Largest loss lubricants are two stroke engines (30%) followed by demoulding agents (25%), greases (17%), and chainsaws (8%). Demoulding agents are used in the building industry to lubricate re-usable moulds so that concrete does not stick to the mould and that on mould removal the concrete is left behind with a smooth surface.

Mould release agents may be based on mineral oil or vegetable oil. The agent must provide an effective barrier between the concrete and the mould, have no adverse effect on either surface, must be safe and easy to apply, and must allow the concrete to slip easily past the mould surface. The first vegetable oil-based release agents were produced in the 1980s but the market share is very limited.

Mould release agents are composed of a base oil, additives, and (in all but bio-based demoulding oils) a solvent. Bio-based products may have a base oil that is either a vegetable oil or a synthetic ester. Vegetable oil bases are generally 2–3 times the price of mineral oil-based products whilst synthetic esters may be 4–5 times the price. However, practical experience of bio-based products indicates a 50% reduction in consumption rate and in some cases consumption is down 7-fold. In addition their use produces no dust, few air bubbles, and a very good surface, i.e. as good as possible, or with better functionality properties (Defrang, 1999).

For construction companies, it is not clear that there are any benefits of changing to bio-products. With a higher unit price and specialist spraying equipment required to reach the low consumption levels where savings can be made, only practical use of the product will indicate its worth. In addition, construction companies may be reluctant to risk structural failure problems where new products are on offer and may prefer to stick with their tried and tested product. At present there appears to be no natural legislation forcing the use of such products though they are eligible for eco-labels.

Main future drivers may relate to health and safety implications from mineral oil vapours compared with vegetable oil vapours, more stringent storage requirements for the petrochemical-based product, and an overall more holistic comparison of the two product types.

At present, approximately 12% of the current mould release agents are made from RRM. Estimates suggest that with government legislation this will increase to 25% of consumption but could reach 50% should legislation force the use of bio-products. Situations where use could be mandatory could be for government buildings, buildings in ESAs, near watercourses, water catchment areas, etc. As vegetable oils with a low-to-medium viscosity are required, rapeseed oil is a suitable base oil for this use.

### 7.3.5 Motor and gear oils

Motor and gear oils represent the largest sector (56%) within the EU lubricants market but less than 0.2% is supplied by RRMs. This small market for lubricants based on vegetable oils is composed of 75% native vegetable oils and 25% synthetic esters, though it is anticipated that these proportions will be reversed in the future because of the functionality improvements of the esters (Bondoli and Igartua, 2000). Theoretically, the different vegetable oils (and their esters) available provide a range of properties sufficient to cover most of the needs of the lubricants sector. Preferred base oils are rapeseed oil and high-oleic sunflower oil (HOSO) with expensive castor oil preferred for certain specialist lubricant requirements. Products based on RRMs exist as alternatives to mineral oil-based lubricants but have difficulty competing (apart from some specialist areas) because of cost and functionality issues. For example, HOSO currently costs approximately £1000/t compared with rapeseed oil at £300–400/t. World production of HOSO stands at only 280 000 t tending to impose a limit on development potential. Modern engines with extended oil-change intervals present severe demands on today's engine lubricants and vegetable oils tend to be more suited to situations of lower technical performance requirements (Schulla, 2002).

Functionality limitations of the vegetable oil-based lubricants relate to the common problems of oxidative stability, lubrication properties at wide temperature range, etc. In addition, purity of the vegetable base oil is particularly critical for the motor and gear oils sector as these oils are recycled within the engine for an extended period of time and any defect can have serious consequences. Efficiency of refining is therefore crucial for this sector particularly in relation to natural constituents (e.g. FFAs), oxidation by-products (e.g. hydroperoxides), and chemical contaminants (e.g. solvent residues and trace metals). Functionality shortcomings can be addressed through additive technology but this may compromise the environmental credentials of the end product. Currently, only eco-labels provide standards which these end products must reach and there is a considerable need for standardisation of environmental criteria for these products to develop.

One interesting approach to overcome functionality limitations of the vegetable-based lubricants is the Plantotronic lubrication system developed by Fuchs, where the lubricant is refreshed with new oil and the waste oil is burned with the diesel fuel.

An additional hurdle that must be achieved for the use of bio-lubricants is the secural of engine warranty from the manufacturers. Currently, there appears to be little in the way of legislation encouraging the use of biodegradable lubricants (other than crankcase oils for tractors and off-highway equipment used close to waterways or on cropped land) and whilst there are clear environmental benefits of using bio-lubricants, there is little pressurising industry to change. It is difficult to foresee this sector becoming a major target for use of RRMs such as rapeseed oil – strategically there are other sectors (e.g. total loss systems) where the environmental benefits would be greater.

### 7.3.6 Metal working fluids

Metal working fluids are used to lubricate machines used in the manufacture of components made from metals. There are many differing manufacturing processes (e.g. cold forming, form rolling, cut or roll topping, machining, etc.) and each may require a different lubricant for optimum performance. The total EU market for metal working fluids currently stands at 338 000 t of which just over 1% is made from renewables. Key to success of vegetable-based metal-working lubricants is functionality and cost. Demands for the lubricants currently used are considerable – in the last 30 years, machine speeds of manufacturing equipment have trebled leading to higher working temperatures. In many cases, machines are operating close to the flash point of the oil that is supposedly cooling them. There are financial factors other than simply oil cost that must be taken into account on comparing mineral-based with vegetable-based lubricants. These include life of the machine, operating and repair costs, filterability of the oil and its operating life, flash point, machine efficiency, oil spillage implications, work place misting, and other health issues from vapours (Cross, 2000). Work by Fuchs and the development of Plantocut oils based on vegetable oils have indicated the potential that can be achieved from vegetable oil-based lubricants (Lea, 2002).

They claim that on one factory site, they have managed to reduce the range of lubricating oils used from 20 mineral oils to 3 grades of biolubes – this clearly simplifies maintenance, reduces lubricating oil stocks required, and reduces the likelihood of error in lubricant application. In addition, being chlorine-free, these oils will satisfy any future environmental legislation. Other advantages of their vegetable-based oil are the volume of waste oil and related disposed costs (disposal cost/litre of some products may exceed purchase price). For example, on one site, disposed volumes were reduced by a third largely through filtration and re-use of the bio-lubricant, in contrast filtration of mineral-based oil may remove some of the additives. They also claim that for deep-hole drilling, a 900% increase in tool life had been achieved (Gallifa, 1999).

Health issues in the work place are improved by the use of vegetable oils as exemplified by rapeseed oil. The flash point of bio-based oils is higher than that for mineral oils. The environmental implications of oil spillage are greatly reduced where the bio-product is used. In addition, misting and associated implications of breathing vapours are reduced with the bio-based oil. Clearly, the use of bio-based oils offers a considerable list of benefits over mineral oils and this sector of the lubricants market shows considerable promise in terms of development. Why therefore has uptake been so poor so far?

There are a number of reasons for this. First, the bio-based lubricants are considerably more expensive, and experience with the product is required before the true comparative cost of the product with mineral oils can be judged. Secondly, bio-lubricants are not compatible with mineral oil-based products, therefore a machine would have to be totally cleaned before it would be possible to transfer to the new products. All filters would have to be changed and all oil lines drained and cleaned. Bio-lubricants therefore are best used from the start of a machine life. Bio-lubricants must not be used where there is a likelihood of the introduction of water, as this would encourage biodegradation. Other commercial factors may deter sales of bio-lubricants. Salesmen are often paid commission on volume of sales, thus moving to a lower volume of fewer products is not attractive to them. In addition, there is an inherent conservatism over lubricant use and machine operations will not change unless there is a clear need to do so.

According to EU figures (Table 7.11) without legislation, use of metal-working lubricants based on renewables will increase fourfold to 20 000 t (1.6% of total use) by 2010 but could increase to 50% of total use should legislation encourage their use. These figures are possibly pessimistic in that the technical, financial, and environmental/health figures relating to metal-working fluids appear stronger than that for other sectors and a greater potential for growth might be expected as a consequence. In terms of legislation, the probable restriction on the use of lubricants free of chlorine and heavy metals together with the other safety aspects (flash point, misting, biodegradability) of the bio-based product all tend to suggest probable expansion of this market sector (Cross, 2002). As with other sectors, EU vegetable oils – particularly rapeseed oil (and their esters) – could readily supply the base oil for these products.

### 7.3.7 Chainsaw oil

Chainsaw oils, or chain bar lubricants, are defined as lubricating oils whose purpose is to reduce friction between the chain and the cutting bar of both manual motor chainsaws and automated harvester machines. A continuous supply of lubricant to the saw chain and guide bar is required for chainsaws. In use, chainsaw lubricants are neither captured nor converted to other forms, but rather are left in the material being cut, on the operators, clothing, or dispersed to the environment (Committee ISO/CD 11679.3, 1996). Being a total loss system (all the lubricant is lost to the environment during use), a high degree of biodegradability is desirable for chainsaw bar lubricants. No problems arising from oxidation of the oil are encountered, due to the rapid use of oil by the saw. Operating temperatures rarely rise significantly above ambient temperatures, and so problems associated with thermal oxidation are irrelevant (Carruthers et al., 1995). However, because chainsaw bar lubricants are used in an outdoor environment, they must function over a wide range of temperature and moisture conditions while protecting the saw chain and guide bar from severe sliding and impact loads (Committee ISO/CD 11679.3, 1996).

In tests of vegetable oil-based lubricants vs mineral oil-based lubricants, it was noted that the vegetable oil appeared to reduce chain and bar wear, relative to the mineral oil product, although it was difficult to quantify (The Forestry Authority/Forestry Commission, 1994). The vegetable oil also appeared to reduce resin build-up on the chain, making maintenance and inspection easier than when using mineral oil (The Forestry Authority/Forestry Commission, 1994). Improved performance of harvesting machinery and reduced machinery wear were also reported by Kytö (1993). However, Anon. (1994) noted that a rapeseed-based chain bar oil reduced cutting blade temperature more than conventional oil and gave greater reduction in friction, but found no difference in blade wear. At present, there is insufficient data to quantify any potential benefits in terms of machinery wear and maintenance.

**Table 7.13** Composition of 1 kg chainsaw bar oil (after Mortensen et al., 1997)

| Component | Rapeseed chainsaw oil (g) | Mineral chainsaw oil (g) |
|---|---|---|
| Rapeseed oil | 924.2 | – |
| Polymethylmethacrylate | 50.0 | 2.0 |
| Di-2-ethylhexyladipate | 25.8 | – |
| Paraffinic mineral base oil | – | 995.5 |
| Polyisobutylene | – | 0.5 |
| Tall oil | – | 2.0 |
| Total | 1000.0 | 1000.0 |

The oil mist from vegetable oils is non-hazardous and non-irritant, and there are low levels of oil mist produced. In contrast, oil mist from mineral oils is irritant and hazardous, and there are high levels of oil mist as mineral oils have higher volatility than vegetable oils (Jőrsmo, 1996). Vegetable oils are free of PAHs and low in potential pollutants such as sulfur-containing compounds (Crawford et al., 1997) and also are biodegradable and a renewable resource. The pressure for increased use of vegetable oils in lubricants, due to both environmental and health and safety considerations, is likely to continue (Crawford et al., 1997).

It is widely reported that the consumption of vegetable-based chainsaw oil is lower than the consumption of mineral chainsaw oil, although the figures vary. A number of sources have reported that the flow rate of lubricating oil on the chainsaw pump can be reduced when vegetable oil is used, due to its superior lubricity and adhesion to the chain. Even when the flow rate of the pump is the same for both mineral and vegetable types of chainsaw oil, however, the consumption of vegetable oil-based lubricant is lower than that of mineral oil-based lubricant (Preskett, 1998).

The composition of 1 kg chainsaw bar oil is given in Table 7.13. Polymethylmethacrylate is an additive commonly used in many lubricants. It acts as a pour point depressant and viscosity index improver (Crawford et al., 1997). Pour point is the ability of the oil to retain fluidity at low temperatures, so ideally the pour point of the oil should be low, so that the oil continues to flow at low temperatures. Polyisobutylene is widely used in lubricants, as a viscosity index improver and also as it improves tackiness (tackifier) (Brown et al., 1997). Tall oil is obtained from the black liquor from paper pulping and is used as a carrier for some additives.

## 7.4 Surfactants

The term surfactants is derived from surface active agents. These substances are generally amphiphilic molecular structures which consist of a hydrophobic part (normally an alkyl chain) and a hydrophilic part (an ionic or strongly polar group). This structure accounts for their property of interfacial activity allowing a bond to develop at the surface of differing compounds. The range of application is considerable and has been classified as follows (ASPA, cited in Carruthers *et al.*, 1999).

(i) *Detergents* (including soap) where surfactants facilitate the removal of stains and dirt, and their dispersion in water in the washing and cleaning process.
(ii) *Dispersing agents* which increase the stability of suspensions of small solid particles in a liquid.
(iii) *Emulsifiers* which facilitate the dispersion of one liquid in another (e.g. oil and water), and thus increase stability.
(iv) *Wetting agents* which further the spreading of a liquid on a solid surface or its rate of penetration into porous materials such as leather, cotton, and paper.
(v) *Foaming and anti-foam agents* which cause or prevent the formation of foam respectively.
(vi) *Solubilisers* which increase the apparent solubility in water of slightly soluble substances.

These functions of surfactants are exploited in a wide range of consumer products and industrial processes.

There are thousands of different surfactants available. They are produced from natural fats and oils or from synthetic raw materials derived from petrochemicals. The most important natural feedstocks are palm kernel oil and coconut oil, though others such as tallow, palm, sunflower, and rapeseed can be used. Total world production of surfactants (including soap) is approximately 15 million tonnes of which 44% is produced in USA, western Europe, and Japan, and the remaining 56% by the rest of the world (Carruthers *et al.*, 1999). Approximately 45% of the world total are surfactant compounds (derived from both oleochemicals and petrochemicals) and 55% are soaps (all derived from oleochemical raw materials).

Raw material choice for surfactant manufacture is very price sensitive. There is some interest in natural raw materials as they are renewable, potentially low cost, of low toxicity, and low irritability. At present, surfactants from natural oils are obtained by processing oils with a high content of lauric ($C_{12}$), myristic ($C_{14}$), or palmitic/stearic ($C_{16}/C_{18}$) acids.

Most products contain a blend of fatty acids, the balance being adjusted according to the product and the price of the raw materials. Medium chain fatty acids ($C_{12-14}$) provide good lather and water solubility whilst longer chain fatty acids ($C_{16-18}$) provide detergency and are kinder to the skin. Although there may be potential to incorporate oleic acid into detergents, its inclusion is currently small and is supplied in the main by fractionated palm oil.

At present, there is limited scope for oilseed rape to supply the surfactant industry. Strategically, for this to happen it would require natural oils to win a greater share of the market from petrochemicals, and oilseed rape would have to compete effectively with tropical oils. The surfactant industry is dominated by petrochemical products largely for price and functionality reasons. With personal care products vegetable oil-based materials are, however, important and can compete with petrochemical products (Carruthers *et al.*, 1995). The public's demand for *natural products* is also encouraging a shift in this direction. However, it is the tropical oils that meet this demand much more than the temperate oilseeds. Yield of tropical oils is higher per unit area, and with lower production costs from these perennial crops, the unit cost is much lower. Volumes of supply are considerable and are set to increase further. These oils meet the specification laid down by the surfactant industry much more closely than does rapeseed oil. The production by Calgene Inc. of a transgenic oilseed rape variety in the mid-1990s with levels in excess of 40% lauric acid in the oil has offered the opportunity for producing the fatty acid via a crop suited to northern Europe. However, the more expensive rapeseed production prevented economic supply of this commodity other than for specialist niche markets, and the lauric rape project was subsequently shelved.

## 7.5 Paints and inks

The paint industry has used natural oils as carriers for coloured pigments for centuries. Paints nowadays may be sold as a solution, a dispersion, or a powder but all forms must dry to leave a solid-coating resistant to different forms of chemical or physical damage. The drying process may be purely due to solvent evaporation but usually involves cross-linking of the binder to form a high molecular weight three-dimensional network (Carr, 2000). Naturally occurring vegetable and mineral oils play an important role in surface coatings due to their versatile reactivity and are constituents of several binder types in many paint products.

Unsaturated vegetable oils react with oxygen on drying to produce unstable hydroperoxides which then decompose to produce cross-linking between molecules. The level of cross-linking depends on the degree of unsaturation of the fatty acids and impacts on the final quality of the coating. Oils such as tung, oiticica, and dehydrated castor oil, all of which have conjugated unsaturation, dry more quickly than those oils without conjugation.

Drying oils (e.g. linseed oil) can be used alone to produce coatings but have fallen from favour because of long drying times, the need for several coats, and because of health issues associated with the volatile aldehydes liberated during drying.

Currently, the largest use of vegetable oils in modern paints is in alkyd resin production. These are oil-modified polyesters which consist of a polyol, a multifunctional acid (phthalic acid or trimellitic acid), and an unsaturated fatty acid. Various other materials (e.g. acrylic, urethane, polyamide, styrene, and modified silicone) may be added to improve specific properties (Carr, 2000).

Choice of oils for use in these paints is determined by their drying characteristics which in turn are determined by the degree of unsaturation. For optimum drying at room temperature, an average of two non-conjugated carbon–carbon double bonds is required (Carr, 2000). The most suitable materials are soybean, tall, linseed, and dehydrated castor oils. Though sunflower oil resembles soybean and tall oils, it is regarded as expensive by paint manufacturers. The opportunities for rapeseed oil are very limited in this sector other than if modified oil varieties are considered. The only other area of potential for rapeseed oil is where the oleic acid can be derivatised with hydroxy or epoxy groups for use in coatings. In this situation, expense of the oil is likely to deter buyers. Conventional inks are largely made up of petroleum-based raw materials (USDA, 1993). Inks consist of dispersions of insoluble colourants or solution of dyes in a varnish or vehicle (Leach and Pierce, 1993). Modern inks also contain additives to improve drying time and adhesion, and also to ensure even flow with no foaming.

Traditionally, linseed oil was the main binding agent used in inks because of its rapid drying qualities. In the US, vegetable oil-based inks using soy oil were first marketed in 1987. By 1993, colour soy ink (approximately 10% inclusion rate) was used by half of the USAs 91 000 newspapers that use colour inks. Newsprint is the most attractive market for substituting conventional inks with vegetable oil-based inks. This is largely because of the ease with which recycled paper can have the pulp de-inked. Research at Western Michigan University has shown that soyoil-based inks can be de-inked more quickly and more clearly than conventional inks, resulting in higher quality recycled paper (USDA, 1993).

In the US, coloured soy inks are well established and are used in a wide range of printing activities (Table 7.14).

In the US, legislation has stimulated demand for soy oil-based inks as federal agencies and the USDA are required to use vegetable-based inks wherever possible.

**Table 7.14** Vegetable oil as a percentage of ink for various ink types

| Ink type | Vegetable oil as % of ink (by weight) |
|---|---|
| Newsprint | over 50 |
| Sheet-fed printing | 20–40 |
| Magazines | 10–15 |
| Business forms | 40 |

Source: Carruthers (1999).

In Europe, an estimated 12 000–15 000 t/year vegetable oil was used for printing inks in 1994 and estimates for 1997 ranged from 25 000–50 000 t/year (Carruthers *et al.*, 1995). Printing processes in which vegetable oils are used include intaglio, letterpress, lithography, screen printing, and dot inkjet printing (Davidson, 1999). Vegetable oils may be used in a number of different forms, including polymerised oils, modified oils (various), alkyds, epoxy esters, and polyamides. Whilst rapeseed oil has found a market within the printing ink industry replacing mineral oils, quantities used are relatively small. The main advantages of vegetable oil-based paints have been their environmental and health benefits – particularly the reduction in VOC emissions together with their ease in paper recycling. Compared to inks based on petroleum distillates, inks formulated with rapeseed methyl esters have given similar results in terms of rheological properties and printability, but drying has been slower (Gandini, 1999). The trend towards faster-running printing machinery and the need for inks to dry more quickly has resulted in vegetable oil content being reduced in certain applications in favour of quick setting systems (Davidson, 1999).

Another application in the printing industry has been the development of vegetable oil-based esters for cleaning printing presses. Here the aim has been to substitute volatile white spirits with non-volatile vegetable-based cleaning agents thereby improving the working environment of the printing press operator.

## 7.6 Polymers

The global market for all polymers is in excess of 125 million tonnes/year and consumption is growing faster than general economic growth (Carruthers *et al.*, 1999). Most polymers are derived from petroleum or natural gas but some incorporate vegetable oil derivatives either as functional additives or reactive ingredients.

Examples of functional additives based on vegetable oil are slip, anti-block, anti-static and plasticising agents, stabilisers, processing aids, and flame retardants. A major example of a slip agent is erucamide derived from high erucic acid rape seed (HEAR) – see Chapter 5. A range of oleochemicals are used as lubricants for plastics including fatty amides, metallic stearates, bisamides, fatty acids, and fatty acid esters (Leonard, 1990). Anti-static agents for protecting packaging materials from static electricity to stop products sticking together may be made from alkyl amines derived from fatty acids. Plasticisers and stabilisers used for PVC manufacture can be made from epoxidised soybean oil and mercaptoethyl oleates, and vegetable oils provide several other plasticisers for plastics used for food packaging and medical uses (Pryde and Rothfus, 1989). Polymerisation results in identical molecules being joined together to form larger molecules with properties different from that of the components. As part of this reaction, reactive ingredients may be attached to the main chain thus further adding to the polymer's characteristics. At present, commercial production of vegetable oil-based polymers is limited to some specialist plastics, e.g. castor oil for printed circuit boards. Research has also concentrated on polyamides (i.e. nylon) and also on deriving polyesters and polyurethanes (via polyols) from vegetable oils. Formation of dicarboxylic acids through *dimerisation* of unsaturated fatty acids produces chemical intermediaries useful for polyester resins and other polymers. Oxidative cleavage of unsaturated fatty acids can produce useful shorter-chain dicarboxylic acids.

Other advances include the development of microbial yeasts for transforming fatty acids to dicarboxylic acids and the study of the range of polymers that could be derived from these acids. Work in the Netherlands is currently looking at a bacterial transformation of vegetable oils to yield PHA (polyhydroxyalkanoate) plastics. Still further work is investigating methods of catalytic hydroxylation to produce a new range of polyols (Carruthers *et al.*, 1995). Another approach is the provision of a direct source of polymers via the genetic modification (GM) of the plant's biochemistry. Whilst this is theoretically possible, it is technically difficult and commercialisation may be discouraged as many countries particularly in Europe show reluctance to embrace GM technology. New opportunities for rapeseed oil within the polymers market do exist. However, the use of vegetable oils is limited currently to certain specialist plastics. Increased use of vegetable oils is possible – particularly in the manufacture of polyamides and polyurethanes – but the big markets are for small ($C_2$) alkene units which are very unlikely to be obtained economically from vegetable oils. It is also important to note that plastics from renewable resources can also be produced from starches and proteins, and plant fibres can also be incorporated into some plastics. Development of starch-based biodegradable plastics is further advanced than vegetable oil-based processes and seems more likely than oilseed-based products to penetrate the market for biodegradable and environmentally benign plastics. In addition, as vegetable oils contain a range of fatty acids and other components, any desired fatty acid for polymer production will *leave behind* a large range of other constituents. In conclusion, the range of plastics or plastic precursors that could be derived from rapeseed oil appears limited. The polyol market appears to be the area with the most potential.

## References

ABI 2003 in Entwistle, G., Walker, K.C., Booth, E.J. and Walker, R.L. (2003). Biodiesel in Europe – outlook and policy issues. *Proc. 11th International Rapeseed Congress,* Copenhagen, July 2003, 608–610.

Anon. (1994) The Forestry Authority/Forestry Commission, Biodegradable Chain Oil, Technical Branch Information Note 1/94.

Anon. (2003) EU Energy and Transport in Figures ISBN 92-894-6244-2. Pub. European Communities, Luxemburg.

ASPA (cited in Carruthers *et al.*, 1999) ASPA. Syndicat national des fabricants d'agents de surface et de produits auxiliares industriels. Paris: ASPA.

Batchelor, S.E., Booth, E.J. and Walker, K.C. (1995) Energy analysis of rape methyl ester (RME) production from winter oilseed rape. *Industrial Crops and products*, **4**, 193–202.

Batel, W. (1980) Pflanzenöl Für die Kraftstoff-und Energieversorgung. *Grundlagen Landtechnik* 30, No. 2.

Battersby (2002) Extending the performance of oilseed rape-based hydraulic fluids. *Conference: Lubricating the Market: the future for bio-based lubricants*, Warwick, May 2002, *ACTIN*.

Betton, C.I. (1993) Relevance of eco-labelling for the oil industry: what will the requirement be for hydrocarbon products? In: *Life Cycle Analysis and Eco-assessment in the Oil Industry* (ed. J. Phipps), pp. 42–46, Proceedings of Conference on 26 November 1992, London: Institute of Petroleum.

Bondoli, P. and Igartua, A. (2000) Lubricants and hydraulic fluids. *CTVO-NET Final Conference Proceedings*, Bonn, Germany, June 2000, pp. 43–54.

Boustead, I. (1997) Eco-profiles of the European Plastics Industry: polycarbonate. Report 13: A report for the Technical and Environmental Centre of the Association of Plastics Manufacturers in Europe (APME), Brussels, September 1997, APME.

Brown, M., Fotheringham, J.D., Hoyes, T.J., Mortier, R.M., Orszulik, S.T., Randles, S.J. and Stroud, P.M. (1997) Synthetic base fluids. Chapter 2. In: *Chemistry and Technology of Lubricants* (eds R.M. Mortier and S.T. Orszulik), 2nd Edition, pp. 34–74, London: Blackie Academic and Professional – An imprint of Chapman & Hall.

Carr, C. (2000) The potential of vegetable oils for paints and coatings – have we just scratched the surface. *CTVO-Net Final Conference*, Bonn, Germany, June 2000, pp. 68–76.

Carruthers, S.P., Marsh, J.S., Turner, P.W., Ellis, F.B., Murphy, D.J., Slabas, T. and Chapman, B.A. (1995) *Industrial Markets for UK-produced Oilseeds*. Research Review No. OS9, HGCA.

Carruthers, S.P., Walker, K.C., Wightman, P.S., Jones, P.J., Eavis, R.M., Tranter, R.B., Bennett, R.M., Little, G.P.J., Harrison Mayfield, L.E., Batchelor, S.E., Rehmann, T. and Miller, F.A. (1999) Cost-benefit analysis, including life-cycle assessment of oils produced from UK-grown crops compared with mineral oils. Final report for MAFF, August 1999.

Committee ISO/CD 11679.3 (1996) Lubricants for chain saws – specification of chain lubricants.

CONCAWE (1979) The environmental impact of refinery effluents. *CONCAWE Report No. 5/79*.

Copan, W.G. and Haycock, R. (1993) Lubricant additives and the environment. CEC publication, May 1993, CEC/93/SP02, *4th International Symposium on the Performance Evaluation of Fuels and Lubricants*. NEC, Birmingham, England, 5–7 May 1993.

Crawford, J., Psaila, A. and Orszulik, S.T. (1997) Miscellaneous additives and vegetable oils. Chapter 6. In: *Chemistry and Technology of Lubricants* (eds R.M. Mortier and S.T. Orszulik), 2nd Edition, pp. 181–202, London: Blackie Academic & Professional – An imprint of Chapman & Hall.

Cross, C. (2000) Renewable resource based products used in machining and farming operations. *CTVO-Net Final Conference*, Bonn, Germany, June 2000, pp. 21–23.

Cross, C. (2002) The use of replenishable resource fluids for product manufacture. *Conference: Lubricating the Market: the future for bio-based lubricants*, Warwick, May 2002, *ACTIN*.

Culshaw, F. and Butler, C. (1992) *A Review of the Potential for Biodiesel as a Transport Fuel*. ETSU-R-71 publication, HMSO, London.

Davidson, J. (1999) New uses for vegetable oils in printing ink formulations. Paper 7 ACTIN Workshop, London, June 1999, Pub. CTVO Net EU Concerted Action.

Defrang, M. (1999) Loss lubricating use of vegetable oil and derivatives in the concrete industry. *CTVO-NET Workshop*, Eibar, Spain, February 1999, 14pp.

Diesel, R. (1895) German patent no. 82168.

Ehrenberg, J. (2002) Current situation and future prospects of EU industry using renewable raw materials. EU Working Group "Renewable Raw Materials". DG Enterprise Unit E.1: Environmental Aspects of Industry Policy Brussels, February 2002.

Ferdinand, C. (ed.) (1992) Non-food uses of agricultural products in Europe. In: *Study written by the Club de Bruxelles* (for conference on 'Biofuels in Europe'). Club de Bruxelles, Bruxelles, Belgium.

Gallifa, R. (1999) Lubricant fluids based on vegetable oils – industrial applications. *CTVO-NET Workshop*, Eibar, Spain, February 1999, 11pp.

Gandini, A. (1999) The use of renewable resources as components of off-set inks. Paper 8 ACTIN Workshop, London, June 1999. Pub. CTVO Net EU Concerted Action.

Goetz, W.A. (1994) Evaluation of biofuels in heavy-duty engines on an engine dynamometer. In: *Proceedings of Bio-oils Symposium.* Saskatchewan Canola Development Commission, Saskatoon, Canada.

Havenith, C. (1993) Pflanzenölmotoren (Vegetable oil engines). In: *Proceedings of the Symposium Kraftstoffe aus Pflanzemől fűr Dieselmotoren* (Fuels from vegetable oils for diesel engines). Technische Akademie Esslingen, Ostfildern, Germany, 14pp.

Hewstone, R.K. (1994) Health, safety and environmental aspects of used crankcase lubricating oils. *The Science of the Total Environment*, **156**, 255–268.

Jőrsmo, M. (1996) Lubricant manufacturers' oil raw material requirments – can natural oils meet them? In: *Lubricants from Oilseeds Seminar*, London, 9 May 1996.

Junek, H. and Mittelbach, M. (1986) *Austrian patent AT 386*, 222 Institute for Organic Chemistry, Hienrichstr. 28, A8010 Graz, Austria.

Kasterine, A. and Batchelor, S. (1998) Biodiesel – environmental impact and potential for niche markets. HGCA Research Review OS12. Pub. HGCA London.

Kőrbitz, W. (1995) Utilization of oil as a biodiesel fuel. *In: Brassica Oilseeds, Production and Utilization* (eds D. Kimber and D.I. McGregor), Pub. CAB International, Cambridge.

Kytő, M. (1993) Biodegradable vegetable oil based lubricants in timber harvesting – a field test. Pub. Technical Research Centre, Finland.

Lea, C. (2002) The versatility of lubricants derived from renewable resources – diverse new applications from production lubricants to novel engine sump oils. *Conference: Lubricating the Market: the future for bio-based lubricants.* Warwick, May 2002, *ACTIN*.

Leach, R.H. and Pierce, R.J. (eds) (1993) *The Printing Ink Manual*. London: Blueprint.

Leonard, E.C. (1990) The marketing and economics of oleochemicals as plastics additives. In: *World Conference on Oleochemicals in the 21st Century* (ed. T.H. Applewhite), Illinois: American Oil Chemists Society.

Mittelbach, M. and Tritthard, P. (1988) Diesel fuel derived from vegetable oils. III Emission tests using methyl esters of used frying oil. *Journal of the American Oil Chemists' Society*, **65**, 7.

Mortensen, B., Hauschild, M., Weidema, B., Nielsen, P., Schmidth, A. and Christensen, B.H. (1997) *Livscyklusvurdering af produkter baseret pa fornybare ravarer. (Lifecycle analysis of products based on renewable resources),* Draft Report IPY, Denmark.

Muntada, L. (1999) The use of vegetable oils in lubricating greases. *CTVO-NET Workshop*, Eibar, Spain, February, 1999, 10pp.

Prankl, H. and Wőrgetter, M. (1995) Standardisation of biodiesel on a European level. *Proceedings 3rd European Motor Biofuels Forum*, Brussels, Belgium.

Preskett, D. (1998) *On the suitability of food-grade cooking oil as a sawchain lubricant – comparative field tests and laboratory analysis.* BSc (Hons) Forestry thesis, University of Wales, Bangor.

Pryde, E.G. and Rothfus, J.A. (1989) Industrial and non-food uses of vegetable oils. In: *Oil Crops of the World: their Breeding and Utilisation* (eds G. Robbelen, R.K. Downey and A. Ashri), New York: McGraw-Hill.

Rasberger, M. (1997) Oxidative degradation and stabilisation of mineral oil based lubricants. Chapter 4. *In: Chemistry and Technology of Lubricants* (eds R.M. Mortier and S.T. Orszulik), 2nd Edition, pp. 98–143, London: Blackie Academic & Professional – An imprint of Chapman & Hall.

Sala, M. (2000) Analytical evaluation of esters based lubricants. *CTVO-NET Workshop Milan*, Italy, February 2000. pp. 7–24.

Sams (1996) *Use of Biofuels under Real World Engine Operation 2nd European Biofuels Forum.* 225–233 Mudge (1997).

Sams, T. and Schindlbauer, H. (1992) Untersuchungen über den Betrieb von Dieselmoteren mit RME (Investigations on running a diesel engine with rapeseed methyl-ester). In: *Proceedings of the Symposium 'RME' – Kraftstoff und Rohstoff*, FICHTE/Technical University, Vienna, Austria, pp. 49–66.

Schäfer, A. (1991) Pflanzenölfettsäure-Methyl-Ester als Dieselkraftstoffe/Mercedes Benz. In: *Proceedings of the Symposium Kraftstoffe aus Pfanzemöl für Dieselmotoren* (Fuels from vegetable oils for diesel engines). Technische Akademie Esslingen, Ostfildern, Germany.

Schulla, P. (2002) High performance lubricants/liquids. *Conference: Lubricating the market: the future for bio-based lubricants*, Warwick, May 2002, *ACTIN*.

Schűtt, H. (1982) Natűrliche Rohstoffe zur Herstellung von Fettalkoholen (Natural raw materials for production of fatty alcohols). In: *Fettalkohole*. Henkel KGAA, Dűsseldorf, Germany, pp. 13–49.

USDA (1993) Industrial uses of agricultural materials: situation and outlook. Commodity Economics Division, Economic Research Service, USDA, June.

Walker, K.C. and Kőrbitz, W. (1994) Rationale and economics of a UK biodiesel industry, Vienna, January. *A report prepared for the British Association for Biofuels and Oils*, Scottish Agricultural College, 581 King Street, Aberdeen AB9 1UD.

Walter, T. (1992) Untersuchungen des Emissionsverhaltens von Nutzfahrzeugmotoren am Prűfstand bei Betrieb mit RME (Investigations on emissions of lorry engines in a bench test using rapeseed methyl ester). Swiss Research Institute, Dűbendorf. *Proceedings of the Symposium "RME" – Draftstoff und Rohstoff*, 49–66, FICHTE/Technical University, Vienna, Getreidemark 9, A-1060 Vienna, Austria.

Zehe, M. (2000) New developments on environmentally friendly greases. *CTVO-NET Workshop*, Milan, Italy, February 2000, pp. 25–33.

# 8 Potential and future prospects for rapeseed oil

Christian Möllers

## 8.1 Introduction

World vegetable oil markets are highly competitive, requiring a steady improvement in oil quality to maintain or increase market share. Competitiveness of the rapeseed crop will depend in the future primarily on general quality improvements, like increasing seed oil and protein content. The objective of modifying oil quality is to develop oils with enhanced nutritional and functional properties, which require, if possible, no further processing for specific end-use markets. Modification of the fatty acid composition for various food, feed and numerous industrial applications is already well on its way. Here, it has to be distinguished whether *new* oil qualities have the potential to become a new commodity, or whether the market prospects are limited to a speciality oil. Increasing the content and modifying the composition of minor bioactive compounds like tocopherols, sterols and carotenoids can be regarded as general quality improvements and hence may also help to improve the value of the crop. In a similar way, reduction of compounds regarded as neutral or undesirable like glucosinolates, cellulose and hemicellulose, tannins, phenolic acid esters, lignins, chlorophyll, etc. will help to improve the general quality and reduce the production and processing costs, e.g. for bleaching and deodourization.

The majority of rapeseed varieties now being used to produce commodity and speciality oils were produced by conventional plant breeding techniques. There was no conscious effort to exclude modern biotechnology as a transformation tool, but efforts to modify oils have been going on for a long time, before genetic engineering became a practical tool to change oilseeds (Krawczyk, 1999). Considering the present achievements and the current knowledge of molecular biology, it has already become clear that, in principle, every biochemical pathway can be modified to improve or diversify the quality of oilseed crops. Now, we are coming to the point where it is necessary to determine which traits are desired, and in what order (Anonymous, 1998). One key point in finding an answer to this question is that a premium needs to be charged on speciality oils to guarantee *return of investment* and to account for the additional costs to keep speciality oils separate from start to finish in the breeding-to-market channels (Krawczyk, 1999). Rapeseed with a novel fatty acid composition or otherwise improved quality must be kept apart, or *identity preserved* (IP), from the commodity quality throughout the production and distribution chains to retain its added value. Maintaining an identity preservation system adds to the cost of speciality oils.

There are few examples of successfully cultivated IP-crops. High erucic acid rapeseed (HEAR) is one of them. It is now grown as an IP-crop in Europe and in the USA/Canada to produce erucic acid for non-edible applications (see Chapter 5). Other more recently developed IP-rapeseed forms include low linolenic and/or high oleic acid oil types. Most farmers grow IP-crops under contract. On one side this guarantees income and reduces risks to farmers, on the other side it provides a unique way to achieve a seamless quality control system and to credibly certify the quality of the harvested crop. A peculiarity of the rapeseed crop is that IP-rapeseed forms in many regions of the world have to be grown in restricted geographical regions. This is due to the possibility of cross-pollination which in many cases may influence the seed oil quality, if the pollen is from a different oil quality genotype. Cross-pollination may be derived from volunteer plants growing in the field or from different rapeseed quality types growing in nearby fields.

The value of an IP-rapeseed crop can be increased by combining, or stacking, traits that add additional value to the product or possible side-products. Trait combinations, in which the same processing technology can be used to produce different products, e.g. distillation can be used to separate erucic acid from other high value fatty acids, provide additional advantage. Similarly, the condensate obtained after oil deodourization of an IP-rapeseed form could also be used as a – possibly enriched – source for the isolation of high value tocopherols, phytosterols or carotenoids. Trait combinations that address similar markets or, for ingredients, present similar acreage demands should also be chosen (Mazur *et al.*, 1999). Increasing the value of side-products includes also the rapeseed meal quality as protein-rich source for feed and food purposes. In some cases, it may be useful to uncouple the production of oil and protein so that the value of the oil is not dependent on the value of the protein. This may be particularly important for production of useful, non-edible fatty acids, which may prove to be unacceptable as post-processing carryover even at low levels in seed proteins used for food and feed (Somerville and Bonetta, 2001).

An intriguing and tempting challenge is converting a speciality oil rapeseed form over to an entire commodity crop. Benefits can be sizable, as high volumes combined with the absence of IP-costs reduce the need for a speciality oil premium. An improved quality and functionality will lead to greater use in current and possibly also in new food and non-food applications. Rapeseed is already a success story in this regard. In the 1970s, the rapeseed commodity oil was high in erucic acid. New cultivars which produced an erucic acid-free oil were introduced as a speciality oil, but very soon became the dominating commodity. In combination with the reduction of the glucosinolate content in the meal in the 1980s, the erucic acid-free rapeseed quality was replaced by the new standard, the now well-known *Canola* or '00'-quality rapeseed.

Development and breeding of new oil qualities requires efficient tools to monitor trait segregation in a breeding programme. Near-infrared reflectance spectroscopy (NIRS) has proven in the past to allow efficient quality analysis of a large number of samples (Velasco, 1999). Nowadays, oil, protein, glucosinolate and moisture content are determined routinely by NIRS in all larger rapeseed-breeding programmes. More recently, this technology is also being used to determine the fatty acid composition of the seed oil (Velasco and Becker, 1998). The big advantage of NIRS is that it allows simultaneous determination of several quality traits in a quick and non-destructive manner. NIRS will remain an important technology for quality breeding and quality control even in the future, with increased diversification of rapeseed oil quality. New developments show that NIRS can also be used to determine the fatty acid composition in single seeds derived from segregating F2-populations. This should also considerably speed up the breeding progress, especially if segregation is complex due to the action of several genes (Velasco et al., 1999).

High expression of a new trait in rapeseed will in many cases require the molecular transfer and expression of several different genes. This complicates further breeding and cultivar development, because of complex segregation of transgenes in progenies. Fortunately, progress is also being made in the size of constructs that can be cloned and transformed into plants. Hence, multiple transgenes can be transformed within one construct and are inherited in the transformed plants as a single mendelian trait (Houmiel et al., 1999).

This chapter summarizes past and current efforts both by traditional seed-breeding methods and by genetic engineering to modify and diversify rapeseed oil quality and will depict possible lines of evolution. Some of the previous chapters of this book, as well as recently published review articles on rapeseed oil quality (Przybylski and Mag, 2002) and molecular genetic modifications of storage lipids (Voelker and Kinney, 2001), complement the present overview.

## 8.2 Oil content and triacylglycerol structure

In some countries, farmers are already paid a premium for above-average oil content. Hence, increasing the oil content is one of the primary breeding aims to maintain and increase the competitiveness of the rapeseed crop. Slow but steady progress in increasing oil content in the future can be expected as a result of traditional seed-breeding methods in combination with the application of the NIRS technology. A larger step forward in increasing the oil content may be achieved through the recently introduced yellow-seeded rapeseed material into breeding programmes (Rakow et al., 1999). Compared to the black-seeded standard varieties the yellow-seeded material has a thinner seed coat, leading to a relatively higher oil and protein content. The black seed coat is rich in tannins which negatively influences the colour and probably other characteristics of the oil (Matthäus, 1998). Although mechanically dehulled seeds of rapeseed have been available for quite a while, very little is known to date on the influence of the black seed coat on the oil quality. Progress in increasing the oil content can also be expected from genetic engineering. However, since oil content is inherited and influenced by a large number of genes, a quantum jump forward in increasing the oil content cannot be expected from the transformation of a single gene. Recent results from these experiments indicate some progress in increasing the oil content, but in many cases results need to be confirmed in field experiments using genotypes with an already high oil content (Roesler *et al.*, 1997; Zou *et al.*, 1997; Jako *et al.*, 2001).

Triacylglycerols are the most abundant lipid class in rapeseed oil. The fatty acid composition of the triacylglycerol is important for their physical characteristics. In many species including rapeseed, it has been found that there is no random distribution of fatty acids at the triacylglycerol backbone. As a general rule, the saturated fatty acids are confined to positions *sn*-1 and *sn*-3, whereas the polyunsaturated fatty acids are located mainly at the *sn*-2 position (Padley *et al.*, 1994; Frentzen, 1998). There is a large body of evidence that the fatty acid distribution results from the respective specificities of the acyltransferases involved in the biosynthesis of triacylglycerols. The transferases responsible for the esterification at *sn*-1 and *sn*-3 positions, glycerol-3-P acyltransferase and diacylglycerol acyltransferase respectively, appear to accept saturated and unsaturated acyl-CoA substrates. In contrast, lysophosphatidic acid acyltransferase (LPAAT), which catalyzes the esterification at *sn*-2 positions, appears to discriminate against saturates in most oilseeds (Frentzen, 1998). As will be shown below, plant breeding and genetic engineering can be used to modify the fatty acid composition and the distribution of the fatty acids at the triacylglycerol backbone. This may result in new triacylglycerol species with modified physical characteristics that make these oils suitable for a variety of specialized applications (Padley *et al.*, 1994).

## 8.3 Modification of the $C_{18}$ fatty acid composition

### *8.3.1 Development of rapeseed with modified linolenic acid (18:3) content*

From a nutritional point of view, the rapeseed oil has an excellent quality because of its high content of the essential ω6-linoleic acid (18:2*n*-6) and ω3-linolenic acid (18:3*n*-3), and because it has the lowest content of saturated fatty acids among vegetable oils. However, due to the high content of polyunsaturated fatty acids, the oil has also only a low oxidative stability, which is a particular drawback when the oil is used for frying and deep frying purposes. Based on induced mutants of the linoleic acid desaturation (Rakow, 1973), attempts have been made to breed oilseed rape cultivars low in linolenic acid. This has resulted in the release of low linolenic (*LL*) spring and winter rapeseed

**Table 8.1** Rapeseed oil qualities with varying $C_{18}$- fatty acid composition (fatty acids in %)

| Oil quality | Palmitic acid 16:0 | Oleic acid 18:1 | Linoleic acid 18:2 | Linolenic acid 18:3 | Erucic acid 22:1 |
|---|---|---|---|---|---|
| '00' resp. Canola | 4 | 60 | 20 | 10 | <2 |
| High erucic acid (HEAR) | 3 | 15 | 13 | 9 | 58* |
| Low linolenic acid (LL) | 4 | 60 | 30 | 2 | <2 |
| High oleic acid (HOAR) | 3 | 84 | 5 | 4 | <2 |
| High oleic acid/low linolenic acid (HOLL) | 3 | 84 | 7 | 2 | <2 |

* Including 8% eicosenoic acid (20:1).

cultivars with less than 3% linolenic acid in the seed oil (Table 8.1). However, at present there is no identity preserved cultivation of *LL*-rapeseed in Europe, and in Canada/USA it is only of minor importance.

The reduction in linolenic acid content led to an increase in linoleic acid, but not to a change in the oleic acid content (Rücker and Röbbelen, 1996). Using gene technological methods, a reduction of linolenic acid content to levels of 2.5% and below has also been achieved (De Bonte and Hitz, 1998). Further reductions in the linolenic acid content of the triacylglycerols could probably be achieved in the future through inhibition of the plastidial linoleic acid desaturase (*fad7*-gene in Fig. 8.1), although it is questionable whether this may affect the viability and vigour of the germinating seedling. On the other hand,

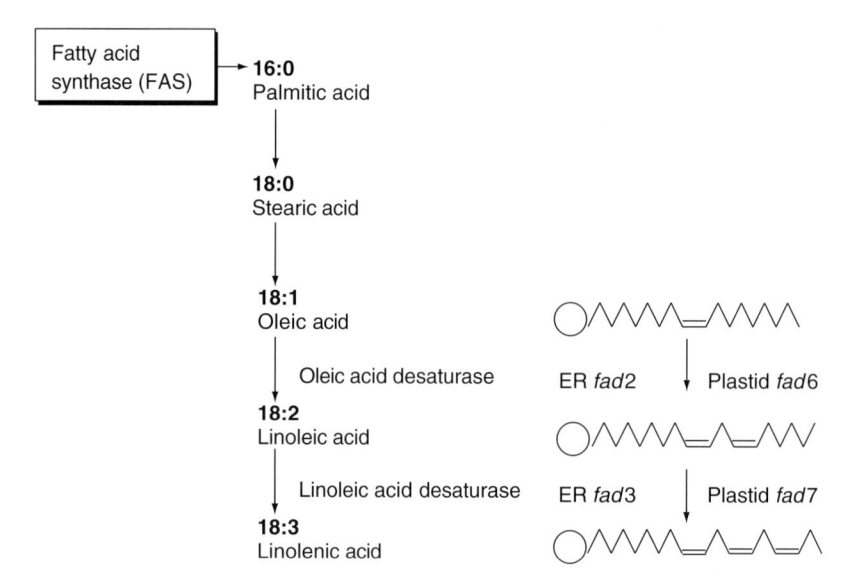

**Figure 8.1** Schematic drawing of the oleic acid desaturation in the plastids and at the endoplasmic reticulum (ER).

the example of sunflower shows that seeds with zero linolenic acid in the triacylglycerols are fully viable. Rapeseed genotypes with a high linoleic acid or linolenic acid content may be useful as a starting material for genetic engineering approaches aimed at the production of ω6- or ω3-long chain polyunsaturated fatty acids respectively (LC-PUFAs). Following overexpression of a ω3-desaturase in rapeseed, an oil with up to 58% α-linolenic acid has been reported (Ursin *et al.*, 2000).

### 8.3.2 *Development of rapeseed with an increased oleic acid (18:1) content*

High oleic (HO)-sunflower, HO-soybean and HO-safflower cultivars have been available for some years and are grown to some extent in Europe, Canada/USA and Australia. More recently, high oleic acid oilseed rape (HOAR) forms are being developed for cultivation in the more temperate regions around the world.

Increases in oleic acid content can be achieved by reducing the activity of oleic acid desaturase, the enzyme which converts oleic acid into linoleic acid in the developing seed (Fig. 8.1). There are two isoforms: one is located in the cytosol at the endoplasmic reticulum (FAD2, oleoyl-phosphatidylcholine Δ12-desaturase, Fig. 8.1), the other is active in the chloroplasts (FAD6, Fig. 8.1). Cytoplasmic desaturation of oleic acid is predominant, however, due to fatty acid exchange between cytoplasma and chloroplasts, it is at present unclear whether some of the linoleic acid found in the triacylglycerides is derived from plastidial desaturation. On the DNA level, two types of *fad2*-alleles, one from the *Brassica oleracea* and the other from the *Brassica rapa* genome, have been cloned from *Brassica napus* (P. Spiekermann, University Hamburg, personal communication). The *fad2*-genes seem to be expressed in a seed-specific manner, whereas the *fad6*-genes are probably constitutively expressed in the whole plant. Mutations in these latter genes can have detrimental effects on vegetative tissue where the correct fatty acid composition is required for normal membrane structure and function (Miquel and Browse, 1994).

Chemically induced mutants have been successfully obtained which raise oleic acid contents up to 80% (Auld *et al.*, 1992; Rücker and Röbbelen, 1995). Further selection in the mutant lines developed by Rücker and Röbbelen (1995), crossing with other high oleic acid genotypes, and continued selection for a high oleic acid content led to the development of genotypes that had 86% oleic acid in the seed oil, as determined after growing the material in the field (Table 8.1; Schierholt and Becker, 2000). These experiments have shown that beside one major gene affecting the oleic acid content, three or more minor genes need to be considered if oleic acid contents of around 85% are to be achieved. This complicates breeding of high oleic acid rapeseed. Fortunately, Schierholt and Becker (2001) found only minor genotype – environment interactions for the HO-trait. The observed high heritability of $h^2=0.99$ indicates that selection of high oleic acid genotypes is very efficient. Environmental effects were significant but small, showing that environmental conditions should not influence the oleic acid content to a larger extent. In further experiments, a positive correlation between oleic acid and oil content (Möllers and Schierholt, 2002), and a negative correlation between oleic acid and yield were found. The increase in oil content (0.6–1%) was not sufficient to compensate for the lower yield (–5%) when the oil yield was calculated (Schierholt, 2001). At present, it is unclear whether the yield reduction is a pleiotropic effect of the HO-mutation or whether it is due to linked genes which are associated with yield reductions. The observed yield reduction of the high oleic acid genotypes was not evident in all tested crosses. Hence, it should be possible to compensate for the yield disadvantage by an increased breeding intensity. Furthermore, other authors did not report a yield drag of the HOAR types (Chungu *et al.*, 2001).

A more detailed molecular analysis of one of the mutants used by Schierholt *et al.* (2000) showed that the *fad2*-allele of the *B. rapa* genome was affected (P. Spiekermann, University Hamburg, personal communication). Further increases in oleic acid content could be expected by combining this material with rapeseed mutants carrying a mutated non-functional *fad2*-allele in the *B. oleracea* genome.

A successful increase in oleic acid content in rapeseed has also been achieved through the application of molecular cosuppression and antisense technology. DeBonte and Hitz (1998) reported the regeneration of transgenic canola plants with 86% oleic acid. Similar results were obtained by Stoutjesdijk *et al.* (1999). These approaches aimed at reducing the activity of the cytoplasmic oleic acid desaturase. Obviously, 85% oleic acid seems to be the threshold for what can be achieved by this approach. It needs to be seen whether an additional seed-specific down-regulation of the plastidial oleic acid desaturase (FAD6) in these genotypes will lead to oleic acid contents of 90% and above.

At present mid-oleic and high oleic acid cultivars with around 75% and more than 80% oleic acid respectively in the seed oil are available in Australia and North America (Abidi *et al.*, 1999; Scarth and McVetty, 1999). Spring and winter rapeseed cultivars with an increased oleic acid content of up to 75% have recently been registered in Europe.

### 8.3.3 Development of high oleic acid/low linolenic acid oilseed rape

High oleic acid/low linolenic acid oilseed rape (HOLL) can be obtained by crossing HO- and LL-rapeseed forms and selection of the desired recombinants in the offspring. Alternatively, rapeseed plants may be transformed with a construct carrying in combination of the *fad2/fad3*-genes under the control of seed-specific promoters to achieve silencing through cosuppression/antisense (DeBonte and Hitz, 1998). Field testing of a doubled haploid population segregating for high oleic acid and low linolenic acid content showed that the combination of the two traits does not lead to an increase in oleic acid content (own unpublished results). Due to an increased number of genes involved in the inheritance of HOLL rapeseed, breeding of this quality type is more complicated. However, it seems that there is an increasing demand for this type of oil (Scarth and McVetty, 1999).

### 8.3.4 Development of high stearic acid (18:0) oilseed rape

Development of a high stearic acid form in rapeseed seems rewarding, since this would be a useful oil source for the production of shortenings and margarine and could reduce the need for hydrogenation of liquid oils or may even render it unnecessary. The advantage of stearic acid in comparison with shorter chain saturated fatty acids is that it apparently does not cause increased blood cholesterol contents. The plastidial stearoyl-acyl carrier protein (ACP) desaturase catalyzes the initial desaturation reaction in fatty acid biosynthesis and thus plays a key role in determining the ratio of total saturated to unsaturated fatty acids in plants. Expression of a seed-specific stearoyl-ACP desaturase antisense construct in *B. napus* led to an increase of stearic acid in the seed oil up to 40% (Knutzon *et al.*, 1992). This increase in stearic acid was accompanied by a decrease in oleic acid, the product of the stearoyl-ACP desaturase, and by a significant increase in linolenic acid. Levels of palmitic acid were reported to be not significantly affected in this material. High stearic acid *B. rapa* plants transformed with the same construct had an oil content reduced by almost a half and, probably as a consequence of this, an increased 18:3 content (up to 25%). Further investigations are necessary to investigate the effects of a high stearic acid content on agronomic characters in *B. napus*. The expression of an acyl-ACP thioesterase of mangosteen (*Garcinia mangostana*) in seeds of *B. napus* led to the accumulation of stearic acid up to 22% in the seed oil (Hawkins and Kridl, 1998). Abidi *et al.* (1999) analysed among others one rapeseed cultivar with 22% stearic acid in the seed oil for minor constituents of the oil, but they did not report any significant differences.

## 8.4 Low saturated fatty acids

In a number of industrialized countries, dietary recommendations limit total fat intake to 30% of the energy and saturated fatty acid intake to 10% of the energy. Consumption of saturated fatty acids with a chain length of 12–16 has been related to an increased blood cholesterol content, especially of the low-density lipoprotein (LDL) cholesterol, and an increased incidence of coronary heart disease. The labelling regulations in the United States and in Canada allow oils with less than 7.1% saturated fatty acid content to be advertised as low in saturated fatty acid. Among all vegetable oils, rapeseed oil contains the lowest content of saturated fatty acids. However, attempts are being made to reduce the saturated fatty acid content in other oil crops, e.g. soybean, to levels comparable to rapeseed oil. Intensive marketing of the low saturated fatty acid attribute has established canola oil as a premium quality vegetable oil in the North American market with a predominant market share (see McVetty and Scarth, 2002). Defence of this market share of canola oil does not only require the maintenance of the low saturated fatty acid content, but needs additional efforts to achieve further reductions in the saturated fatty acid content. In this connection, the Canola Council of Canada has as a mid-term goal the reduction of the saturated fatty acid content to values below 6% and as a long-term goal to values below 3.5%. According to the present US labelling regulations, an oil with 3.5% saturated fatty acids could be identified as *zero saturates* (Rakow and Raney, 2001).

Present breeding approaches to lower the saturated fatty acid content include the resynthesis of rapeseed from its ancestors *B. rapa* and *B. oleracea*, using genotypes with a low saturated fatty acid content, mutagenesis experiments and transgenic approaches (Wong *et al.*, 1995; Rakow and Raney, 2001). Oilseed rape acyl-ACP thioesterase has a strong preference for 18:1-ACP as a substrate although it can also use 16:0-ACP (Slabas *et al.*, 2001) and probably 18:0-ACP. This could explain why only small amounts of 16:0 and 18:0 are exported from the plastid. Identification of new acyl-ACP thioesterase alleles with an even stronger preference for 18:1-ACP could help in breeding rapeseed forms with an even lower saturated fatty acid content. Overexpression of the stearoyl-ACP desaturase gene (Knutzon *et al.*, 1992) could lead to reduced stearic acid contents in the seed oil of rapeseed. Results from transgenic tobacco plants indicate that a reduction in saturated fatty acids (16:0 and 18:0) and a concomitant increase in corresponding monounsaturated fatty acids (MUFA) can be achieved through the expression of a stearoyl coenzyme A desaturase from rat (Hildebrand and Grayburn, 1999). This is a $\Delta$9-desaturase that uses the intermediate CoA-form of plant fatty acids as substrate. A moderate reduction in saturated fatty acid content may also be achieved through breeding for high oleic acid content, since these traits are reported to be negatively correlated (Auld *et al.*, 1992; Möllers and Schierholt, 2002). Near-infrared spectroscopic calibrations have been developed for the saturated fatty acid content in rapeseed (Pallot *et al.*, 1999). Application of these should accelerate cultivar development with reduced saturated fatty acid content in the seed oil.

## 8.5 Medium and short chain fatty acids

Medium and short chain fatty acids (8:0 to 14:0) do not occur in normal rapeseed oil. However, a number of species do have these types of fatty acids in their storage lipids. Thioesterases were identified in these species as the main chain-length determining enzymes in fatty acid biosynthesis. Acyl-ACP thioesterases are plastidial enzymes catalyzing the hydrolysis of acyl-ACPs, resulting in the termination of chain elongation by the fatty acid synthase complex. Knutzon *et al.* (1995) succeeded in cloning the gene encoding for lauric acid (12:0) ACP thioesterase from seeds of the California bay laurel tree (*Umbellularia californica*). Seed-specific expression of this gene in rapeseed resulted in a highly saturated oil with up to 48–59 mol% of lauric acid (Voelker *et al.*, 1996). In these plants, lauric acid was found almost exclusively at the *sn*-1 and *sn*-3 positions of the triacylglycerols, confirming previously found biochemical data that the resident LPAAT in rapeseed does not accept lauric acid as a substrate for the second step of triacylglycerol assembly. In a follow-up approach, a LPAAT from coconut was transferred to high lauric rapeseed, enabling an efficient accumulation of lauric acid at the *sn*-2 position and hence the formation of triacylglycerols with this fatty acid at all three *sn* positions (trilaurate; Knutzon *et al.*, 1999a). The combined expression of the bay tree thioesterase and the coconut LPAAT led to an even higher lauric acid content in the seed oil (up to 53% or 67 mol%). A further increase in the content of medium chain fatty acids beyond those reported above has been obtained through the additional ectopic expression of β-ketoacyl-CoA-synthases with specificities for medium chain fatty acids (Leonard *et al.*, 1998).

The trisaturate type of the trilaurate oil resembles seed oil deposited by natural medium chain producers such as coconut or *Cuphea* species, and it provides further evidence that in the evolution of such oil plants only a very few key enzymes needed to be modified to achieve such drastically different oil compositions and triglyceride species. It has been suggested that *pure* triacyl-glycerides produced from commercial crops could become important feedstocks for a wide range of industries (Shukla and Blicher-Mathiesen, 1993).

Species of the genus *Cuphea* produce seed oils with a wide variety of saturated medium chain fatty acids from 8:0 to 14:0. Jones *et al.* (1995) isolated from *Cuphea hookeriana*, a species with around 50% caprylic acid (8:0) and 10% capric acid (10:0), a thioesterase cDNA (*ChFatB*1). The expression of *ChFatB*1 in seeds of *B. napus* led to the production of an oil with a four to sixfold increased palmitic acid content (6 mol% in untransformed control plants vs 27–34 mol% in transgenic plants). A second acyl-ACP thioesterase cDNA (*ChFatB*2) was isolated from *C. hookeriana* and when expressed in seeds of rapeseed it led to the accumulation of up to 11 mol% 8:0 and 27 mol% 10:0 in the seed oil (Dehesh *et al.*, 1996). *Cuphea lanceolata* was used to clone four thioesterase genes with different substrate specificities (Martini *et al.*, 1995; Töpfer *et al.*, 1995). One of those genes, *ClFatB*4, was responsible for the formation of myristic acid (14:0) in the seed oil of transgenic rapeseed. Rudloff *et al.* (1999) tested this plant material in field trials. They found up to 13.4% myristic acid and an increased palmitic acid content of up to 20%. In single F2-seeds, up to 20% myristic acid was found.

The increased content of medium chain saturated fatty acids was accompanied by a reduction in oleic acid content, whereas the content of linoleic acid and linolenic acid remained fairly constant.

Although considerable success has already been achieved regarding the accumulation of saturated short and medium chain fatty acids in the seed oil of transgenic rapeseed, the results remain in many cases behind the species from which the short and medium chain thioesterase genes were obtained. Additional steps need to be modified to achieve accumulation of high amounts of single, saturated short and medium chain fatty acids (Leonard *et al.*, 1998).

## 8.6  Gamma linolenic acid

There is a considerable interest in using gamma linolenic acid (GLA, 18:3Δ6,9,12) as both a pharmaceutical and a nutraceutical agent. GLA occurs only in a few plant species which are low yielding and not adapted to current agricultural practices. Hence, there is an interest in producing GLA in high-yielding oilseed crops (Huang and Ziboh, 2001). GLA does not occur naturally in rapeseed. However, formation of GLA can be achieved by seed-specific expression of a Δ6-fatty acid desaturase (Fig. 8.2). This Δ6-desaturase uses linoleic acid (18:2Δ9,12) and also α-linolenic acid (18:3Δ9,12,15) as a substrate; however this leads to the formation of stearidonic acid (18:4Δ6,9,12,15). Knutzon *et al.* (1999b) inserted the sequence of the Δ6-desaturase from *Mortierella alpina* – a filamentous fungi – into rapeseed and found around 13% GLA in the seed oil. When they introduced a Δ12-desaturase sequence as a second gene into rapeseed to enhance 18:2 synthesis, they found dramatically increased GLA levels of more than 40% of total fatty acids, only a small amount of stearidonic acid and no 18:2Δ6,9 in the seeds. Three-quarters of

Oleic acid
18:1Δ9
→ Δ6-desaturase → 18:2Δ6,9

Linoleic acid
18:2Δ9,12
→ Δ6-desaturase → Gamma linolenic acid
18:3Δ6,9,12

Linolenic acid
18:3Δ9,12,15
→ Δ6-desaturase → Stearidonic acid
18:4Δ6,9,12,15

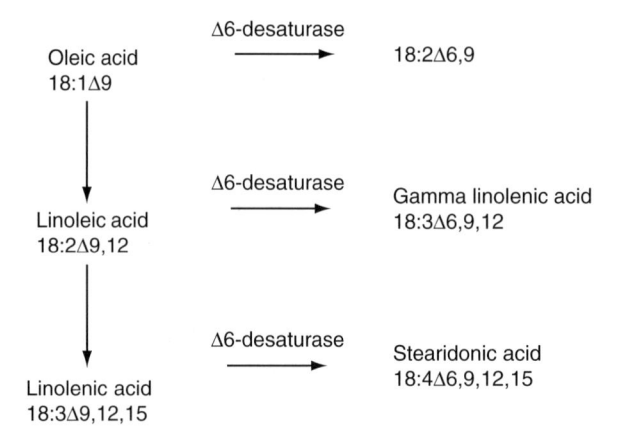

**Figure 8.2** Substrates and products of Δ6-desaturases.

the GLA produced was located at *sn*-1 and *sn*-3 positions of the triacylglycerols. Hong *et al.* (2002) isolated a Δ6-desaturase sequence from the oleaginous fungus *Pythium irregulare* and expressed this under control of the *B. napus* napin promoter in *B. juncea*, which resulted in up to 40% GLA and up to 10% stearidonic acid, small amounts of 18:2Δ6,9 were also detected. GLA was found predominantly in the *sn*-2 position of the triacylglycerols. Even higher amounts of up to 68% GLA and 17% stearidonic acid were reported by Ursin *et al.* (2000).

## 8.7 Long chain polyunsaturated fatty acids

Long chain polyunsaturated fatty acids (LC-PUFA) namely arachidonic acid (AA, 20:4 *n*-6), eicosapentaenoic acid (EPA, 20:5 *n*-3) and docosahexaenoic acid (DHA, 22:5 *n*-3) are valuable commodities. The LC-PUFAs currently consumed are obtained from diminishing natural stocks (cold water sea fish or derived fish oils) or via expensive microbial fermentation. Humans are capable of synthesizing LC-PUFAs from the precursor (*n*-3) and (*n*-6) fatty acids, linolenic acid and linoleic acid respectively (Fig. 8.3). However, this has been reported to work inefficiently and regular consumption of fish as an additional source of LC-PUFAs is recommended. However, it is obvious that it will be difficult to maintain a constant supply from fish in the future. The vast majority of plants including rapeseed do not synthesize LC-PUFAs. Transgenic oilseeds could be one way out of the forthcoming shortage, particularly in view of the fact that a low percentage of LC-PUFAs in daily consumed plant oils would satisfy nutritional requirements. In consideration of their value and the increasing demand, attempts have been started to produce transgenic rapeseed plants that produce AA, EPA and DHA in the seed oil (Abbadi *et al.*, 2001). Figure 8.3 illustrates that starting from linoleic acid and linolenic acid, a number of desaturases and elongases need to be transformed into rapeseed to enable LC-PUFA synthesis. From Fig. 8.3, it is obvious that the concentration of linoleic and linolenic acid is critical for the synthesized amount of AA and EPA/DHA respectively. The low linolenic acid mutant lines mentioned above are useful as starting material for the synthesis of high amounts of AA. Using transgenic rapeseed lines, overexpressing the Δ15-desaturase and having enhanced levels of linolenic acid (Ursin *et al.*, 2000) should result in higher EPA/DHA contents following transformation of the genes necessary for PUFA synthesis.

Recent studies have revealed a novel anaerobic LC-PUFA biosynthetic pathway found in procaryotic and eucaryotic microorganisms that does not require the fatty acid desaturase/elongase system, and instead uses a specialized polyketide synthase (PKS) to produce (20:5 *n*-3) and (22:6 *n*-3) (Metz *et al.*, 2001). Metz *et al.* (2001) isolated a five open-reading frame gene cluster

198          RAPESEED AND CANOLA OIL

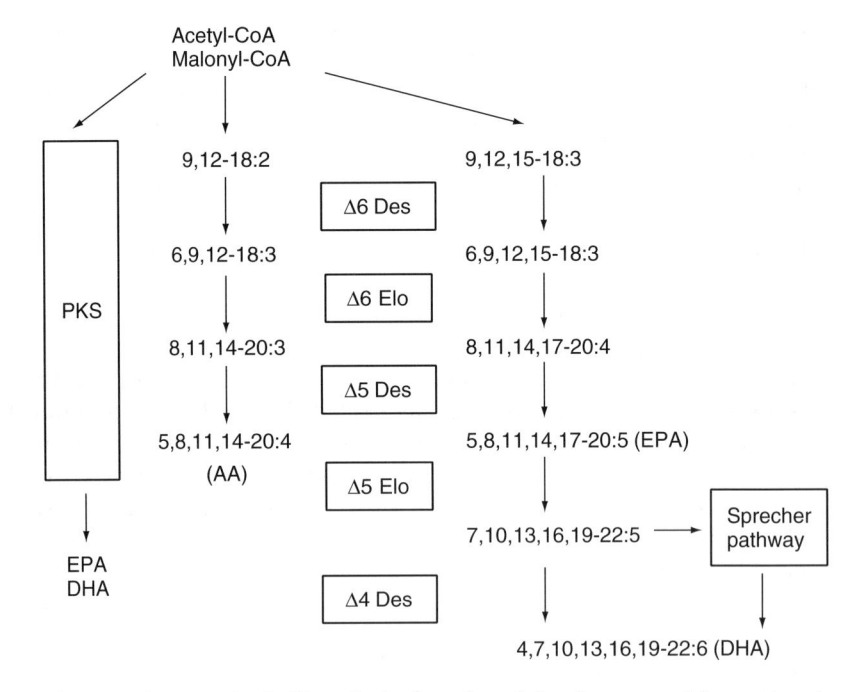

```
                    Acetyl-CoA
                    Malonyl-CoA
              ╱         │         ╲
              │    9,12-18:2      9,12,15-18:3
              │         │              │
              │      [Δ6 Des]          │
              │         │              │
              │    6,9,12-18:3    6,9,12,15-18:3
              │         │              │
    ┌─────┐   │      [Δ6 Elo]          │
    │     │   │         │              │
    │ PKS │   │    8,11,14-20:3   8,11,14,17-20:4
    │     │   │         │              │
    │     │   │      [Δ5 Des]          │
    │     │   │         │              │
    │     │   │   5,8,11,14-20:4  5,8,11,14,17-20:5 (EPA)
    │     │   │       (AA)             │
    │     │   │      [Δ5 Elo]          │
    └─────┘                            │                ┌──────────┐
       │                         7,10,13,16,19-22:5 ──▶ │ Sprecher │
      EPA                                               │ pathway  │
      DHA                     [Δ4 Des]                  └──────────┘
                                                              │
                                                              ▼
                                                4,7,10,13,16,19-22:6 (DHA)
```

**Figure 8.3** Alternative routes for the biosynthesis of very long chain polyunsaturated fatty acids such as eicosapentaenoic acid (EPA), docosahexaenoic acid (DHA) and arachidonic acid (AA). The presence of a Δ4-desaturase needs to be confirmed (adapted from Abbadi *et al.*, 2001). PKS = polyketide synthase; Des = desaturase; Elo = elongase.

from the marine bacteria *Shewanella* and demonstrated that heterologous expression of this open-reading frame gene cluster resulted in LC-PUFA synthesis in *Escherichia coli*. Many technical details still need to be overcome. However, the possibility of using the modular PKS-like system to synthesize DHA in transgenic oilseeds is clearly an attractive approach (Napier, 2002; Wallis *et al.*, 2002).

## 8.8 High erucic acid

Wild-type rapeseed contains erucic acid (22:1) and eicosenoic acid (20:1) as major components in the seed oil. Canola or '00'-quality rapeseed with low contents of erucic acid and eicosenoic acid was derived from these forms by the introduction of spontaneous mutated recessive alleles at the two loci that control the elongation of oleic acid. HEAR cultivars are presently cultivated to a smaller extent in Europe (around 55 000 hectares in 2001/2002) and USA/Canada as an IP-crop. The oil is being used as an environmental friendly and renewable resource for the production of a diverse set of oleochemicals, lubricants and hydraulic oils (Sonntag, 1995; Piazza and Foglia, 2001, see also Chapter 5). Currently available HEAR cultivars contain a maximum of 50% erucic acid and around 8% eicosenoic acid. There is some interest to increase the erucic acid content of the HEAR cultivars, because separation of erucic acid from the remaining fatty acids by distillation is expensive and the by-product fatty acids are sold only at a low price. Since eicosenoic acid is more difficult to separate from erucic acid than the other fatty acids, a reduction of eicosenoic acid content is also desirable.

Attempts are being made by conventional breeding to increase the erucic acid content. This has so far resulted in breeding lines with up to 60% erucic acid and eicosenoic acid. These fatty acids are found in the *sn*-1 and the *sn*-3 positions of the triacylglycerols but are excluded from the *sn*-2 position. This *sn*-2 exclusion limits the erucic acid content together with the eicosenoic acid content to a total of 66% and prevents the synthesis of trierucin. To overcome this bottleneck, the gene for an erucoyl-CoA preferring *sn*-2 acyltransferase (LPAAT) has been cloned from *Limnanthes* spp. and used successfully to alter seed oil *sn*-2 proportions of erucic acid in rapeseed (Lassner *et al.*, 1995; Brough *et al.*, 1996; Weier *et al.*, 1997). However, this achievement was not accompanied by an increase in erucic acid content in the seed oil. In a next step, interest focused on the fatty acid elongation mechanism from oleic acid to eicosenoic acid and then to erucic acid. This elongation is the result of two cycles of a four-step mechanism, in which 18:1-CoA and 20:1-CoA are used as substrates. The first step, the initial condensation reaction of these fatty acids with malonyl-CoA is catalyzed by the ß-ketoacyl-CoA-synthase (KCS). It is believed that this initial reaction is the rate-limiting step (Cassagne *et al.*, 1994). The *fae*1-gene encoding the KCS has been cloned from a range of plant species and has been overexpressed under control of a seed-specific promotor in transgenic HEAR. The results of these experiments showed that there was only a minor increase in erucic acid content (Han *et al.*, 2001; Katavic *et al.*, 2001) Even in combination with the expression of the *Limnanthes* spp. LPAAT, no substantial increase in the erucic acid content has been found (Han *et al.*, 2001). This indicates that there are other bottlenecks in the pathway, such as the pool of oleic acid available for elongation to eicosenoic acid and erucic acid, or the activities of the other three enzymes of the elongase complex, which limit erucic acid synthesis. There is some evidence that the cytosolic pool of available oleoyl-CoA or malonyl-CoA may limit the elongation (Bao *et al.*, 1998; Domergue *et al.*, 1999). Sasongko *et al.* (2003) combined the genes for high erucic acid with those for high oleic acid in order to increase the pool of oleoyl-CoA available for fatty acid elongation. In comparison with the high erucic acid parental line, they found reduced contents of polyunsaturated and saturated fatty acids and increased contents of oleic acid and eicosenoic acid (Table 8.2). However, no significant change in the erucic acid content was

**Table 8.2** Mean fatty acid composition of single F3-seeds ($n=20$) of two selected high erucic, high oleic acid F2-plants and of the parental high erucic acid cv. Maplus (adapted from Sasongko et al., 2003)

| Genotype | 16:0 + 18:0 | 18:1 | 18:2 + 18:3 | 20:1 | 22:1 | MUFA |
|---|---|---|---|---|---|---|
| 6575-1 | 3.2[b] | 29.9[b] | 6.7[a] | 12.4[b] | 46.7[a] | 89.0[b] |
| 6586-3 | 3.6[b] | 28.1[b] | 8.0[a] | 11.1[b] | 47.9[a] | 87.1[b] |
| Maplus | 5.0[a] | 13.9[a] | 25.7[b] | 7.2[a] | 46.7[a] | 67.8[a] |

[a,b] – significantly different at $p < 0.05\%$, Tukey-Kramer.
MUFA = monounsaturated fatty acids (18:1 + 20:1 + 22:1).

observed. The combination of this material with KCS-overexpressing and *Limnanthes* spp. LPAAT-expressing rapeseed genotypes will reveal whether this leads to an enhanced erucic acid synthesis. The KCS-gene from *Arabidopsis* shows a higher substrate specificity for oleic acid than the KCS-gene from rapeseed, which is consistent with the very long chain fatty acid composition of these two plant species; the seed oil of *Arabidopsis* contains predominantly eicosenoic acid, whereas rapeseed oil contains much more erucic acid. A search for new KCS-alleles with altered substrate specificities in different plant species could be of help to achieve a reduction in eicosenoic acid content in HEAR (Millar and Kunst, 1997; Han et al., 2001). Further interest is also focused on overexpression of the cytoplasmic acetyl-CoA carboxylase which is responsible for a sufficient supply with malonyl-CoA for the cytoplasmic chain elongation.

Already in 1993 Warwel suggested the combination of high oleic acid rapeseed with high erucic acid rapeseed to obtain an oil with 50% oleic acid and 50% erucic acid. Although this seems unrealistic at first sight, Kodali et al. (1998) reported the generation of rapeseed plants with an elevated total MUFA content from about 85% to about 90% through the combination of the genes for high oleic acid content with those for high erucic acid content. The high erucic, high oleic acid oil presented in Table 8.2 has an equally high content of MUFAs. Such an oil could provide useful traits for industrial purposes such as lubrication, hydraulic oils, etc. Due to its lower content of PUFAs the oil stability should be improved and the lower content of saturated fatty acids should contribute to improved properties at low ambient temperatures.

## 8.9 Miscellaneous unusual fatty acids

Plants collectively produce many kinds of unusual fatty acids which are potentially valuable for the production of paints, nylon, varnishes, plasticizers, resins, lubricants and polymers. These unusual fatty acids include hydroxy, epoxy and trienoic fatty acids, just to name a few of the more important ones. Ricinoleic acid (12-hydroxyoctadeca-9-enoic acid) is a hydroxylated fatty acid that accumulates in the seeds of castor bean (*Ricinus communis* L.), where it constitutes

85–90% of the fatty acids of the seed neutral lipids. Several species of the genus *Lesquerella* produce as major hydroxy fatty acids either densipolic acid (12-hydroxyoctadeca-9,15-dienoic acid), lesquerolic acid (14-hydroxyeicosa-11-enoic acid) or auricolic acid (14-hydroxyeicosa-11,17-dienoic acid). The epoxy fatty acid vernolic acid (12-epoxyoctadeca-9-enoic acid) is enriched in the seed oils of several *Asteraceae* genera, including *Stokesia*, *Vernonia* and *Crepis*. The seed oil of *Crepis alpina* is made up of about 70% crepenynic acid (9-octadecen-12-ynoic acid), a fatty acid with a carbon–carbon acetylenic (triple) bond.

In vitro studies have shown that the *Ricinus* hydroxylase accepts both oleic acid and eicosenoic acid as substrates for the synthesis of ricinoleic acid and lesquerolic acid respectively. The oleic acid 12-hydroxylase cDNA (*fah*12, Fig. 8.4) was cloned from *Ricinus* and expressed under control of the *B. napus* napin promoter in transgenic *Arabidopsis thaliana* plants (Broun and Somerville, 1997). This resulted in the accumulation of up to 7.8% ricinoleic acid, 6.6% densipolic acid and 2.1% lesquerolic acid. These results are consistent with the previous biochemical evidence that the *Ricinus* hydroxylase utilizes oleic acid and eicosenoic acid as substrates for ricinoleic and lesquerolic acid biosynthesis respectively. Densipolic acid is probably formed by the action of a resident *n*-3-desaturase that can convert ricinoleic acid to densipolic acid (Broun and Somerville, 1997). It is also discussed whether in *Arabidopsis* and also in *Brassica* the *fae*1-gene (see above) can use ricinoleic acid and densipolic acid as substrates for the fatty acid chain elongation to lesquerolic acid and auricolic acid respectively (Fig. 8.4).

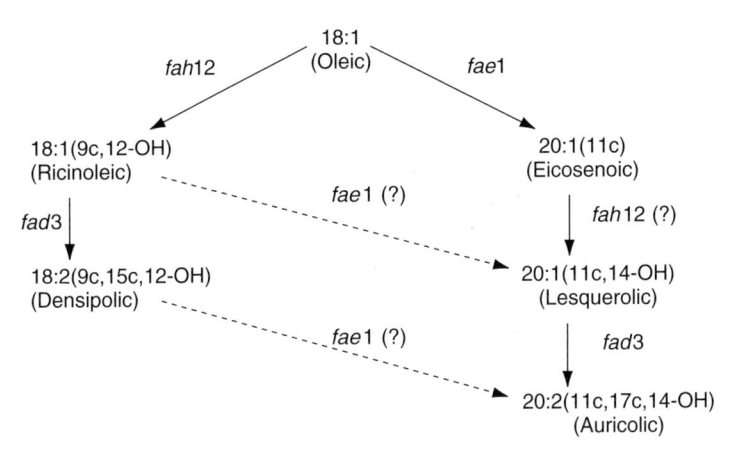

**Figure 8.4** Hypothetical pathway for the synthesis of hydroxylated fatty acids. The name of the gene that is proposed to correspond to each step is shown adjacent to the arrows (Broun and Somerville, 1997). (*fae*1 (?) and *fah*12 (?) refer to hypothetical enzymes which have not yet been confirmed.) Reprinted with permission of the American Society of Plant Biologists (Copyright).

Lee *et al.* (1998) cloned the *Crep*1 acetylenase gene from *Crepis alpina* and expressed it in *Arabidopsis* under control of the seed-specific napin promoter. Total fatty acids from seeds of individual first-generation transgenic plants contained up to 25% (w/w) crepenynic acid. No other acetylenic fatty acids were detected in these plants. Lee *et al.* (1998) also cloned the Δ12-epoxygenase gene from *Crepis palaestina*. The expression of the *Cpal*2 cDNA in transgenic *Arabidopsis* resulted in up to 15% vernolic acid in the seed oil. Similar results should be possible using rapeseed instead of *Arabidopsis* for transformation. Compared to the above mentioned species containing epoxy, hydroxy and acetylenic fatty acids in their seed oil, rapeseed is much more adapted and is a higher yielding crop that would allow a much more economic production of these fatty acids.

## 8.10 Minor bioactive constituents

Minor bioactive constituents in rapeseed oil include polar lipids, tocopherols, sterols and carotenoids. The first three substances are important by-products of the oil-processing industry. Carotenoid content in natural rapeseed is quite low and economical interest may develop with enhanced content achieved through genetic modification. In the future, market profitability of rapeseed oil may depend to a larger extent on its content of minor bioactive constituents. Hence, there is an increasing interest in these compounds.

### 8.10.1 Polar lipids

Mixtures of polar lipids – mainly phospholipids – in combination with other by-products of oil refining, e.g. sterols and tocopherols, available in commercial lecithin preparations and as purified phospho- and glycolipid formulations, find many uses as emulsifiers, stabilizers, antioxidants and as physiologically active compounds in numerous foods, medical and cosmetic applications. The functional properties and thereby the quality of the lecithin are essentially determined by the fatty acid composition and the type of functional head group, but also by the presence of other compounds like phenolic acids and sterols (Schneider, 2001). Abidi *et al.* (1999) have analysed the phospholipid composition in 12 spring rapeseed varieties and found a large variation in the total content of phospholipids as well as in the composition. Phospholipid content varied from 19.6 to 57.1 µg/g oil and phosphatidylcholine content ranged from 28.4 to 48.5%. Analysing a set of 40 winter and spring rapeseed genotypes, Heift *et al.* (2003) found a large variation in the phospholipid and glycolipid content and composition (Table 8.3). They also analysed other quality parameters of lecithin-like sterols, phenolic acids, carbohydrates and colour (green/yellow) and found a large variation in most of these characters.

**Table 8.3** Type and quantity of polar lipid components in rapeseeds (adapted from Heift et al., 2003)

| Component | Mean content (per 100 mg meal of dehulled seeds) | Range (per 100 mg meal of dehulled seeds) |
|---|---|---|
| Phospholipids | 2229 μg | 1371–4497 μg |
| PE | 366 μg | 197–786 μg |
| PA | 43 μg | 7–147 μg |
| PI | 677 μg | 257–1392 μg |
| PC | 1073 μg | 618–2233 μg |
| Glycolipids | 79 μg | 14–124 μg |
| Sterylglycoside | 47 μg | 12–92 μg |
| DGDG | 27 μg | 15–73 μg |
| Cerebroside | 5 μg | 5–24 μg |
| Phenolic acids[1] | 800 μg | 400–1500 μg |
| Sterols | 175 μg | 58–455 μg |
| Carbohydrates[2] | 14 mg | 12–17 mg |
| Green pigments[3] | 5.7 | 1.2–10.8 |
| Yellow pigments[3] | 13.8 | 5.3–27.8 |

[1] Calculated as sinapic acid.
[2] per 100 mg of defatted meal.
[3] colour value.
PA = phosphatidic acid; PC = phosphatidylcholine; PE = phosphatidylethanolamine; PI = phosphatidylinositol; DGDG = Digalactosyldiglyceride.

However, replicated field trials are necessary to confirm the genetic basis of this variation. The preliminary results of Heift et al. (2003) indicate that more variation could be found in a larger germplasm collection and that conventional breeding methods could be applied to develop cultivars with an improved or diversified lecithin quality.

### 8.10.2 Tocopherols

Tocopherols, commonly known as vitamin E, are important natural antioxidants that inhibit fatty acid peroxidation in vegetable oils and act as free radical quencher. They belong to the tocochromanols and exist in four forms ($\alpha$-, $\beta$-, $\gamma$- and $\delta$-tocopherol), which differ in the position and number of methyl groups and in biological effectiveness. Tocopherols are only synthesized by plants as well as some photosynthetic procaryotes. The predominant form in the leaves of higher plants is $\alpha$-tocopherol, whereas seeds often contain more $\gamma$-tocopherol (Demurin et al., 1996). For human and animal utility, $\alpha$-tocopherol has the highest vitamin E activity and has been implicated in a variety of health areas (DellaPenna, 1999; Bramley et al., 2000). $\gamma$-tocopherol is a strong antioxidant for oxidation-sensitive fatty acids in vegetable oil products (Kamal-Eldin and Appelqvist, 1996). Because of the nutritional value and importance for oil stability, tocopherol content in the seed oil recently moved into focus as a value-adding compound.

In rapeseed tocopherol, proportions of 65% γ-tocopherol and 35% α-tocopherol are commonly found in the seed oil. Only very low amounts of δ-tocopherol (<1%) and no β-tocopherol are present. The variation in total tocopherol content in *B. napus* ranges from about 80 to 1000 ppm in the seed oil (Marquard, 1976; Abidi *et al.*, 1999; Dolde *et al.*, 1999; Marwede *et al.*, 2003). Breeding for increased tocopherol content and modified composition has just been started (Leckband *et al.*, 2002). However, genetic progress may be hampered by a strong influence of genotype–environment interactions (Marwede *et al.*, 2003).

Tocopherols contain an aromatic head group, which is derived from homogentisic acid, and a hydrocarbon portion, which arises from phytyldiphosphate. Homogentisic acid is derived from the shikimic acid pathway and phytyldiphosphate is generated from the condensation of four isoprenoid units. The condensation of phytyldiphosphate and homogentisic acid to form 2-methyl-6-phytylplastoquinol is the first committed step in tocopherol biosynthesis and is performed by the homogentisate phytyltransferase. Subsequent cyclization and methylation reactions result in the formation of the four major tocopherols (Fig. 8.5).

The first major breakthrough in the manipulation of seed tocopherol content was achieved by Shintani and DellaPenna (1998). They succeeded in drastically increasing the levels of α-tocopherol in seeds of *Arabidopsis thaliana*, which naturally produces mainly γ-tocopherol (Table 8.4). This was achieved

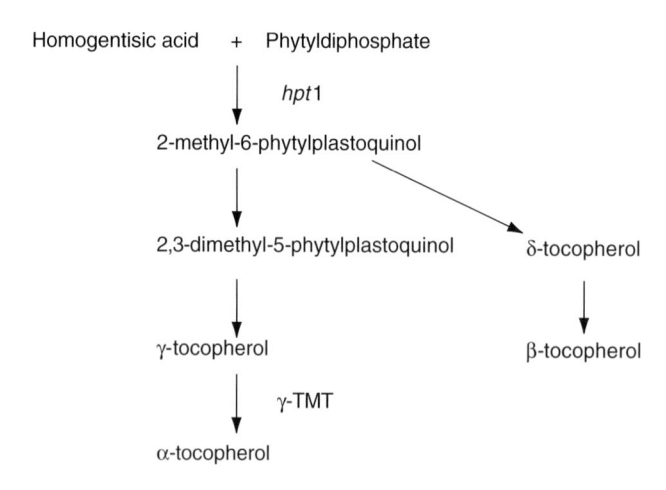

**Figure 8.5** Schematic drawing of the tocopherol's biosynthetic pathway (adapted from Savidge *et al.*, 2002).

Table 8.4 Mean tocopherol content and compositions of segregating T2-seed populations from *Arabidopsis* transformed with the empty vector control pDC3 and with the γ-tocopherol methyltransferase (γ-TMT) overexpression construct pDC3-A.t.g-TMT (adapted from Shintani and DellaPenna, 1998)

| Line | TOC content[1] | α-TOC (%) | β-TOC (%) | γ-TOC (%) | δ-TOC (%) | γ-TMT[2] |
|---|---|---|---|---|---|---|
| pDC3 control[3] | 367 | 1 | 0 | 97 | 2 | 0 |
| pDC3-y-TMT-1 | 361 | 95 | 1 | 4 | 0 | 2.12 |
| pDC3-y-TMT-2 | 383 | 94 | 1 | 6 | 0 | 1.88 |
| pDC3-y-TMT-3 | 340 | 88 | 0.5 | 12 | 0 | 1.71 |

[1] ng per mg of seed; mean of four replicate seed samples.
[2] pmol per mg of protein per hour.
[3] Mean of three independent transformed lines.
TOC = tocopherol.

by overexpression of the γ-tocopherol methyltransferase (TMT), which catalyzes the final step in α-tocopherol synthesis from the precursor γ-tocopherol (Schultz *et al.*, 1985). The modification of the γ-TMT activity in rapeseed through overexpression or antisense technology should be possible without great difficulty.

Savidge *et al.* (2002) reported the isolation and characterization of the homogentisate phytyltransferase *hpt*1-gene from *Arabidopsis*. They also showed that overexpression of this gene in *Arabidopsis* under control of the seed-specific napin promoter led to a twofold increase in seed tocopherol levels compared to untransformed wild types. Selected transformed lines contained in T3-seeds up to 926 ng/mg tocopherol compared to 527 ng/mg in the control plants. The majority of the increase in total tocopherols was caused by an increase in γ-tocopherol. α-Tocopherol did not increase, indicating that γ-TMT is saturated at wild-type levels of γ-tocopherol. Similar results were also found in transgenic *Arabidopsis* lines in which the *hpt*1-gene was expressed under control of the constitutive CaMV 35S promoter through antisense technology. Some of these lines contained up to tenfold lower tocopherol levels, however, only the γ-tocopherol content was significantly affected (Savidge *et al.*, 2002). Savidge *et al.* (2000) speculated that it may not be possible to increase the tocopherol content much beyond the twofold increase achieved in their study without additional manipulation of other, earlier steps in tocopherol biosynthesis. Further experiments have to show whether an overexpression of the *hpt*1-gene in rapeseed will lead to a substantial increase in tocopherol content beyond what can be achieved by classical breeding methods.

Porfirova *et al.* (2002) succeeded in isolating a tocopherol-deficient mutant (*vte*1) from *Arabidopsis thaliana*. Mutant plants completely lacked all the four tocopherol forms and were deficient in tocopherol cyclase activity. The mutant was shown to be defective in the tocopherol cyclase. The reduction in tocopherol cyclase activity was accompanied by the accumulation of its substrate, 2,3-dimethyl-5-phytyl-1,4-hydroquinone (DMPQ). Growth of the *vte*1 mutant, chlorophyll content and photosynthetic quantum yield were similar to those of the wild type under optimal growth conditions and were only slightly affected under photooxidative stress. Obviously, the absence of tocopherol has no large impact on photosynthesis or plant viability, suggesting that other antioxidants can compensate for the loss of tocopherols.

### 8.10.3 Sterols

Sterols are ubiquitous in plant cells, where they serve crucial functions to control the fluidity and permeability of membranes and as precursors to steroid growth regulators such as brassinosteroids. In humans, phytosterols are known inhibitors of cholesterol absorption. Their serum-cholesterol lowering properties have increased the interest in enhancing plant sterol fraction of food items of vegetable origin. Common vegetable oils and plant-based foods contain a large number of sterols belonging to the three major groups: 4-desmethylsterols, 4-mono-methylsterols and 4,4′-dimethylsterols, of which the first group dominates. The sterols may be found as free sterols, steryl esters, steryl glycosides and acylated steryl glycosides. The saturated counterparts (at $\Delta 5$) of campesterol and sitosterol (campestanol and sitostanol) commonly known as stanols, are mainly found as a natural constituent in some cereal lipids. Gordon and Miller (1997) compared ten different vegetable oils and found that rapeseed oil had 6900 mg/kg total sterol content, the second highest content next to corn oil. In rapeseed oil, 36% of the total sterol content was present as steryl esters. So far analysis of genetic variation within the *Brassica* species and breeding for enhanced sterol contents and improved sterol composition has not been realized. Although sterols are moving into the research focus, large-scale screening for genetic variability has not yet been performed, probably due to the sophisticated and expensive analytical procedures required for accurate determination of the different sterols. More recently, a capillary column gas–liquid chromatographic separation of $\Delta 5$-unsaturated and -saturated phytosterols has been reported (Dutta and Normen, 1998), which should allow a comparatively quick and easy, but sufficiently exact determination of sterols in seed samples.

The sterol pathway is divided into two distinct parts (Nes, 2000). First, acetate is converted into squalene epoxide via several steps in the mevalonate pathway. Second, in the first committed step of sterol biosynthesis, squalene epoxide is cyclized to give cycloartenol, which is transformed into end-product sterols in a series of enzyme-catalyzed methylations, demethylations and desaturations (Fig. 8.6). It has been proposed that the C-24 methylation of cycloartenol is a major site of regulation in the sterol biosynthetic pathway (Nes *et al.*, 1991; Chappell *et al.*, 1995; Nes, 2000). Important evidence supporting

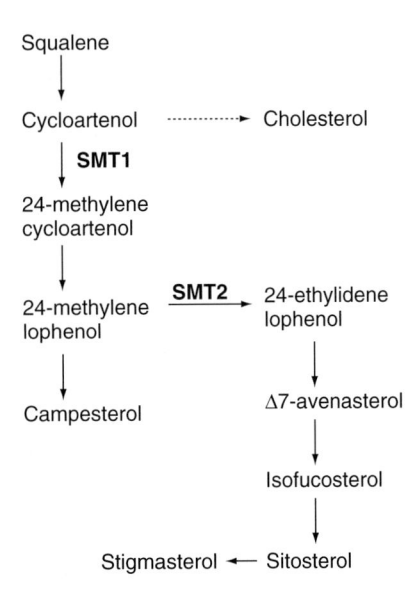

```
              Squalene
                 │
                 ▼
             Cycloartenol ········▶ Cholesterol
                 │ SMT1
                 ▼
             24-methylene
             cycloartenol
                 │
                 ▼
             24-methylene    SMT2    24-ethylidene
             lophenol       ──────▶  lophenol
                 │                      │
                 ▼                      ▼
             Campesterol            Δ7-avenasterol
                                        │
                                        ▼
                                    Isofucosterol
                                        │
                                        ▼
             Stigmasterol ◀── Sitosterol
```

**Figure 8.6** Schematic drawing of sterol biosynthesis (adapted from Holmberg *et al.*, 2002).

this view is that transgenic tobacco (*Nicotiana tabacum*) leaf overexpressing 3-hydroxy-3-methylglutaryl CoA reductase (HMGR) accumulates high levels of cycloartenol, but only relatively moderate levels of 24-alkyl-Δ5-sterols (Chappell *et al.*, 1995; Schaller *et al.*, 1995). The conversion of cycloartenol into 24-methylene cycloartenol is principally catalyzed by an S-adenosyl-L-Met-dependent sterol C-24 methyltransferase type1 (SMT1). First attempts to manipulate the sterol content and composition by genetic engineering have been successful. Overexpression of SMT1 in tobacco under control of a constitutive or seed-specific promoter increased the amount of total sterols in seed tissue by up to 44% (Holmberg *et al.*, 2002). The sterol composition was also perturbed with levels of sitosterol increased by up to 50% and levels of isofucosterol and campesterol increased by up to 80%, whereas levels of cycloartenol and cholesterol were decreased by up to 53 and 34% respectively. Concomitant with the enhanced SMT1 activity an increase of the endogenous HMGR activity was found, indicating a potential regulatory role of SMT1 in seed sterol biosynthesis.

Conversely, sterol C-24 methyltransferase type 2 (SMT2) is mainly responsible for the second methyl addition, but it can also catalyze the first sterol methylation (Schaeffer *et al.*, 2001). Therefore, it was not a surprise that an *Arabidopsis* SMT1 deletion mutant was still able to accumulate alkylated sterols, but at altered levels (Diener *et al.*, 2000). However, the perturbed leaf sterol composition of the SMT1 mutant was associated with several phenotypical abnormalities such as poor growth and fertility, a loss of proper embryo morphogenesis and sensitivity of the root to calcium (Diener *et al.*, 2000). This emphasizes that perturbed sterol composition in vegetative tissue can have severe consequences for plant development.

### 8.10.4 Carotenoids

Carotenoids are a large group of often highly coloured compounds and, in general, are thought to have an antioxidant function. In plants, they are essential for photosynthesis and serve as precursor for the biosynthesis of abscisic acid and gibberellic acid, etc. Certain carotenoids play important roles in human health by serving as precursors for vitamin A synthesis and by possibly reducing the incidence of certain diseases. In photosynthetic tissues of plants, carotenoids are synthesized in the plastids and accumulate in chloroplast membranes. Shewmaker *et al.* (1999) confirmed lutein (30–31 µg/g FW) as the major rapeseed carotenoid in the seeds of two spring rapeseed cultivars. Only negligible amounts of β-carotene (3–5 µg/g FW) and no lycopene and no α-carotene were detectable. Hence, carotenoids occur in rapeseed in concentrations of about ten times lower than that of tocopherols.

The first committed step in carotenoid biosynthesis is the condensation of two geranylgeranyl diphosphate ($C_{20}$) moieties to give phytoene ($C_{40}$, a colourless carotenoid; Fig. 8.7). The gene responsible for this reaction, phytoene synthase, has been cloned from a variety of microorganisms and several plants. It is often assumed that the first committed step in a pathway will be a regulated step. As a consequence, one might predict that enhanced expression of such an enzyme would lead to increased flux through a given pathway. This was done by Shewmaker *et al.* (1999). They expressed a bacterial phytoene synthase (*crtB*) from *Erwinia uredovora* in conjunction with a plastid targeted sequence in rapeseed under control of the seed-specific napin promoter, resulting in a 50-fold increase in carotenoid levels. At 35–40 days post-anthesis, first-generation transgenic T2-seeds accumulating high levels of carotenoids had a very orange appearance instead of a green one. In T2- to T4-seeds derived from different transformation events, more than 1000 µg/g FW and up to 1600 µg/g FW carotenoids were detected. Drastic increases were found for α-carotene (up to 440 µg/g FW), β-carotene (up to 949 µg/g FW) and for the precursor phytoene (up to 430 µg/g FW). Barely no change was observed for lutein and only a slight increase was found for lycopene content (up to 25 µg/g FW). The accumulation of significant amounts of phytoene indicated that enzymes other than the phytoene synthase may be rate limiting as well.

The analysis of other metabolic changes in the transgenic high carotenoid rapeseed revealed about a 5% increased oleic acid content and concomitantly decreased linoleic and linolenic acid contents, which lacks an explanation. More strikingly, low chlorophyll levels were observed in the green developing seeds. At 35 days post-anthesis, total chlorophyll levels were between 51 and 138 μg/g FW in comparison with around 500 μg/g FW total chlorophyll level in the control plants. However, at maturity no difference in chlorophyll content between the high carotenoid transgenic lines and the controls was discernible. Furthermore, an overall decrease in the tocopherol content of up to 50% was found in mature transgenic seeds. This decrease occurred mainly at the expense of γ-tocopherol. In summary, these results indicate that the enhanced carotenoid content occurred at least partly on account of chlorophyll and tocopherol synthesis (Fig. 8.7).

The observed 50-fold increase in carotenoids was possible only because carotenoids normally comprise such a small fraction of the total isoprenoid population. The overall increase in isoprenoid units was only fourfold (Shewmaker *et al.*, 1999). These data as well as that of Savidge *et al.* (2002) on the increase of tocopherol content suggest that there may be a limit to the level to which

$$DMAPP \xrightarrow{+IPP} IPP$$

$$\downarrow$$

Geranylgeranyl diphosphate – $C_{20}$ (×2)

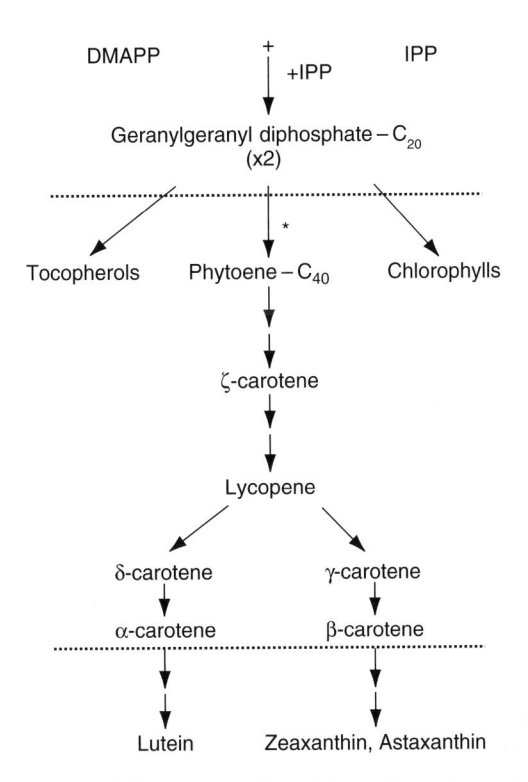

Tocopherols ← Phytoene – $C_{40}$ → Chlorophylls
↓
ζ-carotene
↓
Lycopene
↙ ↘
δ-carotene    γ-carotene
↓             ↓
α-carotene    β-carotene
↓             ↓
Lutein        Zeaxanthin, Astaxanthin

**Figure 8.7** Carotenoid biosynthetic pathway (adapted from Shewmaker *et al.*, 1999). DMAPP = dimethylallyl diphosphate; IPP = isopentenyl diphosphate.
* Phytoene synthase.

isoprenoids can be increased without modifying steps prior to phytoene synthase. There are studies that demonstrate that geranylgeranyl pyrophosphate, a phytol precursor, may be limiting in tocopherol and carotenoid biosynthesis (Furuya et al., 1987).

### 8.10.5 Chlorophyll

The role of chlorophyll in seeds has been a matter of controversy. While maturing *B. napus* seeds have been shown to refix respired $CO_2$ for their metabolic needs, other studies have indicated that photosynthetic processes contribute little to developing seeds. Seeds appear to be dependent on cytosolic processes for adenosine triphosphate (ATP), reducing power and carbon precursors that are required for development and maturation. Normally, chlorophyll is degraded in the maturing seeds (Ward et al., 1994). However, cold temperatures and excess precipitation interfere with the ripening of *Brassica* oilseeds, resulting in a high proportion of green seeds with higher than normal levels of chlorophyll at harvest. Green seeds have affected growers mainly in Canada and in northern Europe (Kimber and McGregor, 1995).

The problem with the green seeds in *Brassica* oilseeds begins when seeds are crushed for vegetable oil production. As much as 60 μg/ml of chlorophyll or four times that of the top-grade seed can leach into the oil, causing oxidation and rancidity and thus reducing the shelf life of the oil (see Tsang et al., 2003). Bleaching technology is used to remove chlorophyll from the oil. However, bleaching is a major expenditure in processing for vegetable oil production and the cost is directly proportional to the extent of chlorophyll contamination. One way to circumvent the green seed problem would be the breeding of a low chlorophyll seed variety.

In green plants, chlorophyll is synthesized from its precursor 5-aminolevulinic acid. Eight 5-aminolevulinic acid molecules are required to form the tetrapyrrole ring of one chlorophyll molecule. 5-aminolevulinic acid is derived from glutamate via a tRNA Glu-mediated pathway. In this pathway, glutamate 1-semialdehyde is converted to 5-aminolevulinic acid by the enzyme glutamate 1-semialdehyde aminotransferase (glutamate 1-semialdehyde 2,1-aminomutase; see Tsang et al., 2003). Tsang et al. (2003) isolated from *B. napus* a cDNA clone encoding glutamate 1-semialdehyde aminotransferase and expressed this under control of the seed-specific napin promoter in transgenic *B. napus* plants. In field experiments with second-generation transgenic T2-plants, transformants showed varying degrees of chlorophyll reduction in the T3-seeds (Table 8.5). Chlorophyll reduction did not appear to have any negative impact on the performance of transgenic plants based on seed yield, seed weight and oil content. Interestingly, enhanced levels of protein content are observed in the seeds of all transgenic lines.

**Table 8.5** Agronomic performance of transgenic *B. napus* lines having a reduced chlorophyll content (Means of three replicates, adapted from Tsang *et al.*, 2003)

| Transgenic lines | Chlorophyll content (mg/kg FW) | Seed yield of 40 plants (g) | FW of 1000 seeds (g) | Seed oil content[1] (%) | Seed protein content[2] (%) |
|---|---|---|---|---|---|
| Untransformed control | 21.7 | 162 | 4.5 | 38 | 48.1 |
| #22 | 16.3** | 170 | 4.5 | 40 | 53.3** |
| #23 | 15.9** | 153 | 4.5 | 39 | 53.7** |
| #29 | 13.9** | 184 | 4.4 | 40 | 51.3** |
| #36 | 15.1** | 179 | 4.6 | 37 | 52.5** |
| #47 | 11.7** | 174 | 4.6 | 39 | 51.8** |
| #55 | 13.1** | 158 | 4.5 | 36 | 52.1** |
| #58 | 17.3* | 159 | 4.5 | 40 | 51.3** |

*, ** Dunnett's test of significance at $p = 0.05$ and 0.01 respectively.
[1] Oil content is reported on a moisture-free basis.
[2] Protein content (Nx = 6.25) is reported on a moisture- and oil-free basis.
FW = Fresh weight.

## 8.11 Conclusions and outlook

There has been considerable progress in understanding and genetic modification of plant lipid biosynthetic pathways. However, with few exceptions, the expression of new biosynthetic capabilities has not led to quantities of the desired compounds acceptable to the chemical industry. The increasing availability of structural and regulatory genes controlling relevant pathways will make it possible in the future to enhance the content of valuable fatty acids or of other minor bioactive compounds.

Genetic engineering allows for other seed quality modifications in addition to those mentioned above. Novel fatty acids may be created, liquid waxes and polyhydroxy alkanoates (PHAs) may be produced in rapeseed oil (e.g. Houmiel *et al.*, 1999; Lardizabal *et al.*, 2000). The extensive diversity in the composition of seed storage fatty acids found in higher plants is synthesized by a family of structurally similar enzymes. Experiments performed by Broun *et al.* (1998) have shown that as few as four amino acid substitutions can convert an oleic acid Δ12-desaturase to a hydroxylase and as few as six result in the conversion of a hydroxylase to a desaturase. Application of site-directed mutagenesis can be used to create new alleles with new enzymatic properties. This has already been used earlier to modify the substrate specificity of an acyl–acyl carrier protein thioesterase (Yuan *et al.*, 1995). The possibility of modifying enzyme specificities extends the tools for creating new *designer oils*, which will lead to an even more diversified spectra of the already known large number of different fatty acids.

## References

Abbadi, A., Domergue, F., Meyer, A., Riedel, K., Sperling, P., Zank, T.K. and Heinz, E. (2001) Transgenic oilseeds as sustainable source of nutritionally relevant $C_{20}$ and $C_{22}$ polyunsaturated fatty acids. *European Journal of Lipid Science and Technology*, **103**, 106–113.

Abidi, S.L., List, G.R. and Rennick, K.A. (1999) Effect of genetic modification on the distribution of minor constituents in Canola oil. *Journal of the American Oil Chemists' Society*, **76**, 463–467.

Anonymous (1998) GMO oils: many are created, few will prosper. *INFORM*, **9**, 671–672.

Auld, D., Heikkinen, M.K., Erickson, D.A., Sernyk, L. and Romero, E. (1992) Rapeseed mutants with reduced levels of polyunsaturated fatty acid levels and increased levels of oleic acid. *Crop Science*, **32**, 657–662.

Bao, X., Pollard, M. and Ohlrogge, J. (1998) The biosynthesis of erucic acid in developing embryos of *Brassica rapa*. *Plant Physiology*, **118**, 183–190.

Bramley, P., Elmadfa, I., Kafatos, A., Kelly, F.J., Manios, Y., Roxborough, H.E., Schuch, W., Sheehy, P.J.A. and Wagner, K.-H. (2000) Review vitamin E. *Journal of the Science of Food and Agriculture*, **80**, 913–938.

Brough, C.L., Coventry, J.M., Christie, W.W., Kroon, J.T.M., Brown, A.R., Barsby, T.L. and Slabas, A.R. (1996) Towards the genetic engineering of triacylglycerols of defined fatty acid composition: major changes in erucic acid content at the *sn*-2 position affected by the introduction of a 1-acyl-*sn*-glycerol-3-phosphate acyltransferase from *Limnanthes douglasii* into oil seed rape. *Molecular Breeding*, **2**, 133–142.

Broun, P. and Somerville, C. (1997) Accumulation of ricinoleic, lesquerolic, and densipolic acids in seeds of transgenic *Arabidopsis* plants that express a fatty acyl hydroxylase cDNA from Castor Bean. *Plant Physiology*, **113**, 933–942.

Broun, P., Shanklin, J., Whittle, E. and Somerville, C. (1998) Catalytic plasticity of fatty acid modification enzymes underlying chemical diversity of plant lipids. *Science*, **282**, 1315–1317.

Cassagne, C., Lessire, R., Bessoule, J.-J., Moreau, P., Creach, A., Schneider, F. and Sturbois, B. (1994) Biosynthesis of very long chain fatty acids in higher plants. *Progress in Lipid Research*, **33**, 55–69.

Chappell, J., Wolf, F., Prouix, J., Cueller, R. and Saunders, C. (1995) Is the reaction catalyzed by 3-hydroxy-3-methylglutaryl coenzyme A reductase a rate limiting step for isoprenoid biosynthesis in plants? *Plant Physiology*, **109**, 1337–1343.

Chungu, C., Murray, B., Thompson, S., Kubik, T., Freeman, S. and Gore, S. (2001) High Oleic and Low Linolenic *Brassica napus*. GCIRC Technical Meeting, 5 June 2001, Poznan, Poland. GCIRC Bulletin, B18. Online at www.cetiom.fr/gcirc (22.05.2003).

DeBonte, L.R. and Hitz, W.D. (1998) Canola oil having increased oleic acid and decreased linolenic acid content. US Patent No. 5850026. Date issued: 15 December.

Dehesh, K., Jones, A., Knutzon, D.S. and Voelker, T.A. (1996) Production of high levels of 8:0 and 10:0 fatty acids in transgenic canola by overexpression of Ch FatB2, a thioesterase cDNA from *Cuphea hookeriana*. *Plant Journal*, **9**, 167–172.

DellaPenna, D. (1999) Nutritional genomics: manipulating plant micronutrients to improve human health. *Science*, **285**, 375–379.

Demurin, Y., Skoric, D. and Karlovic, D. (1996) Genetic variability of tocopherol composition in sunflower seeds as a basis of breeding for improved oil quality. *Plant Breeding*, **115**, 33–36.

Diener, A.C., Li, H., Zhou, W.-X., Whoriskey, W.J., Nes, W.D. and Fink, G.R. (2000) Sterol methyltransferase 1 controls the level of cholesterol in plants. *Plant Cell*, **12**, 853–870.

Dolde, D., Vlahakis, C. and Hazebroek, J. (1999) Tocopherols in breeding lines and effects of planting location, fatty acid composition and temperature during development. *Journal of the American Oil Chemists' Society*, **76**, 349–359.

Domergue, F., Chevalier, S., Santarelli, X., Cassagne, C. and Lessire, R. (1999) Evidence that oleoyl-CoA and ATP-dependent elongations coexist in rapeseed (*Brassica napus* L.). *European Journal of Biochemistry*, **263**, 464–470.

Dutta, P.C. and Normen, L. (1998) Capillary column gas–liquid chromatographic separation of $d^5$-unsaturated and saturated phytosterols. *Journal of Chromatography A*, **816**, 177–184.

Frentzen, M. (1998) Acyltransferases from basic science to modified seed oils. *Fett/Lipid*, **100**, 161–166.

Furuya, T., Yoshikawa, T., Kimura, T. and Kaneko, H. (1987) Production of tocopherols by cell culture of Safflower. *Phytochemistry*, **26**, 2741–2747.

Gordon, M.H. and Miller, L.A.D. (1997) Development of steryl ester analysis for the detection of admixtures of vegetable oils. *Journal of the American Oil Chemists' Society*, **74**, 505–510.

Han, J., Lühs, W., Sonntag, K., Zähringer, U., Borchardt, D.S., Wolter, F.P., Heinz, E. and Frentzen, M. (2001) Functional characterization of ß-ketoacyl-CoA synthase genes from *Brassica napus* L. *Plant Molecular Biology*, **46**, 229–239.

Hawkins, D.J. and Kridl, J.C. (1998) Characterization of acyl-ACP thioesterases of mangosteen (*Garcinia mangostana*) seed and high levels of stearate production in transgenic canola. *Plant Journal*, **13**, 743–752.

Heift, C., Schipmann, K. and Lange, R. (2003) Monitoring on polar lipid composition of rape seeds (*Brassica napus*). In: *Proceedings 11th International Rapeseed Conference*. Copenhagen, Denmark, in press.

Hildebrand, D.F. and Grayburn, S.W. (1999) Fatty acid alteration by a 9 desaturase in transgenic plant tissue. US Patent No. 5,866,789.

Holmberg, N., Harker, M., Gibbard, C.L., Wallace, A.D., Clayton, J.C., Rawlins, S., Hellyer, A. and Safford, R. (2002) Sterol C-24 Methyltransferase Type 1 controls the flux of carbon into sterol biosynthesis in tobacco seed. *Plant Physiology*, **130**, 303–311.

Hong, H., Datla, N., Reed, D.W., Covello, P.S., MacKenzie, S.L. and Qiu, X. (2002) High-level production of γ-linolenic acid in *Brassica juncea* using a Δ6-desaturase from *Pythium irregulare*. *Plant Physiology*, **129**, 354–362.

Houmiel, K.L., Slater, S., Broyles, D., Casagrande, L., Colburn, S., Gonzalez, K., Mitsky, T.A., Reiser, S.E., Shah, D., Taylor, N.B., Tran, M., Valentin, H.E. and Gruys, K.J. (1999) Poly(ß-hydroxybutyrate) production in oilseed leukoplasts of *Brassica napus*. *Planta*, **209**, 547–550.

Huang, Y.-S. and Ziboh, A. (2001) *Gamma-linolenic Acid: Recent Advances in Biotechnology and Clinical Applications*. AOCS Press, Champaign, IL.

Jako, C., Kumar, A., Wei, Y., Zou, J., Barton, D.L., Giblin, E.M., Covello, P.S. and Taylor, D.C. (2001) Seed-specific overexpression of an *Arabidopsis* cDNA encoding a diacylglycerol acyltransferase enhances seed oil content and seed weight. *Plant Physiology*, **126**, 861–874.

Jones, A., Davies, H.M. and Voelker, T.A. (1995) Palmitoyl-acyl carrier protein (ACP) thioesterase and the evolutionary origin of plant acyl-ACP thioesterases. *Plant Cell*, **7**, 359–371.

Kamal-Eldin, A. and Appelqvist, L.-. (1996) The chemistry and antioxidant properties of tocopherols and tocotrienols. *Lipids*, **31**, 671–701.

Katavic, V., Friesen, W., Barton, D.L., Gossen, K.K., Giblin, E.M., Luciw, T., An, J., Zou, J., MacKenzie, S.L., Keller, W.A., Males, D. and Taylor, D.C. (2001) Improving erucic acid content in rapeseed through biotechnology: what can the *Arabidopsis FAE1* and the yeast *SLC1-1* genes contribute? *Crop Science*, **41**, 739–747.

Kimber, D.S. and McGregor, D.I. (1995) The species and their origin, cultivation and world production. In: *Brassica Oilseeds Production and Utilization* (eds D. Kimber and D.I. McGregor), Cab International, Wallingford, Oxon, UK, pp. 1–7.

Knutzon, D.S., Thompson, G.A., Radke, S.E., Johnson, W.B., Knauf, V.C. and Kridl, J.C. (1992) Modification of *Brassica* seed oil by antisense expression of a stearoyl-acyl carrier protein desaturase gene. *Proceedings of the National Academy of Science* (USA), **89**, 2624–2628.

Knutzon, D.S., Lardizabal, K.D., Nelsen, J.S., Bleibaum, J.L., Davies, H.M. and Metz, J.G. (1995) Cloning of a coconut endosperm cDNA encoding a 1-acyl-*sn*-glycerol-3-phosphate acyltransferase that accepts medium-chain-length substrates. *Plant Physiology*, **109**, 999–1006.

Knutzon, D.S., Hayes, T.R., Wyrick, A., Xiang, H., Davies, H.M. and Voelker, T.A. (1999a) Lysophosphatidic acid acyltransferase from coconut endosperm mediates the insertion of laurate at the *sn*-2 position of triacylglycerols in lauric rapeseed oil and can increase total laurate levels. *Plant Physiology*, **120**, 739–746.

Knutzon, D., Chan, G.M., Mukerji, P., Thurmond, J.M., Chaudhary, S. and Huang, Y.-S. (1999b) Genetic engineering of seed oil fatty acid composition. In: *Plant Biotechnology and in Vitro Biology in the Twenty-First Century* (eds A. Altman, M. Ziv and S. Izhar), Kluwer Academic Publishers, Dordrecht, The Netherlands, pp. 575–578.

Kodali, D.R., Fan, Z. and DeBonte, L.R. (1998) Plants, seeds and oils having an elevated total monounsaturated fatty acid content. US Patent No. 6,414,223.

Krawczyk, T. (1999) Edible speciality oils – an unfulfilled promise. *INFORM*, **10**, 552–561.

Lardizabal, K.D., Metz, J.G., Sakamoto, T., Hutton, W.C., Pollard, M.R. and Lassner, M.W. (2000) Purification of a jojoba embryo wax synthase, cloning of its cDNA, and production of high levels of wax in seeds of transgenic *Arabidopsis*. *Plant Physiology*, **122**, 645–655.

Lassner, M.W., Levering, C.K., Davies, H.M. and Knutzon, D.S. (1995) Lysophosphatidic acid acyltransferase from meadowfoam mediates insertion of erucic acid at the *sn*-2 position of triacylglycerol in transgenic rapeseed oil. *Plant Physiology*, **109**, 1389–1394.

Leckband, G., Frauen, M. and Friedt, W. (2002) NAPUS 2000. Rapeseed (*Brassica napus*) breeding for improved human nutrition. *Food Research International*, **35**, 273–278.

Lee, M., Lenman, M., Banas, A., Bafor, M., Singh, S., Schweizer, M., Nilsson, R., Liljenberg, C., Dahlqvist, A., Gummeson, P.-O., Sjödahl, S., Green, A. and Stymne, S. (1998) Identification of non-heme diiron proteins that catalyze triple bond and epoxy group formation. *Science*, **280**, 915–918.

Leonard, J.M., Knapp, S.J. and Slabaugh, M.B. (1998) A Cuphea ß-ketoacyl-ACP synthase shifts the synthesis of fatty acids towards shorter chains in *Arabidopsis* seeds expressing *Cuphea* FatB thioesterases. *Plant Journal*, **13**, 621–628.

Marquard, R. (1976) Der Einfluß von Sorte und Standort sowie einzelner definierter Klimafaktoren auf den Tocopherolgehalt im Raps. *Fette, Seifen, Anstrichmittel*, **78**, 341–346.

Martini, N., Schell, J. and Töpfer, R. (1995) Expression of medium-chain acyl-(ACP) thioesterases in transgenic rapeseed. In: *Rapeseed Today and Tomorrow. Proceedings of 9th International Rapeseed Congress*, 4–7 July 1995, Cambridge, UK, Vol. 2, pp. 461–463.

Marwede, V., Möllers, C., Olejniczak, J. and Becker, H.C. (2003) Genetic variation, genotype x environment interactions and heritabilities of tocopherol content in winter oilseed rape (*Brassica napus* L.). In: *Proceedings 11th International Rapeseed Conference*, Copenhagen, Denmark, 6–10 July 2003, Vol. 1, pp. 212–214.

Matthäus, B. (1998) Effect of dehulling on the composition of antinutritive compounds in various cultivars of rapeseed. *Fett/Lipid*, **100**, 295–301.

Mazur, B., Krebbers, E. and Tingey, S. (1999) Gene discovery and product development for grain quality traits. *Science*, **285**, 372–375.

McVetty, P.B.E. and Scarth, R. (2002) Breeding for improved oil quality in *Brassica* oilseed species. *Journal of Crop Production*, **5**, 345–370.

Metz, J.G., Roessler, P., Facciotti, D., Levering, C., Dittrich, F., Lassner, M., Valentine, R., Lardizabal, K., Domergue, F., Yamada, A., Yazawa, K., Knauf, V. and Browse, J. (2001) Production of polyunsaturated fatty acids by polyketide synthases in both prokaryotes and eukaryotes. *Science*, **293**, 290–293.

Millar, A.A. and Kunst, L. (1997) Very-long-chain fatty acid biosynthesis is controlled through the expression and specificity of the condensing enzyme. *Plant Journal*, **12**, 121–131.

Miquel, M. and Browse, J.A. (1994) High oleate oilseeds fail to develop at low temperature. *Plant Physiology*, **106**, 421–427.

Möllers, C. and Schierholt, A. (2002) Genetic variation of palmitate and oil content in a winter oilseed rape doubled haploid population segregating for oleate content. *Crop Science*, **42**, 379–384.

Napier, J.A. (2002) Plumbing the depths of PUFA biosynthesis: a novel polyketide synthase-like pathway from Marine organisms. *Trends in Plant Science*, **7**, 51–54.

Nes, W.D. (2000) Sterol methyl transferase: enzymology and inhibition. *Biochimica et Biophysica Acta*, **1529**, 63–88.

Nes, W.D., Janssen, G.G., Norton, R.A., Kalinowska, M., Crumely, F.G., Tal, B., Bergentsrahle, A. and Jonsson, L. (1991) Regulation of sterol biosynthesis in sunflower by 24(R,S),25-epiminolanosterol, a novel C-24 methyl transferase inhibitor. *Biochemical and Biophysical Research Communication*, **177**, 566–574.

Padley, F.B., Gunstone, F.D. and Harwood, J.L. (1994) Occurrence and characteristics of oil and fats. In: *The Lipid Handbook* (eds F.D. Gunstone, J.L. Harwood and F.B. Padley), Chapman and Hall, London, pp. 49–170.

Pallot, T.N., Leong, A.S., Allen, J.A., Golder, T.M., Greenwood, C.F. and Golebiowski, T. (1999) Precision of fatty acid analyses using near infrared spectroscopy of whole seed brassicas. In: *New Horizons for an Old Crop. Proceedings of 10th International Rapeseed Congress*, 26–29 Sept. 1999, Canberra, Australia (eds N. Wratten and P.A. Salisbury) (http://www.regional.org.au/au/gcirc/4/573.htm).

Piazza, G.J. and Foglia, T.A. (2001) Rapeseed oil for oleochemical uses. *European Journal of Lipid Science and Technology*, **103**, 405–454.

Porfirova, S., Bergmüller, E., Tropf, S., Lemke, R. and Dörmann, P. (2002) Isolation of an *Arabidopsis* mutant lacking vitamin E and identification of a cyclase essential for all tocopherol biosynthesis. *Proceedings of the National Academy of Science* (USA), **99**, 12495–12500.

Przybylski, R. and Mag, T. (2002) Canola/rapeseed oil. In: *Vegetable oils in Food Technology – Composition, Properties and Uses. Chemistry and Technology of Oils and Fats*, Vol. 5. (ed. F.D. Gunstone), Blackwell Publishing, Abingdon, Oxon, UK, CRC Press, pp. 98–127.

Rakow, G. (1973) Selektion auf Linol- und Linolensäuregehalt in Rapssamen nach mutagener Behandlung. *Zeitschrift für Pflanzenzüchtung*, **69**, 62–82.

Rakow, G. and Raney, P. (2001) Low saturated fat *Brassica napus*. GCIRC Technical Meeting, 5 June 2001, Poznan, Poland. GCIRC Bulletin, B18. Online at www.cetiom.fr/gcirc/ (22.05.03).

Rakow, G., Relf-Eckstein, J., Raney, J.P. and Gugel, R. (1999) Development of high yielding disease resistant, yellow-seeded *Brassica napus*. In: *New Horizons for an Old Crop. Proceedings of 10th International Rapeseed Congress*, 26–29 Sept. 1999, Canberra, Australia [Compact Disc] (eds N. Wratten and P.A. Salisbury) (http://www.regional.org.au/au/gcirc/4/68.htm).

Roesler, K., Shintani, D., Savage, L., Boddupalli, S. and Ohlrogge, J. (1997) Targeting of the *Arabidopsis* homomeric acetylcoenzyme A carboxylase to plastids of rapeseeds. *Plant Physiology*, **113**, 75–81.

Rücker, B. and Röbbelen, G. (1995) Development of high oleic acid rapeseed. In: *Rapeseed Today and Tomorrow. Proceedings of 9th International Rapeseed Congress*, 4–7 July 1995, Cambridge, UK, Vol. 2, pp. 389–391.

Rücker, B. and Röbbelen, G. (1996) Impact of low linolenic acid content on seed yield of winter oilseed rape (*Brassica napus* L.). *Plant Breeding*, **115**, 226–230.

Rudloff, E., Jürgens, H.U., Ruge, B. and Wehling, P. (1999) Selection in transgenic lines of oilseed rape (*Brassica napus* L.) with modified seed oil composition. In: *New Horizons for an Old Crop. Proceedings of 10th International Rapeseed Congress*, 26–29 Sept. 1999, Canberra, Australia (eds N. Wratten and P.A. Salisbury) (http://www.regional.org.au/au/gcirc/4/92.htm).

Sasongko, N.D., Becker, H.C. and Möllers, C. (2003) Increase in erucic acid content in oilseed rape (*Brassica napus* L.) through the combination with genes for high oleic acid: In: *Proceedings 11th International Rapeseed Conference*, Copenhagen, Denmark, 6–10 July 2003, Vol. 1, pp. 229–231.

Savidge, B., Lassner, M.W., Weiss, J.D. and Post-Beittenmiller, D. (2000) Protein and cDNA sequences of *Arabidopsis* dimethylallyltransferase and the uses thereof on altering tocopherol synthesis in plants. Patent Application WO 0063391 A2.

Savidge, B., Weiss, J.D., Wong, Y.-H.H., Lassner, M.W., Mitsky, T.A., Shewmaker, C.K., Post-Beittenmiller, D. and Valentin, H.E. (2002) Isolation and characterization of homogentisate phytyltransferase genes from *Synechocystis* sp. PCC 6803 and *Arabidopsis*. *Plant Physiology*, **129**, 321–332.

Scarth, R. and McVetty, P.M. (1999) Designer oil canola – a review of new food-grade *Brassica* oils with a focus on high oleic, low linolenic types. In: *New Horizons for an Old Crop. Proceedings of*

*10th International Rapeseed Congress*, 26–29 Sept. 1999, Canberra, Australia [Compact Disc] (eds N. Wratten and P.A. Salisbury) (http://www.regional.org.au/au/gcirc/4/57.htm).

Schaeffer, A., Bronner, R., Benveniste, P. and Schaller, H. (2001) The ratio of campesterol to sitosterol that modulates growth in *Arabidopsis* is controlled by sterol methyltransferase 2,1. *Plant Journal*, **25**, 605–615.

Schaller, H., Grausem, B., Benveniste, P., Chye, M.-L., Tan, C.-T., Song, Y.-H. and Chua, N.-H. (1995) Expression of the *Hevea brasiliensis* (H.B.K.) Müll. Arg. 3-hydroxy-3-methylglutaryl-coenzyme A reductase 1 in tobacco results in sterol overproduction. *Plant Physiology*, **109**, 761–770.

Schierholt, A. (2001) Hoher Ölsäuregehalt im Samenöl: genetische Charakterisierung von Mutanten im winterraps (*Brassica napus* L.) – Grundlagen und Ansätze zur züchterischen Entwicklung eines neuen Rapsöltyps. Vorträge für Pflanzenzüchtung, **52**, 129–132.

Schierholt, A. and Becker, H.C. (2000) Entwicklung von Hochölsäure Raps. In: *Öl- und Faserpflanzen* (eds W. Diepenbrock and H. Eißner), UFOP-Schriften 14, Bonn, pp. 319–323.

Schierholt, A. and Becker, H.C. (2001) Environmental variability and heritability of high oleic acid content in winter oilseed rape (*Brassica napus* L.). *Plant Breeding*, **120**, 63–66.

Schierholt, A., Becker, H.C. and Ecke, W. (2000) Mapping a high oleic acid mutation in winter oilseed rape. *Theoretical and Applied Genetics*, **101**, 897–901.

Schneider, M. (2001) Phospholipids for functional food. *European Journal of Lipid Science and Technology*, **103**, 98–101.

Schultz, G., Soll, J., Fiedler, E. and Schulze-Siebert, D. (1985) Synthesis of prenylquinones in chloroplasts. *Physiologia Plantarum*, **64**, 123–129.

Shewmaker, C.K., Sheehy, J.A., Daley, M., Colburn, S. and Ke, D.Y. (1999) Seed-specific overexpression of phytoene synthase: increase in carotenoids and other metabolic effects. *Plant Journal*, **20**, 401–412.

Shintani, D. and DellaPenna, D. (1998) Elevating the vitamin-E content of plants through metabolic engineering. *Science*, **282**, 2098–2100.

Shukla, V.K.S. and Blicher-Mathiesen, U. (1993) Studies in evaluation of unconventional oils from Southeast Asia. *Fat Science Technology*, **95**, 367–369.

Slabas, A.R., Simon, J.W. and Brown, A.P. (2001) Biosynthesis and regulation of fatty acids and triacylglycerides in oil seed rape. Current Status and future needs. *European Journal of Lipid Science and Technology*, **103**, 455–466.

Somerville, C.R. and Bonetta, D. (2001) Plants as factories for technical materials. *Plant Physiology*, **125**, 168–171.

Sonntag, N.O.V. (1995) Industrial utilization of long-chain fatty acids and their derivates. In: *Brassica oilseeds* (eds D. Kimber and D.I. McGregor), CAB International, Wallingford, Oxon, UK, pp. 339–352.

Stoutjesdijk, P.A, Hurlstone, C., Singh, S.P. and Green, A.G. (1999) Genetic manipulation for altered oil quality in *Brassica*. In: *New Horizons for an Old Crop. Proceedings of 10th International Rapeseed Congress*, 26–29 Sept. 1999, Canberra, Australia [Compact Disc] (eds N. Wratten and P.A. Salisbury) (http://www.regional.org.au/au/gcirc/4/190.htm).

Töpfer, R., Martini, N. and Schell, J. (1995) Modification of plant lipid synthesis. *Science*, **268**, 681–686.

Tsang, E.W.T., Yang, J., Chang, Q., Nowak, G., Kolenovsky, A., McGregor, D.I. and Keller, W.A. (2003) Chlorophyll reduction in the seed of *Brassica napus* with a glutamate 1-semialdehyde aminotransferase antisense gene. *Plant Molecular Biology*, **51**, 191–201.

Ursin, V., Knutzon, D., Radtke, S., Thornton, J. and Knauf, V. (2000) Production of beneficial dietary omega-3 and omega-6 fatty acids in transgenic Canola. Abstr. No. 49, 14th International Symposium Plant Lipids, J.L. Harwood and P.J. Quinn (organizers). Cardiff (UK).

Velasco, L. (1999) Analysis of rapeseed quality traits by near-infrared reflectance spectroscopy (NIRS). *Lipid Technology*, **11**, 90–93.

Velasco, L. and Becker, H.C. (1998) Estimating the fatty acid composition of the oil in intact-seed rapeseed by near-infrared reflectance spectroscopy. *Euphytica*, **101**, 221–230.

Velasco, L., Möllers, C. and Becker, H.C. (1999) Estimation of seed weight, oil content, and fatty acid composition in intact single seeds of rapeseed (*Brassica napus* L.) by near-infrared reflectance spectroscopy. *Euphytica*, **106**, 79–85.

Voelker, T.A. and Kinney, A.J. (2001) Variations in the biosynthesis of seed storage lipids. *Annual Review of Plant Physiology and Plant Molecular Biology*, **52**, 335–361.

Voelker, T.A., Hayes, T.R., Cranmer, A.C., Turner, J.C. and Davies, H.M. (1996) Genetic engineering of a quantitative trait: metabolic and genetic parameters influencing the accumulation of laurate in rapeseed. *Plant Journal*, **9**, 229–241.

Wallis, J.G., Watts, J.L. and Browse, J. (2002) Polyunsaturated fatty acid synthesis: what will they think of next? *Trends in Biochemical Sciences*, **27**, 467–473.

Ward, K., Scarth, R., Daun, J.K. and Thorsteinson, C.T. (1994) Characterization of chlorophyll pigments in ripening Canola seed (*Brassica napus*). *Journal of the American Oil Chemists' Society*, **71**, 1327–1331.

Warwel, S. (1993) Transgene Ölsaaten – Züchtungsziele bei Raps aus chemisch-technischer Sicht. *Fat Science Technology*, **9**, 329–333.

Weier, D., Hanke, C., Eickelkamp, A., Lühs, W., Dettendorfer, J., Schaffert, E., Möllers, C., Friedt, W., Wolter, F.P. and Frentzen, M. (1997) Trierucoylglycerol biosynthesis in transgenic plants of rapeseed (*Brassica napus* L.). *Fett/Lipid*, **99**, 160–165.

Wong, R.S.C., Grant, I., Patel, J.D., Parker, J.P.K. and Swanson, E.B. (1995) Edible endogenous vegetable oil from rapeseeds of reduced stearic and palmitic saturated fatty acid content. EP0476093B1, US Patent No. 5,434,283.

Yuan, L., Voelker, T.A. and Hawkins, D.J. (1995) Modification of the substrate specificity of an acyl-acyl carrier protein thioesterase by protein engineering. *Proceedings of the National Academy of Science* (USA), **92**, 10639–10643.

Zou, J., Katavic, V., Giblin, E.M., Barton, D.L., MacKenzie, S.L., Keller, W.A., Hu, X. and Taylor, D.C. (1997) Modification of seed oil content and acyl composition in the *Brassicaceae* by expression of a yeast *sn*-2 acyltransferase gene. *Plant Cell*, **9**, 909–923.

# List of acronyms

| | | | |
|---|---|---|---|
| "00" | double zero rapeseed oil with low erucic acid and low glucosinolate content | HO | high oleic acid |
| | | HOAR | high oleic acid rapeseed |
| | | HOLL | high oleic/low linolenic acid rapeseed |
| α-LNA | α-linolenic acid; 18:3 $n$-3 | | |
| γ-TMT | γ-tocopherol methyltransferase | HOSO | high-oleic sunflower oil |
| AA | arachidonic acid; 20:4 $n$-6 | HPLC | high performance liquid chromatography |
| ABI | Austrian Biofuel Institute | | |
| ACP | acyl carrier protein | IHD | ischemic heart disease |
| AHA | American Heart Association | IP | identity preserved |
| ALD | adrenoleukodystrophy | IPP | isopentenyl pyrophosphate |
| AOM | active oxygen method | IR | infrared |
| ASPA | Agents de surface et da produits auxiliaries industriels | IV | iodine value |
| | | KCS | β-ketoacyl-CoA-synthase |
| ATP | adenosine triphosphate | LA | linoleic acid |
| BHA | butylated hydroxyanisole | LC | long chain |
| BHT | butylated hydroxytoluene | LDL | low density lipoprotein |
| BOPP | biaxially orientated polypropylene | LDPE | low density polyethylene |
| CFPP | cold-filter plugging point | LEAR | low erucic acid rapeseed |
| CHD | coronary heart disease | LERO | low erucic rapeseed oil |
| CVD | cardiovascular disease | LL | low linolenic acid |
| DGDG | digalactosyldiglyceride | LLDPE | linear low density polyethylene |
| DHA | docosahexaenoic acid; 22:6 $n$-3 | L-LNA | low-linolenic acid |
| DMAPP | dimethylallyl diphosphate | LPAAT | lysophosphatidic acid acyl transferase |
| DMPQ | 2,3-dimethyl-5-phytyl-1,4-hydroquinone | | |
| | | MDA | malondialdehyde |
| EAHF | environmentally acceptable hydraulic fluids | MI | myocardial infarction |
| | | MRFIT | Multiple Risk Factor Intervention Trial |
| EPA | eicosapentaenoic acid; 20:5 $n$-3 | | |
| ESA | environmentally sensitive area | MT | metric tonnes |
| ESR | electron-spin resonance | MUFA | monounsaturated fatty acids |
| EU | European Union | NIR | near infrared |
| FDA | Food and Drug Administration (USA) | NIRS | near-infrared reflectance spectroscopy |
| FFA | free fatty acid | NMR | nuclear magnetic resonance |
| FT-IR | Fourier-transform infrared | $NO_x$ | total oxides of nitrogen |
| FT-Raman | Fourier-transform Raman | OSI | oil stability index |
| FW | fresh weight | ox-LDL | oxidized LDL |
| GC | gas chromatography | PA | phosphatidic acid |
| GLA | gamma linolenic acid | PAH | polycyclic aromatic hydrocarbon |
| GLC | gas-liquid chromatography | PC | phosphatidylcholine |
| HAC | hydrophilic and amphiphilic compounds | PE | phosphatidylethanolamine |
| | | PEG | polyethylene glycol |
| HCO | hydrogenated canola oil | $PGF_{2α}$ | prostaglandin $F_{2α}$ |
| HDL | high density lipoprotein | PHA | polyhydroxy alkanoate |
| HEAR | high erucic acid rapeseed | PI | phosphatidylinositol |
| HERO | high erucic rapeseed oil | PKS | polyketide synthase |
| HMGR | 3-hydroxy-3-methylglutaryl CoA reductase | Pm | particulate matter |
| | | pNMR | pulsed nuclear magnetic resonance |

# LIST OF ACRONYMS

| | | | |
|---|---|---|---|
| PP | polypropylene | SV | saponification value |
| PUFA | polyunsaturated fatty acid | t/ha | tonne(s) per hectare |
| PV | peroxide value | TAG | triacylglycerols |
| PVC | polyvinyl chloride | TBA | thiobarbituric acid |
| RBD | refined, bleached and deodorized | TBARS | thiobarbituric acid reactive substances |
| RI | refractive index | | |
| RME | rapeseed oil methyl esters | TBHQ | tert-butyl hydroquinone |
| RR | relative risks | TMP | trimethylolpropane |
| RRM | renewable raw material | TOC | tocopherols |
| SMT1 | sterol C-24 methyltransferase type1 | USDA | US Department of Agriculture |
| SMT2 | sterol-C24 methyltransferase type2 | UV | ultraviolet |
| | | VOC | volatile organic compounds |
| $sn$ | stereospecific numbering | ZDDP | zinc di-(2-ethylhexyl)-dithiophosphate |
| $So_x$ | total oxides of sulfur | | |

# Index

adrenoleukodystrophy  126
agronomy  1, 115, 154
ALD  126
alkali-refined oils  66
alkali refining  27, 29
anisidine value  85
antioxidants  83
*Aspergillus niger*  33
autoxidation  81

baking fats  133
biodiesel  154
    economics  163
    feedstocks  155
    market opportunities  166
    production  158, 159, 162–165, 166, 167
    specification  157
biorefining  32, 33
biosynthetic pathways  190, 196, 198
*Brassica carinata*  117
*Brassica napus*  111
*Brassica napus* agronomic performance  188
*Brassica* varieties  2

calorific value  156
canola oil *see* rapeseed oil
canola trademark  2
carbon dioxide for extraction  26
cardiac arrhythmia  147
cardiovascular disease  134
carotenoids in rapeseed oil  66, 207–209
cetane number  156
chainsaw oil  176
chemical properties  79
chlorophyll  68, 69, 210, 211
cholesterol  135, 140
clot formation  146
cloudpoint  156
cold test  92
composition of crude oil  26
conditioning  20
consumption, 13, 14
*Crambe abyssinica*  115, 116

Crop establishment  3
crystal structure  93, 133
crystallisation  121

degumming  27, 28
dehulling  19
density  87, 156
deodourisation  27, 31
desaturase  190, 196
desolventising  25
diseases *see* oil seed rape
double-low rapeseed oil  2

eicosanoids  143
electron-spin resonance  86
emissions  160, 161
energy ratios  162
epoxidised oils  181
epoxy acids in rapeseed oil  202
erucamide  126
erucic acid
    producers  124
    users  124
    uses  126, 127
exhaust emission *see* emissions
exports *see* oil seed rape and rapeseed oil
extraction  17, 18

fatty acids  38, 40, 43, 44, 47, 49, 52, 54,
    112, 156, 190, 191
fertiliser *see* oil seed rape
fire point  91
flaking  19
flash point  91, 156
food uses  132
fractional distillation  120
frying oils  133
fuel characteristics  160
fuel efficiency  160

gamma linolenic acid in rapeseed oil  196
gas chromatogram  44
gear oils  173

# INDEX

genetic modification 84, 128
glucosinolate 2, 20
glycolipids 67, 68
greases 171

harvesting *see* oil seed rape
HEAR
    agronomy 113
    downstream processing 120, 122
    fatty acid composition 112
    genetic modification 128
    meal quality 123
    methanolysis 119
    mineral content 124
    processing 118, 120
    production 114
    seed varieties 114
    splitting 118, 119
heat of combustion 91
heat of crystallization 91
heat of fussion 90
hexane 23
high-erucic acid oil *see* HEAR
high-erucic acid rapeseed oil 38, 45, 62, 198
high-lauric acid rapeseed oil 45, 54, 56
high-oleic acid rapeseed oil 190
high-oleic/low-linolenic acid rapeseed oil 192
high-stearic acid rapeseed oil 54, 55, 59, 193
hydraulic fluids 169, 170, 171
hydrogenation 97
hydroxy acids in rapeseed oil 200, 201
hyperlipidemic subjects 137

imports *see* oil seed rape and rapeseed oil
inks 179
interesterification 101
interfacial tension 89
iodine value 80, 156
isopropanol for solvent extraction 26

lipid peroxidation 140
lipoproteins 135
Lorenzo's oil 126
low-erucic acid rapeseed oil 2
low-linolenic acid rapeseed oil 45, 52, 53, 189
low-saturated acid rapeseed oil 193
lubricants 167
lunaria 112
Lyon diet heart study 147

margarine 132
market opportunities
    biodiesel 166
    hydraulic fluids 171
    lubricants 167
mayonnaise 132
meadowfoam 112
meal–fatty acid profile 123
mechanical extraction 21, 22
medium and short chain acids in rapeseed oil 194
melting behaviour 93
metal working fluids 174
metals 71
methanolysis 119
methyl esters
    *see also* biodiesel
minerals 71, 124
motor oils 173
mould release agents 172
mustard seed oil 38, 40, 116
myrosinase 20

non-food uses of rapeseed oil 154
normolipidemic subjects 135
nutritional properties 131, 134

oil content 188
oil seed rape 15
    crop establishment 3
    diseases 9
    fertiliser requirement 6
    harvesting 10
    pests 8
    seed production 10
    trade in seeds 11, 12
    weeds 7
    yields 12
Oil Stability Index 85
oleic acid desaturase 190
oxidative stability 81

paints 179
peroxide value 85
pests *see* oil seed rape
phospholipids 67, 68
photo-oxidation 82
physical properties 87, 156
physical refining 27, 29, 66
phytosterols 140
plasma cholesterol levels 140
plasma phospholipids 143

# INDEX

platelet aggregation  143
platelet phospholipids  143
polar lipids  67, 68, 202, 203
polarized light microscopy  97
polyketide synthesis  198
polymers  181
polymorphism  93, 133
pressing  22
production *see* biodiesel, erucic acid, HEAR, oil seed rape, and rapeseed oil
prospects for rapeseed oil  186
prostacyclin  143
PUFA in rapeseed oil  197

rapeseed methyl esters *see* biodiesel
rapeseed oil
    carotenoids  66, 208
    chemical properties  79
    chlorophyll  68, 69, 210, 211
    consumption  13, 14
    development  37, 41
    double low  2
    fatty acids  38, 40, 43, 45, 47, 48, 52, 54, 190, 191
    future prospects  186
    gamma linolenic acid  196
    glycolipids  67, 68
    high-erucic acid  38, 45, 62, 198
    high-lauric acid  45, 54, 56
    high-oleic acid  190
    high-oleic and low linolenic acid  192
    high-stearic acid  54, 55, 59, 193
    hydroxy acids  200
    long chain PUFA  197
    low-erucic acid  2
    low glucosinolate  2
    low-linolenic acid  45, 52, 53, 58
    low-saturated acids  193
    medium and short chain acids  194
    minerals  71
    modified linolenic acid content  189
    oil content  188
    phospholipids  67, 68
    physical properties  87
    polar lipids  67, 68, 202, 203
    production levels  11
    specialty types  38, 41
    sterols  59, 206
    sulfur containing compounds  46, 70
    tocopherols  63, 203, 205
    trade in oil  11, 13
    triacylglycerols  48, 50, 53, 54, 55, 56, 189
    waxes  66
refining  23, 27
refractive index  90
regiospecific analysis  57, 58, 59

salad oils  132
saponification value  80
sediment from bottled oil  47
seed cleaning  17
seed production  11
seed varieties  114
sensory analysis  86
settling  23
short spacings  96
shortening  133
smoke point  91
solid fat index  101
solubility  92
solvent extraction  23, 26
solvent recovery  24
solvent removal  25
specification for biodiesel  157
spectroscopic properties  92
splitting  118, 119
sterol esters  63, 64
sterols in rapeseed oil  59, 62, 64, 206, 207
sulfur-containing compounds  46, 70
supercritical carbon dioxide  26
surface tension  89
surfactants  177

TBA test  85
tempering  18
thrombogenesis  142
toasting  25
tocopherol  32, 63, 65, 66, 203, 204
trade *see* oil seed rape and rapeseed oil
triacylglycerols  48, 50, 53, 54, 55, 56, 188, 189

users of erucic acid  124
uses of erucic acid  126, 127

viscosity  88, 156
vitamin E  32

waxes  66
weeds  7
winterisation  27, 31

yields  12